Lagomorphs

A Volume in the Series
Mammals of Russia and
Adjacent Regions

Lagomorphs

V.E. Sokolov, E.Yu. Ivanitskaya
V.V. Gruzdev, V.G. Heptner

Scientific Editors

Robert S. Hoffmann
Series Scientific Editor
Smithsonian Institution, Washington, DC

Andrew T. Smith
Volume Scientific Editor
Arizona State University, Tempe, AZ

A Volume in the Series
Mammals of Russia and
Adjacent Regions

Translated from Russian

Science Publishers

Enfield (NH)   Jersey   Plymouth

Science Publishers   *www.scipub.net*
234 May Street
Post Office Box 699
Enfield, New Hampshire 03748
United States of America

General enquiries   :   *info@scipub.net*
Editorial enquiries :   *editor@scipub.net*
Sales enquiries   :   *sales@scipub.net*

Published by Science Publishers, Enfield, NH, USA
An imprint of Edenbridge Ltd., British Channel Islands
Printed in India

ISBN: 978-1-57808-522-4

CIP data will be provided on request

Translation of:
Mlekopitaiushchie Rossii i
Sopredel'nykh Regionov: Zaitseobraznye
Nauka Publishers, Moscow, 1994

UDC 599.333

This book is a continuation of the series of monographs on different orders of mammals, initiated by V.G. Heptner in 1961 by Vysshya Shkola Publishers. This volume is devoted to the description of the order Lagomorpha, which is represented by two extant families—Leporidae and Ochotonidae.

The book is aimed at mammalogists, ecologists, zoogeographers and game specialists.

Illustrations 58, tables 57, bibliography of 13 pages.

Reviewers
V.G. Borkhvardt, Doctor of Biological Sciences
G.V. Shlyakhtin, Doctor of Biological Sciences

FOREWORD

The present book is a continuation of the multi-volume monograph *Mammals of the Soviet Union*, the first four volumes of which were published between 1961 and 1976 under the editorship of V.G. Heptner and N.P. Naumov by Vysshaya Shkola Publishers.

Since the publication of monographs of S.I. Ognev (1940) and A.A. Gureev (1964) much new data have appeared on the systematics and biology of lagomorphs, which have been included mainly in regional compendia on mammals. M.A. Erbaeva (1980) published the monograph entitled *Pishchukhi Kainozoya* [Pikas in the Cenozoic] on the systematics of pikas in the world fauna. However, this monograph does not contain much information on their biology. Similarly, biological descriptions of species are given briefly in the book "*Rabbits, Hares* and *Pikas: Status Survey and Conservation Action Plans*" (Smith et al., 1990).

All descriptions of groups and species are given according to the scheme followed in the preceding volumes of the series. Palaeontological data on Recent species and genera are limited to just brief mentions, since detailed information on fossil pikas is available in the monograph of M. A. Erbaeva (1980) and on hares—in the book of A.A. Gureev (1964).

All synonyms are given a restricted treatment according to the follwing principle: only the proper synonyms are given; of the generic synonyms, only those are given, in which the type of genus is a recent species, notwithstanding whether or not this species is present in our fauna. In the species synonymy, only the valid names of Recent forms appear. More complete lists of synonyms, including also the invalid names for the territory of the former USSR, were recently published by I.Ya. Pavlinov and O.L. Rossolimo (1987).

For widely distributed species, description of the range is given in a generalized form: only its boundaries are described. For species with narrow ranges, as far as possible, all known points of occurance are given. Unlike preceding volumes, all references to the sources of information are given directly in the text while describing their distribution.

---

*Page number of the original Russian text—General Editor.

viii

Range maps have been compiled from published data, and also from the collections of the Zoological Museum of Moscow University (ZM MGU), Zoological Institute of the Russian Academy of Sciences (ZIN), Biological Institute of the Siberian Department of the Russian Academy of Sciences (BINS), Institute of Zoology of the Ukrainian Academy of Sciences (UkrZIN) and Kiev State University (KGU), Institute of Zoology of Kazakhstan Academy of Sciences (IZK), and Kazakh State University (KazGU), Institute of Zoology and Parasitology, Academy of Sciences of Tadzhikistan (IZIP TadzhAN), Institute of Zoology and Parasitology, Academy of Sciences of Uzbekistan (UzbZIN), Irkutsk State University (IGU), Saratov State University (SGU), Institute of Plant and Animal Ecology, Russian Academy of Sciences (IE'RZh)*, Microbe Institute (Saratov), Antiplague Institute of Siberia and the Far East (Irkutsk), and others. We express deep gratitude to colleques from these institutes for their help in the preparation of range maps.

In this book we have included unpublished data on the biology and distribution of pikas, which were very kindly made available by N.A. Formozov, D.G. Derviz and V.V. Labzin.

N.A. Rybakova (IE'ME'Zh) actively participated in the preparation of the manuscript. She prepared the range maps of all species of lagomorphs and wrote the rough draft of the text on the geographic distribution of species. The authors also express great gratitude to G.I. Shenbrot, G.V. Shlyakhtin and G.G. Aksenova for their valuable comments and help in the work on the manuscript.

---

*Abbreviations are used in the text while citing the ownership of collections of specimens.

CONTENTS

ORDER OF LAGOMORPHS
ORDO LAGOMORPHA BRANDT, 1855

The majority of members of the order of lagomorphs have large ears, relatively long hind legs, and a short or externally inconspicuous tail. The body size varies from 120 to 600 mm (or more). The fore limbs have 5 digits and the the hind limbs, 4. The underside of the foot is covered with a dense brush of hairs.

The skull varies in form and size. The lateral surface of the maxillary bones has an ethmoid structure or has one broad orifice. The incisor sockets are long and broad. The infraorbital foramen is small. The bony palate is short.

The maxilla in lagomorphs has two pairs of incisors of different sizes. The smaller incisors (oval or rounded in form) are situated behind large ones, which always have a furrow on the anterior surface. The number of true molars varies from 5/5 to 6/5; premolars always 3/2. The crowns of molars are high and prismatic; their masticatory surface has sharp enamel margins.

The molars on the mandible, besides the first (P_3), consist of two conids—paraconid and protoconid. The anterior conid of the lower molars is higher than the posterior. The distance between the molars on the maxilla is less than that on the mandible. Rami of the mandible in the incisor part are firmly united or fused. The articular fossae on the zygomatic process of the temporal bone are shallow and relatively broad in the transverse direction.

Key for Identification of Families of the Order Lagomorpha

1(2). Ears short, round. Tail inconspicuous. Sides of maxillary bones with large foramina. Nasal bones anteriorly broadened. Supraorbital processes absent. Upper anterior incisors with deep notch on the cutting surface. Molars 5/5... Family **Ochotonidae.**

2(1). Ears long, often pointed at tip. Tail short but conspicuous. Sides of maxillary bones with ethmoid structure. Nasal bones posteriorly broadened. Supraorbital processes present. Upper anterior incisors without notch on cutting surface. Molars 6/5 .... Family **Leporidae.**

FAMILY OF PIKAS
FAMILIA OCHOTONIDAE THOMAS, 1897

Sizes small, smallest in the order. The body length varies from 120 to 280 mm; weight from 170 to 240 g.

Externally pikas are similar to small rabbits, but with shorter limbs and small ears. The limbs are relatively short; the hind legs just slightly (not more than 20-25%) surpass the length of fore legs. The fore limbs have 5 digits; the hind limbs, 4. The tail is inconspicuous. The ears are short; their length in some species slightly more than half the length of the head. The tip of the ears is rounded. The eyes are medium-sized.

The hair coat is thick, dense and soft. Its color is reddish or brown to gray, usually lighter on the abdominal side. Long vibrssae are characteristic, in some species surpassing a third of the body length.

6 Two or three pairs of teats. The scrotum is absent, and the os penis very small.

The length of the intestine surpasses the body length by 10 times and more, the length of its small intestine constitutes more than half the length of the entire intestine; the caecum is long, more than 1.5 times that of the body length. The inner surface of the large intestine is morphologically differentiated into several sections; the abundance of lymphoid tissue is characteristic (Naumova, 1981).

The skull is elongated and strongly compressed in the interorbital region. The parietal bones do not have supraorbital foramina. The interparietal bone is retained and is relatively larger in size than in hares. The orbit is often round and considerably open above. The zygomatic bone, behind the zygomatic process of the squamosal bone, is extended as a long process, almost reaching the auditory opening. The bony palate is short; moreover, the length of section of the bony palate forming it is more than the width of the palatine process of the maxillary bone. The interpterygoid space is narrow and shallow. The tympanic bulla is relatively large and spongy in structure. A special articular fossa is absent on the zygomatic process of the squamosal bone.

Dental formula i 2/1, c 0/0, p 3/2, m 2/3 = 26. The third* upper

---

*second—wrong in Russian original—Editor.

premolar is smaller than the fourth* premolar and the latter is similar to the molar. The third upper molar is absent. The upper front incisors have a deep V-shaped groove on the anterior surface, clearly dividing the distal end of the tooth into outer and inner lobes, the outer lobe being appreciably broader and longer than the inner. The vertebrae in the thoracic section (17, sometimes 16), unlike the corresponding vertebrae of hares, have a short neural spine. The lumbar vertebrae (5, sometimes 4) are thick, with a very short neural spine and transverse processes. Caudal vertebrae number 7 or 8.

The hand is articulated mainly with the ulna, which is twice as thick as the radius. As in hares, the pubic symphysis is absent in the pelvis.

The diploid complement of chromosomes varies from 38 in *Ochotona pallasi* to 68 in *O. pusilla*, *O. collaris* and *O. princeps*.

Pikas live in semidesert, steppe, stony taluse, rock outcrop and montane meadow habitats at different altitudes—from sea level to 4-6 thousand meters above sea level. Pikas living in the open plains and in the foothills dig burrows, and those dwelling in stony biotopes use hollows between stones for shelter. They live singly or in colonies. They are active in the daytime, although sometimes they are active at night. In the majority of species, the vocal signal is well developed, particularly in the event of danger.

Pikas feed on various plants. Storage of food is a characteristic; moreover, many species of pika have a unique method of drying and storing food—hiding it under rock ledges or piling it in a stack. Coprophagy plays an important role in the nutrition of pikas.

The breeding period is greatly protracted and does not coincide in time in different species or throughout the range of the same species. Gestation lasts for about 30 days. The litter consists of 2-7 young. There are one to three litters in a year. The young are born helpless. In some species (large-eared, red, and Altai pikas), the newborns are densely covered with hair; in others (little, Afghan, and Pallas's pikas), the young are born naked.

Present-day pikas live in the mountains of North America, from the southeastern half of Alaska and the Canadian province of Yukon in the north to the central regions of the states of California, Colorado, and New Mexico in the south; extending as far east as the southwest of the province of Alberta and central regions of the states of Montana and Wyoming. In the Old World pikas extend from the southern Urals (Orenburg and Saratov regions) in the west to the Gulf of Anadyr, Hokkaido, and the Korean Peninsula in the east; as far north as the the Arctic Coast; and in the south to the northern parts of Iran, Afghanistan, Pakistan, India and Burma.

---

*third upper premolar cannot be smaller than the fourth premolar, as there are only 3 upper premolars—Editor.

7 The most ancient members of the pika famiy are known from the Oligocene of Asia (Sinolagomyinae). By the Early Miocene the members of this subfamily lived in Europe, Africa and North America, but became extinct everywhere by the end of the Miocene. The subfamily Lagomyinae started to grow from the Early Miocene, and it contained 8 genera. Pikas reached their peak in the Pliocene (4 genera, 22 species), and by the Early Pleistocene the members of 3 genera had become extinct. Thus, from the Pleistocene, the family of pikas is represented by only one genus—*Ochotona*. The earliest records of the members of the genus (*O. lagrelli* Schlosser, 1924) relate to the Late Miocene (Inner Mongolia). In the Pleiocene, the range of *Ochotona* covered Europe, Asia and North America. Paleontological data confirm the fairly large diversity of size of the Pliocene pikas (Erbaeva, 1988). In the Pleistocene, small forms of pikas were predominant. In Europe at this time, pikas of the *pusilla*-group were widely distributed; and North America contained the species which, probably, may be considered ancestoral to present-day *O. collaris* and *O. princeps*. In Asia, many of the Pleistocene pikas became extinct by the Holocene, and this part of the Palaearctic at the present is inhabited by relatively younger species. For the present-day species of pikas, the most ancient records relate to the Late Pliocene (*O. pusilla*, *O. thibetana*); *O. rufescens* and *O. daurica* are known from Pleistocene deposits (Erbaeva, 1988). It is customary to consider Central Asia as the center of origin of pikas (Gureev, 1964; Dawson, 1967). V.S. Bazhanov holds a different point of view. He assumes that the area of origin of pikas lies in Eurasia, probably in its western part. Pikas do not have a substantial economic significance. Some species have epidemiological significance. Individul species may inflict damage to forestry by destroying saplings, or to agriculture by damaging pastures. In the famiily Ochotonidae there is one present-day genus—pikas.

Genus of Pikas
Genus *Ochotona* Link, 1775*

1773. *Lepus* Pallas. Reise dur verschiedene Provinzen des Russischen Reichs. Bd. II. S. 701-702.

1775*. *Ochotona* Link. *Beitrage zur Naturgeschichte.* Bd. 1, Theil 2. S. 74.

1799. *Pika* Lacèpéde, Tableau des divisions der Mammiferes. Paris. p. 9.

1800. *Lagomys* G. Cuvier. *Lecons d'anatomie compares.* V. 1, tabl. 1. p. 1.

1867. *Ogotoma* Gray. Notes on the skulls of hares and pikas in British Museum. Annals and Magazine of Natur. History. V. XX. p. 220.

---

*1975 in Russian original—Translator.

1904. *Conothoa* Lyon. Classification of the hares and their allies. *Smiths. Miscellan. Collections.* V. 45(1). part 3-4. p. 438.

1941. *Lagotona* Kretzoi. Weitere Beitrage zur Fauna von Gombas-zog. *Ann. Mus. Natur. Hung. pars. Miner. Geol. Paleont.* Bd. 34. S. 112

1988. *Buchneria* Erbajeva. Pishchukhi kainozoya [Pikas in the Cenozoic]. Moscow: Nauka, S. 170.

The head is relatively large. The ears are of different sizes. Dermal outgrowths present on the inner surface of ears. Underside of planta is covered with quite long and dense hairs. In some species the digital pads are naked. The hair coat consists of directed, guard, and downy hairs. In pikas the labial vibrissae are well developed, but in different species they differ somewhat in length. For rock-dwellers (*O. macrotis, O. rutila, O. kamensis, O. erythrotis, O. collaris*) the longest vibrissae are characteristic; in the dwellers of plains and montane steppes (*O. pusilla, O. daurica, O. curzoniae*), the vibrissae are shortest (Fedosenko, 1974).

In many species of pikas: Altai, northern, red, Daurian, little, Afghan, Tibetan (*O. thibetana*), American (*O. princeps*), Pallas's and Thomas's pikas specific paired dermal glands are present in the cheek region (Orlov, 1983). The glandular area in the Altai pika is oval in form and lies below the base of the pinna, nearer the neck. The animals mark the territory occupied by them by rubbing the side of the cheek on the substrate (stones, etc.). The functional activity of the cervical gland is not directly associated with reproduction. Pikas have prepucial and anal glands.

In the skull, the facial section is short and the cranial section is enlarged. One large foramen is present on the side of the maxillary bone (maxillary foramen). In some species the incisor fossa is completely or partially separated by the protrusions of the intermaxillary bones. The nasal bones are broadened anteriorly. In some species there are two foramina on the frontal bones. The zygomatic bone has a long process almost reaching the tympanic bulla. The mandible is low in the dental section. The angular process is small. The articular process is well developed.

The upper outer incisors have a median furrow on the anterior surface and a triangular notch on the cutting margin. On the upper molars (except P^2 and P^3) there are outer and inner entrant folds, the inner fold being deeper. The lower molars (except P_3 and M_3) are formed by two prisms joined by cement. These prisms in transverse section are rhombic in form. The upper second premolar tooth is small, with one anterior entrant fold, the upper third molar has a deep C-shaped antero-external entrant fold and a small inner fold (Fig. 1a). The lower third premolar comprises two sections and has several entrant folds (3 or more) (Fig. 1b). The lower third molar tooth is small and tubular.

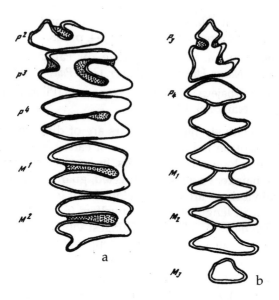

Fig. 1. Molar teeth of pikas (*Ochotona alpina*).

The seventh cervical vertebrae in pikas does not have the arterial canal in the transverse process. The radius is thinner than the ulna or of the same thickness. The pelvis is narrow and weak. The ilium is long and narrow; the processes in the anterior part of these bones are directed sideward.

Ecology, distribution and paleontology—in the description of the family. The number of species in the family is not definitely established. A.A. Gureev (1964) recognized the existence of 15 species: *O. alpina, O. curzoniae, O. daurica, O. erythrotis, O. kamensis, O. kozlovi, O. ladacensis, O. nepalensis, O. pricei* (=*pallasi*), *O. pusilla, O. roylei, O. rufescens, O. rutila, O. thomasi,* and *O. thibetana.* According to the data of V.E. Sokolov (1977), there are 18 species in the genus *Ochotona*—besides the species listed above *O. hyperborea, O. collaris,* and *O. princeps* are also added here. In the compendium *Mlekopotayushchie mira* [Mammals of the World] (Honacki et al., 1982) the existence of 20 species is recognized: *O. alpina, O. daurica, O. curzoniae, O. collaris, O. erythrotis, O. hyperborea, O. kamensis, O. kozlovi, O. ladacensis, O. lama, O. macrotis, O. nepalensis, O. pallasi, O. princeps, O. pusilla, O. roylei, O. rufescens, O. rutila, O. thibetana,* and *O. thomasi.* M.A. Erbaeva (1988) recognizes the existence of 21 species—*O. himalayana* and *O. angdawi* are added to the above list and *O. nepalensis* is relegated to the synonymy of *O. roylei.* Some researchers segregate *O. gloveri* as a species separate from *O. erythrotis,* and *O. cansus* from *O. thibetana* (Feng and Zeng, 1985). In 1986, a new species *O. iliensis* was described from northeastern China (Boro-Khoro Range) (Li and Ma, 1986), which, in the opinion of M.A. Erbaeva (1980), is a large form of *O. macrotis*.

There is no unanimity of view regarding the number and composition of supraspecific groups of pikas. S.I. Ognev (1940) proposed to divide them into three subgenera.

Subgenus *Pika* with *O. alpina*, *O. hyperborea*, *O. princeps*, and *O. collaris*.

Subgenus *Ochotona* with *O. daurica*, *O. curzoniae*, *O. ladacensis*, *O. rufescens*, and *O. pricei* (= *pallasi*).

Subgenus *Conothoa* with *O. roylei*, *O. macrotis*, *O. rutila* and *O. erythrotis*.

Moreover, pikas of the *Pusilla* group were separated with *O. pusilla* and *O. thibetana*, which, in the opinion of S.I. Ognev, should also be segregated in a separate subgenus.

A.I. Argiropulo (1948) recognized only two subgenera in the genus *Ochotona*; moreover, in the type subgenus he included all the species, which were combined in groups:

"*alpina*" group with *O. alpina* (with *O. hyperborea*), *O. princeps*, *O. collaris*, and *O. schisticeps*;

"*roylei*" group with *O. roylei* (including *O. macrotis*), *O. erythrotis*, *O. kamensis*, and *O. gloveri*;

"*pusilla-thibetana*" group with *O. pusilla*, *O. thibetana*, *O. forresti*, and *O. thomasi*;

"*daurica-curzoniae*" group with *O. daurica*, *O. curzoniae*, *O. rufescens*, *O. ladacensis*, and *O. pallasi*.

O. kozlovi was included in a separate subgenus *Tibetolagus*.

A.A. Gureev (1964) combined the present-day pikas not in subgenera but in groups, in essence conserving the system of A.I. Argiropulo (1948).

"*roylei*" group—*O. roylei* (including *O. macrotis*), *O. erythrotis*, *O. kamensis*, *O. nepalensis*, and *O. rutila*;

"*daurica*" group—*O. daurica*, *O. curzoniae*, *O. kozlovi*, *O. rufescens*, *O. ladacensis*, and *O. pallasi*;

"*alpina*" group—*O. alpina* (including *O. hyperborea*, *O. collaris*, and *O. princeps*);

"*thibetana*" group—*O. thibetana*, *O. pusilla*, and *O. thomasi*.

The comparative analysis of karyotypes of nine species of pikas made it possible for N.N. Vorontsov and E.Yu. Ivanitskaya (1973) to identify the following supraspecific groups:

1. *O. pusilla*, *O. princeps*, *O. collaris*;
2. *O. macrotis*, *O. rutila*;

8

3. *O. rufescens, O. daurica;*
4. *O. alpina, O. hyperborea, O. pallasi.*

M.A. Erbaeva (1988), based on the study of teeth and some characteristics of the skull of fossil and recent species of pikas, proposed a new classification of Ochotonidae, in which the present-day species of pikas are divided into seven subgenera:

Subgenus *Lagotona* Kretzoi, 1941 (*O. pusilla*);

10   Subgenus *Pika* Lacèpéde, 1799 (*O. alpina, O. hyperborea, O. princeps, O. collaris, O. pallasi*);

Subgenus *Ochotona* Link, 1795 (*O. daurica, O. curzoniae, O. rufescens*);

Subgenus *Conothoa* Lyon, 1904 (*O. roylei, O. macrotis, O. lama, O. himalayana, O. angdawi*);

Subgenus *Buchneria* Erbajeva, 1988 (*O. erythrotis, O. kamensis, O. rutila, O. ladacensis*);

Subgenus *Tibetolagus* Argyropulo, 1948 (*O. kozlovi*);

Subgenus *Argyrotona* Rekovetz in litt. (*O. thibetana, O. thomasi*).

A critical analysis of the available supraspecific systems of the genus *Ochotona* makes it possible to conclude about the affinity of pikas inhabiting the territory of Russia, Kazakhstan, Russian Central Asia, and Mongolia to four subgenera:

1. Subgenus *Lagotona* with *O. pusilla;*
2. Subgenus *Conothoa* with *O. macrotis* and *O. rutila;*
3. Subgenus *Ochotona* with *O. daurica* and *O. rufescens;*
4. Subgenus *Pika* with *O. alpina, O. hyperborea,* and *O. pallasi.*

The range of the genus coincides with the range of the family.

Economic significance of pikas—in the description of the family.

Key for Identification of Species of the Genus *Ochotona*

1(6). Incisor fossa partitioned by entrant prominences of intermaxillary bones, covering their palatine processes from above.

2(3). Dimensions small, length of foot less than 28 mm, condylobasal length of skull less than 41 mm, height of occipital part of skull less than 12 mm, maximum length of tympanic bulla less than 13.5 mm .... Northern pika, **Ochotona hyberborea.**

3(2). Dimensions large; length of foot more than 28 mm, condylobasal length of skull more than 41 mm, height of occipital part of skull more than 11.5 mm; maximum length of tympanic bulla more than 13.5 mm.

4(5). Planatae with dark brown hairs. In summer, back reddish-brown, belly yellowish-gray; winter fur lighter and grayer; interorbital distance more than 4 mm .... Altai pika, **Ochotona alpina.**

5(4). Planatae covered with lighter hairs. In summer, hair coat lighter, yellowish-gray, in winter—pale gray. Interorbital distance less than 4.5 mm .... Pallas's pika, **Ochotona pallasi.**

6(1). Incisor fossa not partitioned by protuberances of intermaxillary bones; if protuberances present, then hiatus between them retained.

7(10). Hiatus present between protuberances of intermaxillary bones entering incisor foramen.

8(9). Dimensions relatively large: body length more than 200 mm, condylobasal length of skull more than 40 mm. In summer, color of hair coat reddish-rufous, back often with ochreous tinge; head, hind part of body rufous. Hairs on plantae lighter .... Red pika, **Ochotona rutila.**

9(8). Dimensions smaller: body length less than 200 mm, condylobasal length of skull less than 44 mm. Color of hair coat dark brown in summer (somewhat lighter in winter) with small longitudinal light-colored streaks, Hairs on plantae dark .... Little pika, **Ochotona pusilla.**

10(7). Inner protuberances of intermaxillary bones absent.

11(14). Interorbital distance less than 5 mm, vibrissae small—less than 60 mm. Length of ear less than 25 mm. Frontal bones with crest above orbits, without foramina. ·

12(13). Sides of neck with rusty-reddish or reddish-brown spot. Margin of ears without white border. Majority of vibrissae dark brown, not less than 50 mm long. Color of hair coat in summer ochreous-gray with brown bloom on back, gray on belly. Condylobasal length of skull more than 43.5 mm, length of teeth of upper row not less than 9 mm .... Afghan pika, **Ochotona rufescens.**

13(12). Sides of neck without spot. Margin of ears with white border. Majority of vibrissae white, less than 50 mm long. Color of hair coat in summer ocherous-yellow or light ochreous, in winter same as in summer. Condylobasal length of skull less than 44.0 mm, length of tooth [upper] row less than 9 mm .... Daurian pika, **Ochotona daurica.**

14(11). Interorbital distance broad—more than 4.5 mm. Vibrissae long— more than 60 mm. Length of ear more than 25 mm. Frontal bones flat between orbits, without crest on edges .... Large-eared pika, **Ochotona macrotis.**

Little (Steppe) Pika
Ochotona (Lagotona) pusilla Pallas, 1768

1768. *Lepus pusillus* Pallas. Descriptio Leporids pusilli, Novi Commentarii Academiae Petropolitanae. T. XII p. 531-538. Volga area steppes, Orenburg region, Samara, river Buzulik (Ognev, 1940: 104).

1932. *Ochotona pusilla angustifrons* Argyropulo. Tr. Zool. In-ta Akad. Nauk SSSR, 1, p. 55. Kazakhstan, Karaganda region, Karkaralinsk district, "Dzhamcha River".

Diagnosis

Body less than 210 mm long. Condylobasal length of the skull less than 40 mm. Hiatus present between protuberances of intermaxillary bones; projecting in the incisor fosso. Color of hair coat in summer dark gray or brownish-gray (in winter somewhat lighter) with small longitudinal lighter streaks. Hairs on plantae long and dark. Vibrissae short—less than 50 mm long.

Description

The body length of adult animals 153-210 mm. The minimal body length of males is 158 mm; of females, 153 mm; the maximal body length of males is 198 mm; of females, 210 mm. Foot length 25-36 mm; in males 25-36 mm, in females 25-33 mm. Ear length 17-22 mm; in males 17-21 mm; in females 17-22 mm. Body weight of males in April to June—from 95 to 203 g, of females—from 105 to 277 g.

The color of hair coat in summer is dark gray or brownish-gray with longitudinal light ripples on the back, on the sides sometimes with an ochreous bloom. The belly is grayish-white. The upper lip and chin are ochreous-yellowish or ochreous-brown. The front side of the pinna and the outer margin of the inner surface are black-brown. The margin of the ear has a light border (Fig. 2). The upper side of the wrist and the foot are brownish-gray. Plantae have long hairs covering the digital pads, ochreous-brown or gray in color; plantae of the hind paws are usually darker than those of fore paws. In winter the hair coat is much longer (twice) and denser. Its color in general is the same as in summer but somewhat lighter. The vibrissae are up to 40 mm long.

The claws are relatively short and thin. By the beginning of winter the protuberances from the claws and plates on the claws of front legs are broadened at tip; then become rounded and in width surpass the claw itself (Fedo-Senko, 1974). Such an occurrence can be considered as

11

11

Fig. 2. Little pika *Ochotona pusilla* (Photo by V.V. Kucheruk).

12

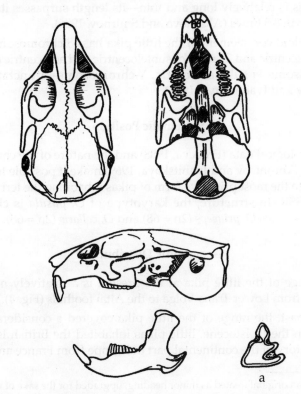

a

Fig. 3. Skull of the little pika *Ochotona pusilla*. *a*—third lower premolar tooth.

12

adaptation to digging the frozen ground. In spring these protuberances are worn out, decrease in size and later disappear.

The skull of the little pika is massive, relatively broad and bulged in the cranial section; the facial section is longer than the cranial (Fig. 3). The condylobasal length is 32.5-39 mm. The preorbital foramen is triangular in form and occupies half the length of the lateral surface of the rostrum. The nasal bones are weakly expanded in the anterior part. Foramina on the frontal bones are absent. The orbits are round. The zygomatic processes are parallel are to each other. The tympanic bullae are large. The incisor foramen is partitioned by the protrusions of the intermaxillary bones

14 incompletely. The mandible is quite massive and high; its incisor section is short. The articular process is weakly expanded in the anteroposterior direction and is less bent backward (cf. Fig. 3).

The third upper premolar tooth is triangular in form. The anterior section of the posterior segment is much narrower than its posterior section; the anterior segment is small, round or rhombic, with a rounded tip; the hiatus between the anterior and posterior segments is broad (cf. Fig. 3a).

Os penis is relatively long and thin—its length surpasses its thickness by more than 8.5 times (Aksenova and Smirnov, 1986).

The diploid complement of the little pika has 68 chromosomes: 8 pairs of submetacentric and 25 pairs of subtelocentric and acrocentric autosomes, X-chromosome is submetacentric, Y-chromosome—punctate element (Vorontsov and Ivanitskaya, 1973).

Systematic Position

Paleontological data (Erbaeva, 1988) and the nature of the chromosomal apparatus (Vorontsov and Ivanitskaya, 1969) make it possible to relate the little pika to the most primitive form of pikas inhabiting the territory of the former USSR. In structure, the karyotype of *O. pusilla* is close to the American species *O. princeps* (2n = 68) and *O. collaris* (2n = 68).

Geographic Distribution*

The range of the little pika at the present is a relatively narrow belt stretching from Lower Trans-Volga to the Altai foothills (Fig. 4).

In the past, the range of the little pika covered a considerably larger territory. In the Pleistocene, little pikas inhabited the British Islands and in the territory of the continental part of Europe from France and Belgium

---

*In Russian original treated as minor heading, upgraded for the sake of uniformity— General Editor.

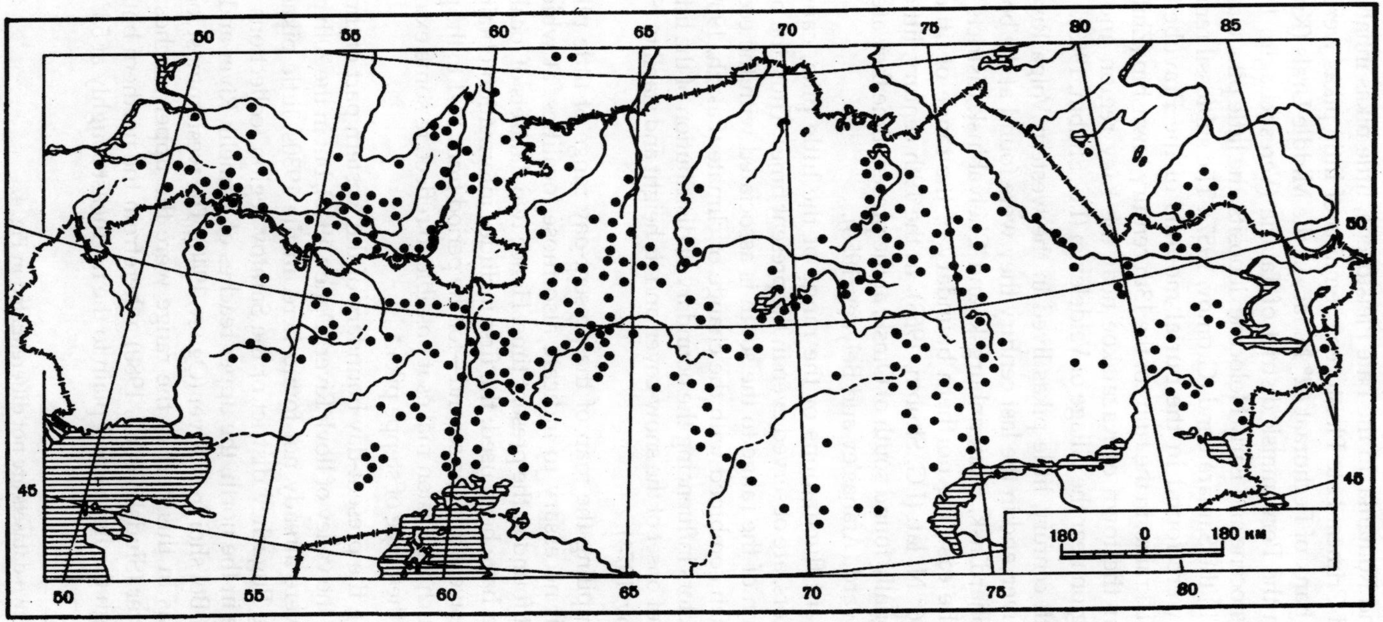

Fig. 4. Localities of the little pika.

to Poland and Romania. In the Late Pleistocene, little pikas inhabited the Ukraine and Crimea. Late Pleistocene remains of little pikas were found on the right bank of Pechoraat 62° N. lat., in the Middle Urals (Kuzmina, 1965) and in the Pervomaisk district of Tataria (Vereshchagin, 1953); in the Late Pleistocene and Early Holocene deposits, the little pika was found in the caves in the Samara bend (Gromov, 1957). The subfossil remains of little pikas were found in the burial mounds of the Pravoberezhnoi [Rightbank] Ukraine, in the ruins of the 13th century township (Zhitomirsk region), near the town of Kazatskoe near the Kiev region and in the Voronezh region near the village of Veidelevka (I.G. Shubin, 1963).

In the 18th century, little pikas lived in the western Volga River area near Ilovli River, and in the last century they were found at the border of the former Shardinsk, Ekaterinburgsk, and Chelyabinsk districts (55°30' N. lat.); in the east, the northern boundary of the range of little pikas reached to 56° N. lat. (I.G. Shubin, 1963). In the 19th century, little pikas were occasionally found south of Omsk; at the present time they are found much southward (Afanas'ev and Belyaev, 1953).

Insignificant fluctuations of the range of the little pika, caused by different factors, are observed even in the present times. Thus, the present-day expansion of the range to the south is associated with the economic activity of man combined with the change of climate (Pakizh, 1969). The important factors influencing the boundary of distribution of the little pika are: the time of onset of the snow-cover, and its height and stability (Shubin, 1965; Smirnov, 1974).

While compiling the map of the present-day range of little pikas we considered it necessary to indicate also those localities, at which this species is not found at the present time. Hence two datasets of localities of the little pika have been used: the first includes information obtained up to 1940, inclusively, and the second—for the period from 1941 to the present (cf. Fig. 4). Such a division makes it possible to reflect, to some extent, the dynamics of the range of steppe pika.*

In the past, the present-day boundary of the western part of the range passed along the valley of Ilovli River (Ognev, 1940), but in the 20th century little pikas were already not found here. In the 1930s, little pikas were caught in the Pugachev district of the Saratov region (collections of ZM MGU); while in the north, the upper reaches of Kamelik River and in the flood plain of Bol'shoi Irgiz River (Ognev, 1940). The westernmost localities for little pikas in this part of the range were the upper reaches of the Kamenka River (Sludskii et al., 1980), wherefrom the northern boundary passes through the Irgiz flood plain to the northeast roughly at 53° N. lat.

*Note: these two datasets are not differentiated in Fig. 4.

(Ognev, 1940). Further, the northern boundary of distribution of the little pika almost precisely coincides with the southern boundary of the meadow steppe, passing Orenburg region through the former Buguruslan district (collections of ZM MGU), the upper reaches of the Salmysh River at Obshchii Syrt* (Ognev, 1940) and descended in the south to the foothills of the southern Urals, passing along the mouth of the Salmysh River (V.A. Popov, 1960). Farther east of the northern part of Obshchii Syrt, little pikas were found very rarely (Ognev, 1940). In Bashkiria, little pikas were caught in 1938 at the mouth of the Bolshaya Uzala River (collections of ZIN).

In the southern Urals, the northern boundary of the distribution of the little pika passed through the upper reaches of the Belaya River through Abzelilovsk district of Bashkiria and the northern part of the Bashkirian preserve (collections of ZM MGU) along the montane forest parts of the Kraka Range (Kirikov, 1952) and crossed the eastern slope of Ural-tau following the boundary of forest-steppe and steppe zones to the upper reaches of the Ural River (Sludskii et al., 1980). In the 18th century, little pikas were numerous between the Urals and Uem.

At the present time, the northern boundary in this region passes along the line of the Dzabyk—Karagaisk pine forest. Apparently the town of Uiskoe should be considered as the northernmost locality of the little pika here (Pavlinin and Shvarts, 1957). From Uiskoe the northern boundary passes along the foothills of the southern Trans-Urals in the southeast through Varna station and the village of Denisovka to the west of Kustanai region (Pavlinin and Shvarts, 1957; Sludskii et al., 1980). From Denisovka (the present Ordzhonikidze) the boundary of distribution of the little pika passes to the northeast passing roughly 50 km west of Kustanai and descends in the southeast passing 8 km northeast of the village of Shalakterek—the northernmost locality of the little pika in the north of the Turgai depression. Eastward, roughly on the same latitude, in 1963-64, little pikas were found in the flood plain of the Ishim River (collections of Geography Faculty of MGU). Along the flood plain of the Ishim River, the boundary descends in the south and passes roughly 30 km west of the confluence of Ishim and Tersakkon rivers. Eastward, the northern boundary passes along the Kokpekta River (Shubin, 1962), through the town of Shortanda and vicinity of Turgai (collections of BIN). In the northeast of Tselinograd and southeast of Kokchetav regions, the northern boundary of the little pika ascends along the Sileti River, not reaching the southern tip of Lake Siletitengiz and farther in the southeast to the Irtysh flood-plain, almost coinciding with the southern boundary of the herb-turf-cereal meadow habitat. Along the right bank of Irtysh River it turns south-east,

*Common watershed upland—General Editor.

where little pikas were recorded in the vicinity of the villages of Lebyazh'e, Agachi, Krivinki, Bolshoi Akazhar and Semiyarskoe (Shubin, 1965; Pakizh, 1969; Sludskii et al., 1980). On the left bank of the Irtysh River, little pikas were found on talus and in the foothill steppes of Semeitau, 50 km south of Semipalatinsk and in the southern foothills of the Kalbinsk Altai (Kuznetsov, 1932).

Through the steppes of the Kalbinsk Altai, the boundary of distribution of the little pika ascends in the north to the western foothills of the Altai, passes through the village of the Steklyanka (Steklyanskii) in Sempalatinsk region and the town of Shemonaikha in eastern Kazakhstan (Afanas'ev and Belyaev, 1953; Pakizh, 1969) and leaves the Altai territory. There, little pikas were found near Zmeinogorsk, in Krasnoshchekovsk and Charyshsk districts (Kuznetsov, 1932; Bazhanov, 1955; collections of ZM MGU). The extreme northeastern place of habitation of little pikas is apparently the vicinity of the village of Novoshipunovo in Krasno-shchekovsk district (collections of the Geography Faculty of MGU).

Later the boundary of distribution of the little pika descends south and through the southern foothills of the Kalbinsk Altai and passes along the Zaisan basin along Bukon'sk sands to the southeast in the foothills of the Saur Range. The extreme eastern locality of the little pika lies in the vicinity of the village of Maikapchagai in eastern Kazakhstan (Sludskii et al., 1980). Calls of little pikas were recorded on the southern slope of the Saur Range near the village of Chagan-Obo quite high in the mountains (Kuznetsov, 1932).

Along the southwestern foothills of the Saur Range, the distributional boundary of the little pika proceeds to the foothills of the western Tarbagatai, where little pikas were reported only from the town of Podgorne (Kuznetsov, 1932; Bazhanov, 1955). The extreme southeastern localities of the little pika lie in the Semipalatinsk region in the vicinity of the village of Bakhta (Kuznetsov 1948). Through the village of Urdzhar the southern boundary passes to the northwest up to the village of Ayaguz and farther— to the foothills of the Chingiztau Range in the region of Tersapryk and Zhorga rivers and slopes of Koksengir low-hills (Kuznetsov, 1932; Bibikov and Stogov, 1963; Sludskii et al., 1980; collections of the Geography Faculty of MGU). Bending around the eastern Kazakh plains, the boundary of distribution of the little pika passes along the gentle slopes of the Kyzyl-Rai and Mai-Tas mountains and descends to Lake Balkhash (Afanes'ev and Varagushin, 1939; collections of ZM MGU and ZIN). On reaching Kounrad the southern boundary passes to the west toward the northeastern edges of Betpak-Dala desert (Afanas'ev and Varagushin, 1939; Ognev, 1940). The easternmost locality of the little pikas in the northern part of Betpak-Dala is the area of the village of Mointa, and the southernmost, the Kogashik and Chulakturanga tracts (Ismagilov, 1961). Following farther along the western part of Betpak-Dala, the southern boundary passes along the Dzhetykonursk sands and along the Sarisu valley. The westernmost locality of little pikas within limits of Betpak-Dala lies in the area of the confluence of the Sarisu and Karakengir rivers (Ismagilov, 1961).

From the western tip of Betpak-Dala, the southern distributional boundary of the little pika passes to the northwest and approaches 30 km south of the village of Karsakpai, Ulutau mountains, farther—to the north bending around the hilly plains of the true deserts of the Turan lowlands and the northwestern Aral Sea area (Kuznetsov, 1948; Bazhanov, 1955; Sludskii et al., 1980). Along the Uly-Dzhilanchik River and its tributaries, the boundary ascends to Amangel'dyisk district of Turgai region, approaching 19 km south of the settlement of Urpek (Bazhanov, 1955; collections of the Geography Faculty of MGU). Further the southern boundary passes through the Akkum sands and bending around from south of the Turgai lowland descends along the flood plain of Turgai River near the village of Taup (roughly 10 km south of the confluence of Irgiz and Turgai river) and farther along the flat-hill plain passes to the south (Garbuzov and Shilov, 1963; collections of the Geography Faculty of MGU). In the northern Aral Sea area the easternmost locality of little pikas lies 55 km northeast of Aralsk (Garbuzov and Shilov, 1963). Along the low-knolly sands of the northern Aral Sea area, the southern boundary passes to the west crossing through the weakly fixed sands of Bol'shie [Greater] and Malye [Lesser] Barsuki. Along the basins and slopes of ravines in the Bol'shie Barsuki, little pikas penetrate Ushtagan outlier, Naiza-Kisken tract, Akkuma and the eastern edge of sands in the area of the Altapan spring (Garbuzov and Shilov, 1963). The southernmost point of occurrence of little pikas lies at 46°20' N. lat. (Sludskii et al., 1980).

Bending around the Aral Sea from the north, the southern boundary of distribution of the little pika passes farther to the west to the southern tip of the Chushkakul mountains and ascending in the north along the valley of the Chagan River meets the western boundary. Here little pikas were found at the Akbulak spring, in the environs of Karabulak and in the valley of Mailisai River (Garbuzov and Shilov, 1963; Sludskii et al., 1980). Later through the Chuskel'sk mountains the boundary proceeds to the Mugodzhary mountain. There steppe pikas were reported on the water divide plateau between the Srednyaya [Middle] Emba and its left tributary Azhaksa, in the vicinity of Kotr-Tas, Berchogur and Mugodzharskaya station, and also in the upper reaches of Kaul'dzhur River (Garbuzov and Shilov, 1963; Sludskii et al., 1980). From here the boundary turns back to the northwest, and approaching 30-50 km north of Emba station, passes along the dry turfy-grassy steppes west of the Aktyubinsk region (Sludskii et al., 1980). From Kandagach (presently Oktyabr'sk) station, the boundary runs to the west passing through the upper reaches of the Temir River and along the right bank area of the Uil River (Dubrovskii, 1963; Sludskii et al., 1980). Along the Taisaigam sands, little pikas enter the Gur'evsk region in the north, from where the boundary of distribution turns to the north and passes through the meander of the Uil River south of its confluence with Kiyl River (Dubrovskii, 1963). Later through the Koguzyakkum low-knolly sands the boundary passes to the west of Uralsk (Sludskii et al., 1980). In the 1930's, little pikas were caught 20 km west of the village of Kushum and in the vicinity of the villages of Chapov Kamenskii and Teplovsk districts of Uralsk region (collections of ZM MGU).

17

The internal structure of the range of the little pika is quite complex (cf. Fig. 4), which is explained mainly by the sporadic nature of distribution of this species.

Outside of Russia and Kazakhstan, reliable finds of the little pika are absent. However, in the opinion of some researchers (Sludskii et al., 1980), little pikas enter Xinjiang from the Saur Range.

Geographic Variation

Within limits of the present-day range of the little pika it is possible to separate two subspecies.

1. *O. pusilla pusilla* Pallas, 1768. European little pika.

Small form—body length 153-182 mm, condylobasal length of skull 32.5-38.3 mm, width of tympanic bulla 7.5-10.0 mm, zygomatic width 18.0-20.1 mm (Table 1). The third upper premolar tooth of triangular form.

In summer pelage brown tones predominate with distinct dark brown or black tips of hairs.

Distributed to the west of the Ural River.

2. *O. pusilla angustifrons* Argyropulo, 1932. Siberian little pika.

Large form—body length 182-210 mm, condylobasal length of skull 36.1-39.0 mm, width of tympanic bulla 8.5-11.0 mm, zygomatic width 19.0-21.4 mm. P_3 trapezoid in form.

Pelage color is somewhat lighter than in the type subspecies. Pale gray ripple is more distinct, which is masked on the back by black tips of hairs.

Distributed to the east of the Ural River.

Biology

Habitats. Little pikas inhabit mainly the semidesert zone and almost do not enter the steppe and desert zones. Their habitats are determined mainly by the presence of shrubby vegetation. Notwithstanding the fact that these pikas are dwellers of mesophilic habitats and live sometimes alongside water bodies (Bazhanov, 1955; Ismagilov, 1961), everywhere they avoid settling on very wet soils (Sludskii et al., 1980). Extremely dry conditions are also not favorable for their living. Hence in deserts little pikas enter only along the azonal elements of the terrain, colonizing willow thickets and birch at the bottom of basins in undulating sands. Since in winter little pikas are active under the snow, fairly deep and stable snow cover is necessary for them, which also affects the nature of distribution of this species.

Table 1.1. Dimensions of skull (in mm) of adult little pikas *O. pusilla pusilla* and *O. p. angustifrons*

	O. p. pusilla			O. p. angustifrons		
	min.	M	max.	min.	M	max.
Condylobasal length	32.5	35.6	38.3	36.1	37.78	39.0
Length of rostrum	13.0	14.3	15.3	14.0	14.97	15.9
Length of upper tooth row	7.0	7.7	8.2	7.7	8.17	8.9
Length of orbit	10.9	11.5	12.1	11.1	11.8	12.5
Length of tympanic bulla	11.0	11.77	13.0	12.1	12.76	13.8
Width of tympabic bulla	7.5	9.0	10.0	8.5	9.99	13.8
Interorbital width	3.8	4.31	5.0	3.9	4.33	5.0
Zygomatic width	18.0	19.19	20.1	19.0	20.21	21.4
Maximum width of cranium	15.0	16.41	18.0	15.9	16.92	18.2
Width in region of tympanic bulla	17.8	19.15	20.1	18.8	20.46	21.6
Postorbital width	12.0	12.83	13.6	11.8	12.92	14.0
Width of upper part of base of rostrum	5.8	6.29	7.2	6.0	6.58	7.1
Occipital height	10.0	10.77	11.8	10.3	11.12	12.0
Height in frontal region	5.5	6.3	7.1	5.9	6.54	7.1
Height of rostrum	13.0	13.77	14.6	12.0	14.25	15.1

Some researchers explain the extinction of little pikas over a considerable part of their range by the destruction of steppe shrubs by man (Kirikov, 1952; Dubrovskii, 1963). However, such an explanation should hardly be considered correct, since the development of Kazakhstan steppes by man has not led to the extinction of little pikas in this territory, while in the Ukraine, within limits of the earlier range of this species, even now places are conserved with shrubby vegetation (I.G. Shubin, 1963). Probably, shrinking of the range of the little pika in its western part occurred as a result of the increase of general humidity in this region (I.G. Shubin, 1963).

At the southern tip of the Urals, the favorite places of habitation of little pikas are thickets of Russian almond and ground cherry. In montane-forest terrain little pikas are rarely found. In the Yuzhnyi Kraka mountain massif, pikas live in sparse forest on slopes of southern exposures, which is a shrubby-stony steppe with occasional pines. Little pikas are common in shrubby-stony steppe of the low-hilly area on the leftbank of the Sakmara River and in the Sakmara-Ural interfluve. Pikas are found in the montane forest-steppes of the Kyrkta and Bolshoi Irendyk ranges, and also in the broad-leaved upland forest steppe of Shaitan-Tau. Little pikas often colonize forest edges of chopped oak forests, and if there is shrubby underbrush of steppe-cherry and Russian almond, they enter deep inside the chopped forests (Kirikov, 1952).

Little pikas are most numerous in Aktyubinsk dry steppes with thickets of spirea. In the middle and lower part of the Tabysai valley, they usually live in willow thickets and tamarisk, and in the Mugodzhary—in spirea groves. South of Mugodzhary little pikas are quite common along the edges of the Bolshie Barsuki sands, where they live in basins with reeds and willows (Garbuzov and Shilov, 1963).

In the west of Kazakhstan in the Ilek-Irgiz interfluve, little pikas most often colonize dense thickets of steppe shrubs on slopes of upper flood plain terraces and along the drainage basins. Less often they are found in willow groves along river flood plains and in shrubby steppe of water divides. In the northern steppes they remain on the edges of dry-bottom forests. In areas with low rounded isolated hills, pikas are found not only in the drainage trough, but sometimes in rocks at the peaks of mud volcanos. East of the Irgiz River, where river valleys and drainage troughs are rare, little pikas live in the shrubby steppe. There, they are more frequent than in other regions, colonize cemeteries, meadows and in steppe under isolated high shrubs of wormwood [*Artemisia procera*—Translator] (Dubrovskii, 1963). Northeast of the Turgai region near Lake Shoindykol', little pikas are found in thickets of wild almond along the flood plains of drying steppe rivers (Sludskii et al., 1980).

In the Kazakh upland (in the vicinity of Kyzyl-Rai, Karkaralinsk and in the mountains near Akotogai), little pikas colonize predominantly dry areas in river valleys. In the high-poulation years, pikas colonize all habitats suitable for life: thickets of Tatarian honeysuckle, spirea, rose, pea shrub and willow. They are found even on the edges of aspen-birch chopped forests, and also among wormwoods and cereals 1-2 km away from shrubs. In low-population years, pikas survive mainly in thickets of Tatarian honeysuckle (Sludskii et al., 1980).

In the northeastern part of central Kazakhstan, little pikas prefer to colonize mountain slopes and dry valleys with soft ground and shrubs. In plains between mud domes, pikas are very rare. Here they dig burrows under solitary shrubs of spirea or among thinned-out scrub of this shrub. In the central part of the Karaganda region, little pikas are found in the rubbly areas on the slopes of mud domes and in valleys between mud domes. Sometimes pikas inhabit alfalfa fields. But pikas do not live in sheep's fescue-feather grass flat steppe, on highly saline soil and unirrigated fields (Beme, 1952).

In the Irtysh flood plain, steppe pikas more often choose to inhabit steppe areas on secondary terraces, ridges, bed embankments with thickets of rose, shengil*, buckthorn, and spirea. Less often they colonize thickets of pea shrub and willow, and on coastal slopes pikas live in shrub thickets and weeds. Sometimes pikas were caught in areas with sandy and sandy-loam soils in thickets of wormwood, wild rye and other plants (Sludskii et al., 1980). In the flood plains of the Ulenta and Shiderta rivers, little pikas are found in willow groves along the river-beds, and in short-grass meadows— among weeds and in shrub thickets (Pakizh, 1969).

In the Zaisan basin (Bukon'sk sands), little pikas are most numerous on sandy barhans and juniper thickets alternating with thickets of spirea. In 1967, in these places, there was a high population of little pikas; all the same, they were not found in the river flood plains and in depressions with meadow herbage—probably because of the high humidity of these places (Sludskii et al., 1980).

In the zone of sympatry with Pallas's pika, the favorite habitats for which are the rocks and piles of stones, little pikas avoid colonizing these habitats. Thus, in the environs of Mt. Bektauta in the northern Lake Balkhash area, both species live together, but little pikas settle only under thickets of spirea and on the periphery of birch clearings, and Pallas's pikas—under isolated stones or in rock recesses.

Population. In northeast Kazakhstan, in relation to landscape ecology it is possible to identify three main types of colonies of little pikas: shrubby steppe, mountain-steppe and flood plain (Pakizh, 1969). Colonies of the first type in the steppe rounded hill areas in the Ulenta-Shiderta-Irtysh interfluves are situated sporadically, sometimes tens of kilometers from each other, depending on the differences in the shrub thickets or aspen forests. There, the population of little pikas is not high and its density reaches 10 individuals per hectare. The mountain-steppe type of habitations are found in montane areas with rounded hills and along slopes of the Shiderta and Ulenta rivers. There, little pikas mainly colonize intermontane steppe, but

---

*Local name of shrub—Translator.

once again in thickets of rose, buckthorn and barberry. Typical colonies are in individual pancake rocks with rose shrubs around. The population of pikas is higher here than in the steppe, and the density may reach 60 individuals per hectare. On rocky riverbanks, little pikas live among stony blocks, where bushes of willow, rose, spirea, buckthorn, etc., grow in small groups. Where shrubs with herbaceous plants form pyramidal clumps, the density of little pikas in the fall months may reach 70-80 individuals per hectare. The density of little pikas in the Irtysh flood plain constitutes, on average, 5-9 individuals per hectare in the south and 0.05-0.1 per hectare in the north (Pakizh, 1969).

In the northern Lake Balkhash area on stony mountain slopes and in dry valleys with soft ground and shrub thickets, the population of little pikas is usually high—the density of occupied burrows here may reach 32 per hectare. In humid valleys, little pikas live singly or, as a rule, under shrubs of rose, spirea, honeysuckle—here, on the average, there are 12-17 occupied burrows per ha. In the valleys of streams and rivulets, little pikas colonize dry places in thickets of rose; here, there may be 13-24 occupied burrows per ha (Andrushko, 1952).

In western Kazakhstan (Aktyubinsk region) in pea shrub thickets on slopes of upper flood plain terraces, the population density of little pikas is the highest for this region—40-50 individuals per hectare, and in river valleys, as in the east of the range, pikas are found far less often and, on the average, it is 1 individual per 3 ha (Dubrovskii, 1963).

Shelters. The burrows of little pikas are temporary and permanent. Temporary burrows, as a rule, are dug by adult females under solitary thickets. Here they hide in the event of danger. Temporary burrows have one or two entrances. The length of their tunnels is short (not more than 1-2 m), and they do not have brood chambers. The number of temporary burrows increases at the end of summer, when they are actively dug by young pikas not participating in reproduction (Sludskii et al., 1980).

The permanent or brood burrows are built by adult animals and the majority of reproducing females. In the west and northwest of the range, these burrows are situated at a depth of 20-50 cm (sometimes to 70-100 cm) and have from 1 to 5 entrances (Kirikov, 1952; Dubrovskii, 1963). In the southwestern part of the central Kazakhstan rounded hill area, brood-burrows of little pikas are more branched, and the number of entrances to them reaches 14 (Smirnov, 1974). The diameter of entrance is 6-7 cm. In the permanent burrow, one, less often two, brood chambers are made, the length of which is usually 16-18 cm and with a width of 10-12 cm. Brood chambers may be made in the broadened part of the tunnel or at the end of the side burrows (Fig. 5). The bottom of brood chambers is laid with leaves of sheep's fescue, feather grass or other grasses. The total length of

19

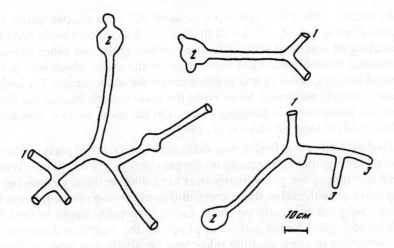

Fig. 5. Scheme of burrows (plan) of little pikas in the north of Aktyubinsk
region. 1—entrance; 2—brood chambers; 3—side burrows
(from Dubrovskii, 1963).

tunnels of a permanent burrow reaches 2-3 m and covers an area of 1.0-
1.5 m². In the permanent and temporary burrows, usually there are 2-3 dead
ends (Dubrovskii, 1963). In the western part of the range, little pikas
construct lavatories in the burrows, but in the east of the range lavatories
are situated outside the burrow. Such a difference is linked with the fact
that, in the western part of the range, little pikas spend a large part of their
time in burrows (Kirikov, 1952).

The permanent brood burrows of little pikas are well masked, near them
the animals do not gnaw the grass and do not make earth mounds. In stony
habitats, for example near Lake Shondykol' and on Mt. Kokshetau, as also
in the southern Urals, little pikas construct nests in rock recesses and rock
piles (Kirikov, 1952).

In Kazakhstan and the southern Urals, little pikas often use for residence
abandoned burrows of hamsters, water voles, Pallas's pikas and mole
voles, broadening the narrow tunnels of the earlier hosts. Occasionally,
little pikas find shelters for rest and sleep among shrubs under the previous
year's shreds of grasses or use dry grass to prepare a nest under ledges of
large rocks. In the trans-Urals, pikas can build shelters in stacks where
they make a nest chamber (Kirikov, 1952, 1955; Dubrovskii, 1959). The
majority of young animals do not dig burrows, and it happens that in some
areas more animals are caught than the occupied burrows present there.

21 On soft chernozems and loessic sandy soils, little pikas dig burrows at
places where the ground water is not close to the surface (Sludskii et al.,
1980).

In winter, little pikas construct nests of different shapes under the snow—from spherical with a wall thickness of 2-4 cm to a simple bedding consisting of leaves of sheep's fescue, feather grass and other grasses. Numerous tunnels issue from the nest under the snow, which lead to the place of foraging, lavatory and to entrances on the snow surface. The length of such tunnels may reach 50 m. From the main tunnels branch out short passages terminating in foraging places. On the snow surface pass short paths 30-60 cm long (Sludskii et al., 1980).

Feeding. Little pikas feed almost exclusively on the aerial parts of plants. In early spring, they feed mainly on the previous year's dry plants. When green sprouts appear, pikas change over to feeding on them. Grasses begin vegetative growth earlier than others, and in the course of one-two weeks pikas feed predominantly on them. Later, little pikas begin to feed on wormwood, giant fennel and other plants. In the Aktyubinsk region, in the beginning of summer, little pikas feed on alfalfa and Russian thistle (Dubrovskii, 1959).

In little pikas the storage of food for winter begins at different times in different parts of the range. In the west of the range, they begin to collect grass in piles only in August (Kirikov, 1952; Dubrovskii, 1959). In the forest-steppe zone at the southern tip of the Urals, the storage process starts from the middle of August to the middle of October (Kirikov, 1952). In central Kazakhstan, stocks of little pikas begin to be spotted in July to August (Bazhanov, 1955). In northeast Kazakhstan, pikas begin to stock food in the second half of June (Pakizh, 1969). In the vicinity of Karaganda and the railroad station of Bosag, the beginning of stocking food is observed in the first half of July (Beme, 1952; Sludskii et al., 1980). In the Irtysh flood plain in 1960, in mid-July piles made by pikas were absent, but in the following year at this very place these piles began to appear from June 1 and by the end of June were found everywhere. The earlier start of food storage in 1961 was associated with the dry and earlier summer as compared to 1960 (Sludskii, et al., 1980).

Little pikas living in dry places and in uplands begin storage 30-40 days earlier than those living in more humid habitats. In some years, for example, in unusually dry 1974, little pikas did not store food at all (Sludskii et al., 1980).

The young non-reproducing pikas are the first to start storing food, while others begin to participate in the collection of plants after the breeding season. Young dispersing pikas, that could not still find a nest for themselves, did not store food at all (Sludskii et al., 1980).

In little pikas the mass of haypiles varies from 300 g to 7 kg, and the height usually does not exceed 50 cm (Sludskii et al., 1980; Smirnov, 1974). In the south of the range, stacks made by little pikas, on average, are somewhat larger than in the north, although in northeast Kazakhstan a haypile weighing 10.5 kg was found (Pakizh, 1969).

Haypiles are most often made on the reclining trunks of small trees or shrubs, on bed of twigs and on dry remnants of previous year's grass. In mountains, the haypiles are sometimes made under rock ledges and in recesses between rocks (Pakizh, 1969).

Little pikas usually collect plants not farther than 20-30 m from the haypile. The haypiles are sometimes located above the entrances of burrows, but more often they are built in shrubs at a distance of 11-15 m from the entrance of the burrow (Dubrovskii, 1959). The data of L.B. Beme (1952) on the distant routes of pikas for plants for haypiles are not confirmed by other researchers (Sludskii et al., 1980). These animals usually carry plants not longer than 20-30 cm and place them atop the haypile. The new layer of grass is piled only when the earlier layer becomes dry. The dry hay preserves well and decays little. During rains water flows out rapidly from the stack. According to the observations of Beme (1952), little pikas begin to use hay from the haypile from its inner and lower part where the dried grass has good qualities.

The species composition of plants of haypiles of little pikas is extremely diverse but, in general, herbs predominate in the reserves. In western Kazakhstan, the stored vegetation of little pikas is comprised of 56 species of plants (Dubrovskii, 1959). In all haypiles one (less often two) species of plants predominates, which constitutes more than 50% (sometimes to 90%) the volume of the haypile; 10-40% of the haypile is constituted by 2-4 plant species dominant in the given locality; the remaining plants occur as individual plants in the haypile. Throughout the range of the little pika, grasses are rarely found in their haypiles. In the vicinity of Mt. Ulutau, in six investigated haypiles, up to 80-90% by volume, consisted of spirea [Spirea hypericifolia], sophora [Sophora alopecuroides], common licorice and oriental dodartia. In the vicinity of Lake Shondykol' in Turgai region, in 30 investigated haypiles, up to 70-90% consisted of Russian almond, which was the dominant species here (Sludskii et al., 1980). The plants more frequent than others in the stocks of little pikas are spirea, wormwood, Russian almond, and giant fennel. Possibly, in little pikas there exists individual variation in the preference for one or the other species of plant. Thus in four neighboring colonies of little pikas in the Aktybinsk steppes, different plants predominated in different haypiles (Dubrovskii, 1959). The list of plants stored by little pikas in different parts of the range is given in the works of A.M. Andrushko (1952), S.V. Kirikov (1952), V.S. Bazhanov (1955), Yu.A. Dubrovskii (1959) and A.A. Sludskii et al. (1980).

In winter, little pikas feed on plants from haypiles. Moreover, they feed on branches of shrubs and grass under the snow. The animals make long tunnels under the snow, which stretch 30-40 m. Usually there are 2-3 such tunnels connected to each other. Side tunnels may branch out from the main tunnel terminating in blind ends—through these tunnels these animals reach the places of their foraging. If herbaceous plants are not sufficient for pikas, they feed on the bark or twigs of spirea, Russian almond, rose, willow, European aspen (*Populus*) and other trees and shrubs (Kirikov, 1952).

Little pikas eat 120-170 g of grass or 20-25 g of hay each day (Kirikov, 1952). The fore gut of little pikas is longer than the hind gut (64%), which confirms the less manifest granivory of this species, as compared to other pikas, for example, Pallas's pika (Ismagilov, 1961).

Little pikas obtain the moisture necessary for them from green food. At the peak of summer, when plants die out, little pikas possibly migrate to more favorable places.

Activity and behavior. Little pikas can be active at night as well as during the day. The peaks of activity change according to the season. At the end of spring and in summer, calls of little pikas can be heard any time of the day, but they call more actively at dusk and in the first half of night (Afanas'ev and Varagushin, 1939). On hot days pikas do not come out of their shelters for a long time. In fall with the onset of cold, the activity of pikas drops considerably; their calls are produced only at night. When the dew falls and during rains, the activity of pikas likewise falls; however in warm rainy weather little pikas can be fully active (Sludskii et al., 1980).

During the period of reproduction the males are most active and relocate to a considerable distance, while females remain in the same place. There are strong fights between males in this period, and many scars can be seen on their body (Shubin, 1965).

During the period of nursing young ones with milk, the female comes close to them only when they begin feeding. Feeding continues for 5-10 min and is repeated after 3-5 hr (according to observations in vivaria) (Shubin, 1975). The female does not warm even the newborns, but the temperature of the nest is regulated by changing the thickness of the upper layer of grass over the nest. In cold weather the female increases the thickness of the nest cover to 5-6 cm, and the young ones huddle together. With the rise of the air temperature, the female removes some of the grass, as a result of which the temperature in the nest is maintained relatively constant at 30-35°C.

Intensive vocal signalling is characteristic of little pikas. Loud cries of little pikas with a metallic tone resemble the sound combination "chok-chok-chok ..." or "chiu-chiu-chiu," which is repeated several times in a row at different intervals. These cries are produced by adults as well as the current year's brood. In the period of reproduction, vocal signalling of

27

little pikas becomes more intensive; it creates a mating song which, unlike usual calls, is more sonorous, sharp and polyphonic. Most mating songs can be heard at sunset and in the early part of the night before the dewfall, but at the peak of reproduction little pikas sing also during the day. Reproducing females sing very rarely; their songs differ somewhat from the songs of males. Pregnant females do not produce calls. During singing, pikas sit on their hind legs and the fore legs are stretched forward and do not touch the ground. Pikas often squeek, sticking out their neck from the burrow. Sounds produced by little pikas can be heard at distances up to 2-3 km. As the enemy approaches little pikas hide silently and do not issue a danger call. At the end of summer and in places with low population, the pikas produce sounds more often than in areas with high density (Smirnov, 1976, 1988; Sludskii et al., 1980).

Area of habitation. Little pikas lead a settled mode of life and their seasonal translocation does not exceed some tens of meters. Only in years of a rise in population can they migrate 3-4 km. Relatively young animals can settle far from the parental colony. Dispersing young pikas were found in the open steppe at distance of 8-10 km from the nearest colony.

Individual territories of female little pikas occupy an area of 80-100 m², but females usually do not leave the nest farther than 3-4 m. In habitats with sparse shrub cover, the area of an individual territory is larger and reaches 200-300 m². During the reproductive period females live singly, occupying the most favorable area for them. They retain the central part of their area and are aggressive in relation to others. When the reproductive period terminates, aggressiveness in the behavior of little pikas disappears—they use the same paths and collectively store hay. The area of the territory of a male is more than that of a female and may cover 3-4 territories of females, which are often situated hundreds of meters from each other. In continuous or winding thickets, the individual territories of little pikas may be overlapping (Shubin, 1965; Sludskii et al., 1980).

In the area of habitation of little pikas, there is a network of beaten paths 4-5 cm wide and 3-4 cm deep. These paths lead to feeding areas, haypiles and temporary burrows. In the areas of little pikas stamped (sometimes 2-3 cm deep) feeding areas of size from 8 × 10 to 15 × 17 cm are found. One or, less often, two paths leads to such areas. The main paths have a length of 20-30 m, sometimes 40-60 m. Their lateral branches have a length of 2-3 to 6-7 m. In compact colonies of little pikas, paths criss-cross each other, often stretching several hundreds of meters (Sludskii et al., 1980).

Reproduction. In Kazakhstan little pikas begin to reproduce in the first-second decade* of April, and the start of reproduction in them is very

*Ten-day period—Translator.

28

synchronous. Moreover, the first wave of reproduction smoothly blends with the second, as a result of which in little pikas during reproduction there is a persistently high percentage of gestating females, with the exception of short periods when there are large-scale births (Fig. 6). In the southern Urals the start of reproduction of little pikas also occurs in April (Kirikov, 1952).

Duration of the reproductive period is different in different years and depends apparently on the weather conditions. For example, in 1959 at Bosaga station, spermatogenesis in males already stopped in the second half of June and gestating females were not found. In the following year, at the same place, gestating females were caught till the end of July (Fig. 6). In the western part of the range, gestating females were caught till August (Kirikov, 1952).

Data are available to show that little pikas produce one brood in a year (Pakizh, 1969). But over the large part of the range during the reproductive period, little pikas succeed most often to produce three broods each. In individual years a fourth or even a fifth brood is possible. Thus in 1960 near the town of Semiyarskoe, even in the second decade of July, reproducing females were found; apparently, in this year many adult females carried not less than four broods (Shubin, 1965). Females of the first, second and sometimes even third brood of current year participate in reproduction. On the whole, the intensity of reproduction of little pikas (number of broods and the extent of participation of little pikas in reproduction) varies considerably from year to year. The number of the

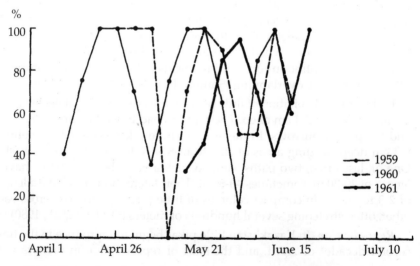

Fig. 6. Dynamics of reproduction (ratio of pregnant females in %) of little pikas in north Kazakhstan in 1959-1961 (from Shubin, 1965).

current year's individuals participating in reproduction positively correlates with the duration of reproductive period of adult individuals. (Shubin, 1965).

The number of embryos in a female varies from 3 to 13 (Table 1.2). The maximum number of embryos (average 10) is found in the second and, in case of more prolonged reproductive period, in the third brood. For the reproductive period, one female can bear up to 27-36 young (Sludskii et al., 1980).

Young females of a current year's brood become sexually mature at the age of 4-5 weeks and may bear a little before the onset of fall. In the same year of birth, sometimes even females from the second brood also reproduce. In 1959, in central Kazakhstan, first growing females to become pregnant began to be found on July 5; in 1960 on July 8; in 1961, on May 21. Surviving (or resident) females could bear one or even two broods. The number of embryos in young females during a second pregnancy is somewhat more than during the first (Shubin, 1965).

Surviving (or resident) males in their first summer do not participate in reproduction. They become sexually mature by the following spring.

Embryo mortality is insignificant in little pikas. The share of females with resorbed embryos does not surpass 19.7% (usually 5-10%). Most often, resorbed embryos are found in young and emaciated females. Of the total number of embryos, 1.7% are resorbed (Shubin, 1965; Sludskii et al., 1980).

Growth and development. Gestation in little pikas lasts probably 22-24 days (Shubin, 1965; Sludskii et al., 1980). The young of little pikas are born naked, blind, with undeveloped (0.5 mm) nails and cutting lower incisors; they weigh 6-7 g. On the second day after birth dense soft hairs, 0.5 mm long on the back and sides appear, they are black, but whitish on the belly. In 3-day olds, the length of lower incisors reaches 1 mm and the upper ones begin to cut. By the fourth day the hair coat becomes lighter, acquiring a gray tone; by this time the ear passage opens. At the age of five days, young ones weigh 12-15 g. On the eighth day their eyes open and molar teeth are cut, and the length of hairs reaches 4-5 mm (Shubin, 1965).

In females lactation continues for 20-21 days. In captivity, young ones fed on mother's milk for 23-24 days. However, from the age of 20 days, little pikas begin to feed on grass. After a changeover of young pikas to feeding on green food, their weight increases more intensively. In the first 10 days of feeding the weight of juveniles increases by 12 g, and by the 20th day—by 25 g; that is, by this age the weight of little pikas, on the average, reaches 44 g (Shubin, 1965).

Three to four days after the large-scale appearance of juveniles on the land surface, they begin their active dispersal. At this time young ones can be found at a distance of 100 m from the territory of a female. Juvenile

Table 1.2. Number of embryos in little pikas in Kazakhstan (from Sludskii et al., 1980)

Pregnancy	Number of females examined	Number of females with given number of embryos											Average number of embryos
		3	4	5	6	7	8	9	10	11	12	13	
Coast of Lake Shondykol', 1958													
First	9	–	–	–	–	1	4	2	–	2	–	–	8.8
Vicinity of Bosaga station, 1959													
First	22	–	–	–	–	3	10	7	1	–	1	–	8.5
Second	46	–	–	1	1	1	1	1	15	16	5	4	10.7
Third	27	–	2	4	2	4	9	1	5	–	–	–	7.4
Vicinity of Bosaga station, 1960													
First	14	–	–	–	–	6	1	7	–	–	–	–	8.1
Second	29	–	–	–	–	–	2	7	11	5	2	2	10.1
Third	7	–	–	–	–	–	1	–	2	4	–	–	10.5
Vicinity of town of Semiyarskoe, 1961													
Third	54	1	–	1	2	6	9	10	15	5	4	1	9.1
Fourth	27	1	2	2	4	3	8	3	4	–	–	–	7.3

mortality in little pikas is very high. Roughly a fourth of the newborns survive to the age of one month (Sludskii et al. 1980).

26   *Molt.* In little pikas spring molt begins in the second half of May. However, in some animals, especially in lactating females and old individuals, it begins 15-20 days later and extends to the middle of July. Molting starts from the back, sides and the head. The last to molt is the hind part of the body (Shubin, 1965, 1972; Sludskii et al., 1980).

Fall molt starts immediately on termination of the spring molt—in the end of July—and continues to the middle of October. The middle part of the back molts first, followed by the sides and nape, and later the sacrum and head; the last to molt are the neck, belly and legs (Shubin, 1972; Sludskii et al., 1980). Winter hairs initially do not differ in length from summer hair, but later additional long hairs grow, and the first winter hairs that appear also become longer.

Juvenile molting occurs in current year's brood at the age of 20-25 days and coincides with their large-scale dispersal; the change of juvenile fur begins from the lower part of sides and the nape, followed by the back, head, sacrum, middle part of the belly and the limbs. In many young animals juvenile fur is completely replaced, and the second molt begins. It also often starts from the sides and spreads later to the back and other body parts. In this case only new hairs grow and old ones are shed. This second molt occurs at very different periods (Shubin, 1972).

Sex and age composition of population. The male to female ratio in embryos and newborns of little pikas is close to 1:1. However, in the first year of their life this ratio changes in favor of an increase in the number of males in the population (Table 1.3). Apparently, the mortality in young of the year female little pikas is higher than in males because of their earlier sexual maturity and feeding of young ones. However, more males die in the period of the rut, which leads to a fluctuation of the sex composition of population of little pikas from season to season (Shubin, 1968).

The ratio of wintered and young of the year individuals in the population of little pikas changes from year to year in the period of their

Table 1.3. Sex ratio of adult little pikas in different years (Shubin, 1968)

Year	Males		Females	
	Number	%	Number	%
1958	17	53.1	15	46.9
1959	179	58.0	130	42.0
1960	64	50.0	63	50.0
1961	164	55.8	131	44.2

25

large-scale appearance in the open. There are also changes in the ratio of young from different broods—in high-population years, the size of the first brood increases. Moreover, the number of young ones in the populations of little pikas depends on the number of small rodents, since with the lowering of their number increases the probability of extermination of young pikas by predators (Shubin, 1966). On the average, in populations of little pikas, in the period of appearance of young ones, grown-ups are not more than wintered; however, by the end of reproduction, young of the year may make up to 93-96% of the population. Thus, the ratio of restoration of population of little pikas is very high and constitutes 1.5-2 years.

Dynamics of population. The population of little pikas is subject to considerable fluctuation. Potentially the population of little pikas may increase 15-30 fold by fall, in comparison with the preceding spring. However, as a result of considerable mortality of young (including also due to predators), the population of little pikas increases only 8-10 fold by the fall. By the following spring, even in relatively favorable years, the population of little pikas decreases 7-9 fold, in comparison with the preceding fall, which is linked with different weather factors and predator stress.

In severe winters with shallow snow cover, the decrease in population of little pikas can be most significant—tens, even hundreds of times, as, for example, was found in Kazakhstan near Shoindykol' and Tengiz lakes in 1956, at Bosaga station in 1960, and in the vicinity of Semiyarskoe town in the Irtysh flood plain in 1962 (Sludskii et al., 1980).

In the vicinity of Bosaga station in 1959, by the end of the reproductive period the population of little pikas was exceptionally high. Over an area of 1 ha in thickets of honeysuckle 57-103 individuals were caught (average 87.9), but in the spring of 1960 the population of pikas decreased sharply and in April, over an area of 1 ha of the same habitat, 3-14 individuals lived (average 8.2). The first half of winter in these places had low snow levels and was icy. The population of pikas decreased more than 100 fold. By 1962 this population was restored, but after the severe winter of 1962/63 pikas have almost completely disappeared from the territory. In 1966 pikas were few here, but by 1970 their population increased somewhat; it increased particularly appreciably in 1971. In 1972 little pikas in this region were quite numerous, but in 1973 they again became rare (Smirnov, 1974; Sludskii et al., 1980).

Enemies. The main enemies of little pikas are predatory mammals, for whom it is easier than birds to catch pikas in burrows and shrub thickets. In the Kazakhstan upland, at Bosaga station in 1953-54, the remains of little pikas in the feces of predators were: foxes—8.6-10.3%, Siberian polecats—3.4-6.9%, corsac fox—3.9% (Shubin, 1962a). In Betpakdala the contribution of little pikas in the diet of red foxes was 2.3-4.0% (n = 230), and in the diet of wolves (from analysis of 859 feces)—2.2%. In spring and summer, after the migration of saiga deer, the share of little pikas in the diet of wolves increased to 4.2-6.6% (Sludskii, 1962; Sludskii et al., 1981). In high-population years, little pikas can be the main dietary object of predators. Thus in 1961, in the Irtysh flood plain in Semyarskoe town, little pikas constituted 47.3% in the feces of corsac foxes (n = 262); while in the feces of Siberian polecats, 9.1% (Sludskii et al., 1980, 1982). In the Irtysh flood plain, ermines also feed on little pikas (Pakizh, 1969). Red and corsac foxes can exterminate quite a large number of young of little pikas. In May to June 1960, at the Bosaga station during the period of feeding young ones, more than half of the burrows were dug down to their nest chambers by red foxes and corsac foxes; subsequently, it was not possible to catch a single young pika in this area. Interestingly, in the preceding year in this very region, predators dug up not more than 4-6% of the burrows (Shubin, 1966).

Predaceous birds rarely attack little pikas; in regurgitates of pale harriers the remains of little pikas constituted just 2.1-2.8% (Shubin, 1962a); while in regurgitates of eagle owl, 8-12% (Sludskii et al., 1980).

Apparently, snakes can also attack little pikas. In the stomach of an arrow snake caught at Basaga station, two young little pikas were found (Sludskii et al., 1980).

Competitors. In the southern part of the Kazakh upland, Pallas's pikas live as neighbors of little pikas and are their main competitors. During an increase in number, Pallas's pikas disperse from their typical habitats—rocks and rock dumps—to shrubby biotopes, dislodging little pikas from there (Smirnov, 1974; Sludskii et al., 1980).

Occasionally, common hamsters settle down in habitats of little pikas. In such a case, little pikas remain not closer than 150-200 m from the burrows of hamsters. Possibly hamsters actively chase away pikas (Sludskii et al., 1980).

In the northern Sempalatinsk region during periods of rising water level, when the thickets of coastal vegetation become inundated, the dwellers of these habitats—water voles—settle down in shrubs, entirely dislodging little pikas (Sludskii et al., 1980).

In their turn, little pikas act oppresively on the populations of some rodents—narrow-skulled and northern red-back voles, striped and common field mice (Sludskii et al., 1980).

Diseases and parasites. So far, infectious diseases and epizootics have not been observed in little pikas (Sludskii et al., 1980). In the Kazakh uplands, the common parasites of little pikas are the larvae of sub-cutaneous bot flies, *Oestromyia leporina* (Grunin, 1962). Among ixodid mites,

34

large-scale parasites of pikas in the Kazakh upland are the larvae and nymphs of *Ixodes laguri, Dermacentor marginatus,* and *Rhipicephalus pumilio.* Less often *Haemophysalis warburtoni, H. humidiana, Dermacentor pictus,* and *Rhipicephalus rossicus* are found (Ushakova and Buslaeva, 1962; Sludskii et al., 1980). In northeast Kazakhstan in 1963-1965 and in summer of 1967, larvae of nymphs of the tick *Dermacentor marginatus* were found in 42 of 62 pikas. Occasionally gamasid mites *Haemolaelaps glasgovi* and *Elaelaps stabularis* were found on little pikas (Pakizh, 1969).

In central Kazakhstan, fleas—*Amphalius runatus, Ceratophyllus insertus, Frontopsylla elata popovi, (Ctenophyllus (Ochotonobius) bondari* (Mikulin, 1956)—were found in small numbers on little pikas. In the Shiderta valley in northeast Kazakhstan, little pikas showed 7 species of fleas: *Amphalius runatus, Ceratophyllus tesquorum, Frontopsylla elata, Ctenophyllus bondari, Ctenophthalmus assimilis, C. brevatus* and *Neopsylla pleskei.* Among the named species *F. elata*—a flea with wide range of hosts—is most common (Pakizh, 1969).

In little pikas from the Kazakh upland, 8 species of helminths have been found: *Schizorchis altaica, Diuterinotaenis spasskyi, Hydatigera* sp., *Cephaluris andreievi, Dermatoxys schumakovitschi, Labiostomum vesicularis, Nematodirus aspinosus* and *Trichocephalus* sp. (Gvozdev, 1962).

Practical Significance

Little pikas are food of many fur animals: foxes, ermines, Siberian polecats and weasels.

Little pikas cannot cause any damage to pastures, since they mainly live among shrub-thickets, where they feed on plants not used by cattle. Sometimes in fall, they can live in stacks, feeding on and polluting hay (Kirikov, 1952; Sludskii et al., 1980). Domestic sheep and Pamir argali may feed on the hay cured by little pikas (Pakizh, 1969).

Little pikas possibly also have some epidemiological significance, since they feed on larvae, nymphs and ticks, as also other parasites.

Little pikas are not used as an object of commerce—their skin, despite having dense long hairs, is not durable and wears away fast.

Long-Eared Pika
Ochotona (Conothoa) macrotis Gunther, 1875

1875. *Lagomys macrotis* Gunther. Description of some leporini animals from Central Asia. Ann. and Mag. Natur. Hist. Ser. 4, Vol. 16, No. 93,. p. 228-231, China (Xinjiang), Kunlun Range, Duba River.

1914. *Ochotona sacana* Thomas. On small mammals from Djarkent. Ann.
and Mag. of Natur. Hist., Vol. 13, p. 572, Kirgizia, Issya-Kul' region,
Prizheval'sk.

1952. *Ochotona wollastoni* Thomas et Hinton. The mammals of the 1921
Mount Everest expedition. Ann. Mag. of Natur. Hist., Vol. 9, p. 184-
186, Nepal, Mt. Everest.

Diagnosis

Skull bulged, medium-sized—condylobasal length up to 4.5 mm.
Interorbital distance more than 4.5 mm. Incisor fossa not partitioned by
protuberances of the intermaxillary bones. Frontal bones with two
foramina. Ears large—length of ears 32.5 mm. Vibrissae long—up to 83
mm. Color of summer fur on back grayish-brown with ochreous tinge. Belly
white with pale yellow tinge. Winter fur with dense long hairs, on back
bright gray, sometimes with pale yellow tinge; belly white. A small white
spot present behind ears. Ears covered with hairs.

Description

The body length of adult animals varies from 160 to 225 mm. On average,
the body length of females from the Tien Shan is 208 mm; that of males,
224 mm (Zimina, 1962); females from the Pamirs and Badakhshan are on
average 179 mm long, and males—181.8 mm (Davydov, 1974). The hind
foot length varies from 28 to 34 mm (average 31 mm). The length of ears—
from 24.5 to 32.8 mm (average 29 mm). The weight of adult long-eared pikas
from the Tien Shan varies from 230 to 300 g (Bernshtein, 1970), and from
the Pamirs—129 to 235 g (Davydov, 1974).

In summer the hair coat of long-eared pikas is smooth, not very dense
and relatively short. The length of guard hairs on the back varies from 18
to 25 mm. The back is grayish-brown in color with an admixture of
ochreous tones. On the head, cheeks, and the anterior part of the back and
sides, the ochreous tone is more strongly manifested. The sides and paws
are lighter than the back. A transverse stripe of pale reddish-brown color
crosses the occiput. The belly is dirty white, sometimes with a yellowish
tone. Ears are grayish-brown. The winter coat is very dense, long and furry.
The length of hairs on the back reaches 35 mm. The back is pale gray with
a pale yellow tone. The upper part of the head and shoulders has a reddish-
brown tone. Cheeks are pale gray in color. Fur is lighter on the sides and
paws than on the back. The belly is white, ochreous-white or ochreous-
gray. Ears are whitish; their anterior margin is dark ash in color.

Individual variation in color is characteristic of long-eared pikas; this
relates particularly to the degree of manifestation of ochreous tones on the

upper part of the body. The summer fur of long-eared pikas inhabiting upland watershed in the Tien Shan is more dull than in pikas living in talus in the forest belt of the Terskei-Alatau, where complete and partial melanistic forms are found (Bernshtein, 1970).

The young animals are brownish till their first molt and are more monotone than the adults.

Digits have well developed corneous pads and relatively short pointed claws. The underside of the paws is covered with a dense tuft of short stiff hairs.

The skull is medium-sized; its condylobasal length varies from 41.1 to 43.9 mm; zygomatic width—21.9 to 23.3 mm; interorbital width—4.6 to 6

29

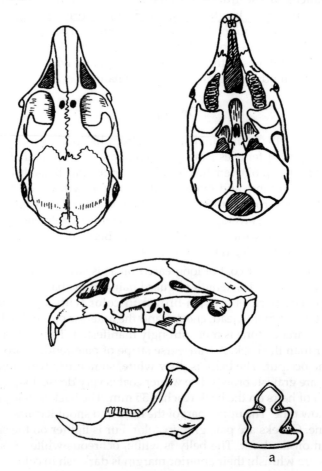

Fig. 7. Skull of the long-eared pika *Ochotona macrotis*.
a—third lower premolar tooth.

mm. The preorbital foramen is large and triangular in form. The lateral margins of the nasal bones are parallel to each other. Frontal bones always have equal-sized foramina. Zygomatic arches are somewhat broadened in the middle. Tympanic bullae are medium-sized. The ascending ramus of the mandible is not high. The angular notch is shallow (Fig. 7). The third lower premolar tooth has a quite large anterior segment. The posterior segment is broad, with different degree of manifestation of the inner fold (Fig. 7a).

Os penis is relatively short and thick—its length surpasses its thickness by not more than 8 times (Aksenova and Smirnov, 1986).

The diploid complement has 62 chromosomes: 5 pairs of submetacentrics, 6 pairs of subtelocentrics and 19 pairs of acrocentrics; the X-chromosome is submetacentric and the Y-chromosome is the smallest acrocentric (Vorontsov and Ivanitskaya, 1973).

Systematic Position

The majority of taxonomists consider the long-eared pika as a separate species. A.A. Gureev (1964) and, following him, Kuznetsov (1965), Yanushevikh et al. (1972), Corbet (1978), Sludskii et al. (1980) consider O. macrotis as a subspecies of O. roylei.

O. macrotis differs from O. roylei in having longer ears and tympanic bullae (Ellerman and Morrison-Scott, 1951), a longer and narrower rostrum, more bulged skull (Abe, 1971; Feng and Zheng, 1985), and more pubescent inner surface of the ears (Feng and Zheng, 1985). Adults of O. roylei are more brightly colored than O. macrotis (Argiropulo, 1932) and do not have foramina in their frontal bones (Abe, 1971). On average, O. macrotis is larger than O. roylei.

Differences have also been noted in the habitat affinity and behavior of these species: O. macrotis is adapted to conditions of low temperatures and aridity, while O. roylei prefers to colonize southern slopes covered with forest. O. macrotis does not store food, whereas O. roylei makes reserves of grass among stones (Feng and Zheng, 1985). In Nepal (region of Mt. Everest) these species are sympatric, have different habitat affinity, and differ in their timing of daytime activity, as well as in social structure (Kawamichi, 1971).

32   It should be noted that, in the book Mammal Species of the World (Honacki et al., 1982), where separate status of O. macrotis as a species is recognized, an error has crept in the description of the range boundary of O. roylei. From this description, it follows that both species of pika occur in the territory of Russian Central Asia.

38

Geographic Distribution

The long-eared pika has quite an extensive range (Fig. 8a), and the Pamirs, Badakhshan and Tien Shan are its northern limits of distribution.

In Tadzhikistan the eastern boundary of the range of long-eared pikas passes along the western spurs of the Sarykol'sk Range in Gornyi [Montane] Badakhshan (Fig. 9). In this region the extreme northeastern place to encounter long-eared pikas is the northern coast of Lake Karakul (Ognev, 1940). From here the eastern boundary of the range of long-eared pikas turns southeast to the northern coasts of the Shorkul and Rangkul lakes and bypassing them from the south proceeds southwest, reaching the environs of Chichekta biological station on the eastern spurs of the Muzkol Range (Ognev, 1940; Davydov, 1974). Individual colonies of long-eared pikas are found along the Oksu River to its upper reaches (Davydov, 1974). The extreme southeastern range of long-eared pikas in Tadjhikstan—is in the environs of Lake Kyzyl-Rabat (Rozanov, 1935; Minin, 1938). Here the southeastern bundary of the range of long-eared pikas meets the southern and passes to the west along the northern slopes of the Vakhansk and Yuzhno [Southern]-Alichursk ranges (Davydov, 1974). Colonies of long-eared pikas have been observed in the valley of Lake Sassyk-Kol' to the south of the Alichurskaya valley and in the Khargush Pass, on the western spurs of the Yuzhno [southern] Alichursk Range in the region of the Kokbai Pass, in the middle and lower reaches of the Mats River, and in the valley of Kok-Bai River (Rozanov, 1935; Minin, 1938; Davydov, 1974). On the northern slopes of the Shakhdarinsk Range, long-eared pikas are found in the valleys of the Dzhavshangoz and Bodomdara rivers (Minin, 1938); here the southern boundary of the range of long-eared pikas turns north, joins the western and passes along the western spurs of the Ishkashimsk Range (Davydov, 1974). In 1937, long-eared pikas were caught in the environs of Khorog at the mouth of the Shakhdara River; and in 1957 and 1962 in the valley of the Dzhizedvar River (a tributary of Bartang) and in the environs of Kishlak* of Vir on the Gunta (Davydovv, 1974).

The northwestern boundary of the range of the long-eared pika in Tadzhikistan passes along the left bank of the Bartang River. Here, long-eared pikas were found in the environs of Lake Sarezhskoe (Davydov, 1974; collections of ZIN). The extreme northwestern regions of occurrence of long-

*In Central Asia, a winter pasture or a winter village—Translator.

Fig. 8. Range of: (a) Altai and (b) long-eared pikas. 1—Schematic range according to S.I. Ognev (1940); 2—locality points of Altai pika in Mongolia (Sokolov and Orlov, 1980); 3—schematic range of the long-eared pika according to Corbet (1978); 4—our data.

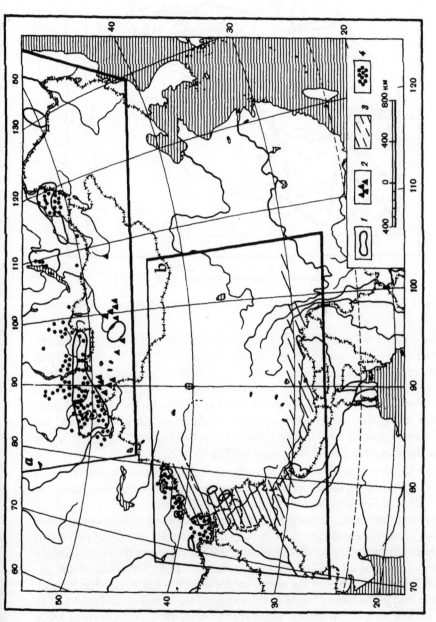

Fig. 8. See page 38 bottom for caption.

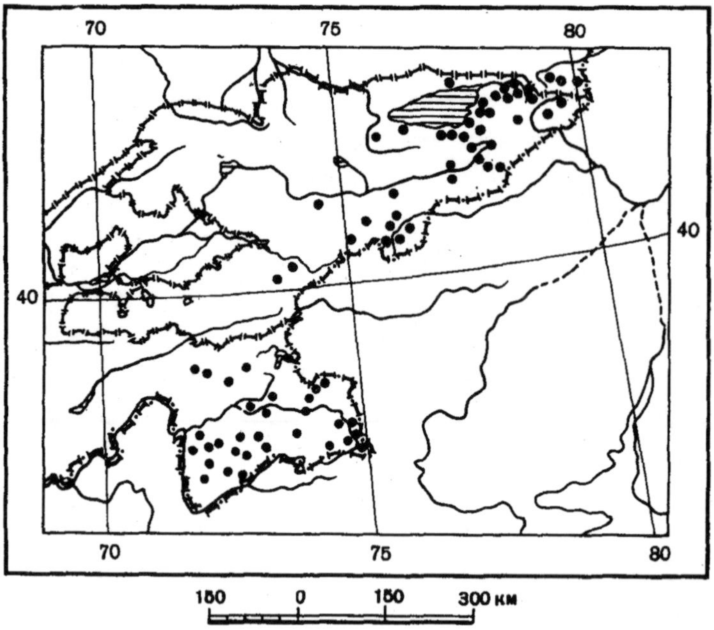

32

Fig. 9. Locality of long-eared pikas in Russian Central Asia and Kazakhstan.

eared pikas in the territory of Tadzhikistan lie in the region of the Fortambek glacier in the Akademiya Nauk Range (Rozanov, 1935; Davydov, 1974). Eastward, long-eared pikas were caught in the valley of the Tanymas glacier in the region of Lake Karakul (Minin, 1938). The northeastern boundary of the range of long-eared pikas within limits of Tadjhikistan passes along the western edge of the Akademiya Nauk Range and the Fedchenko glacier, and in the valley of the Muksu and Kyzyl-Su rivers joins the Kirgistan part of the range (cf. Fig. 9).

The southernmost localities of long-eared pikas in the Kirgistan part of the range were reported on the southeastern spurs of the Alai Range in the valley of Terek-Su River (Argiropulo, 1932; Yanushevich et al., 1972). Following farther in the northeast, the range boundary of long-eared pikas passes along the eastern regions of Narynsk through Lake Chatyr-Kel' and farther—along valleys of the Terek and Ak-Sai rivers on the southwestern spurs of the Kak-Shaal-Too Range (Dzhoiloev, 1959; Yanushaevich et al., 1972). Northeastward (on the northern slopes of the Kak-Shaal-Too Range), long-eared pikas were caught in the valley of the Bordkoldoi River and in the middle fork of the Bedel' River (Bernshtein, 1970; collections of IZK). On the northeastern slopes of the Kak-Shaal-Too Range, long-eared pikas were caught on the Sary-Dzhas syrts (Yanushkevich et al., 1972).

The northeastern edge of the range of long-eared pikas lies in the territory of the Naryngol'sk district of the Alma-Ata region. From the Bayangol River the northern boundary of this part of the range turns west and follows along the northeastern spurs of the Terskei Alatau Range along the valleys of Malyi [Lesser] and Bolshoi [Greater] Kokpak and Tekes rivers (Bernshtein, 1970; collections of ZM MGU). Later the boundary turns southwest and passes along the territory of the Issyk Kul region through the upper reaches of the Tyup River on to the southeastern slopes of the Kungei Alatau, bending about Issyk Kul from the south. Long-eared pikas were reported on the northern slopes of the Terskei Alatau Range near Przhevalsk, in Dzhety-Oguz ravines, in the basins of the Chong-Kyzyl-Su River, and in the Barskaunsk and Turgen' ravines (Kuznetsov, 1948b; Zimina, 1962; Bernshtein, 1970; Yanushkevich et al., 1972). Proceeding further west, the northern boundary of the range of long-eared pikas passes along the northeast of the Narynkol'sk district through the upper reaches of the Kel'deke (Kel'ukek?) River (Zimina, 1962). The extreme northwestern range of long-eared pikas lies along the northeastern spurs of the Kirgiz Range—in the Tyuuk ravine and at the Issyk-Atinsk forest resthouse (Yanushkevich et al., 1972). From here the boundary turns southwest and passes through the Kargalyk Pass, connecting Toguz-Tokrau with the valley of the Alabuga (Alabuki) River and extending farther south along the valley of the Taldyk (Tandyk) River in the Alaisk Range (Argiropulo, 1932).

The internal structure of the range of the long-eared pika is complex (cf. Fig. 9) and is caused by the sporadic distribution of this species. However, mention must be made of the insufficiently complete study of the character of distribution of the long-eared pika, particularly in the Pamirs and at its border with Badakhshan.

Outside of Russian Central Asia and Kazakhstan, long-eared pikas inhabit Afghanistan (Vakhan Range), India (Ladakh, Gilgit, valley of Kishanganga in Kashmir), southern Chinese Turkestan; in China (Xinjiang—in regions bordering Tibet—Qinghai, Tibet); in Nepal (Gosain Kund—central part of Nepal—in the environs of Mt. Everest) (cf. Fig. 8b).

Geographic Variation

Long-eared pikas inhabiting the Pamirs and Tien Shan practically do not differ from each other either in metric characters or in the color of fur. Thus, in the territory of Russian Central Asia and Kazakhstan lives one subspecies of long-eared pika—*O. macrotis macrotis*, which is distributed also in Xinjiang (Kun'lun Range). In Tibet and in the Himalayas inhabits one more subspecies of long-eared pika—*O. macrotis wollastoni* Thomas et Hinton, 1922, which differs from the nominate [type] form in the color of fur and larger ears (Erbaeva, 1988). *O. sacana* Thomas, 1914, described from the vicinity of Przhevalsk, is conspecific or the nominate [type] form of *O. macrotis*.

Biology

Habitat. In the Tien Shan the altitudinal limit of distribution of long-eared pikas lies in the range of 1,800-4,000 m above sea level—from the forest-meadow-steppe belt to the alpine, inclusive (Yanushkevich et al., 1972; Sludskii et al., 1980). In Badakhshan, they are found at altitudes from 2,200 to 3,500 m above sea level, and in the Pamirs—at altitudes from 3,600 to 5,600 m above sea level (Abdusalymov, 1962; Davydov, 1974). In the Himalayas the upper limit of distribution of this species is 6,000 m above sea level (Ellerman and Morrison-Scott, 1951).

The long-eared pika is a typical rock-dweller. In the Pamirs, its populations are located mainly in stationary large detrital mounds, most often in their lower third. In interdetrital mounds, as well as in open meadow-steppe and desert habitats of the high mountains, long-eared pikas are not found. In Badakhshan, long-eared pikas live on slopes of both southern as well as northern exposures. Stony mounds in the wood-shrub belt are usually surrounded by shrubs and occasional trees, which together with herbaceous vegetation serve as the food base for pikas. In the subalpine zone along edges of mounds occur rare shrubs and herbaceous vegetation, and in the alpine zone exceptionally sparse herbaceous vegetation (Davydov, 1974).

In Kazakhstan, on the northern and northeastern slopes of the Terskei Alatau [Range] , there are several types of habitats of long-eared pikas. In forest meadow-steppe belt, pikas may inhabit stone mounds covering sometimes several kilometers. Such mounds are situated usually at the foot of rocky slopes, they are partly turfaceous, and on the edges have well-defined vegetation. In this altitudinal belt, long-eared pikas readily colonize small (300-1,000 m^2) nonturfaceous mounds situated mostly on slopes of northern exposure. However, on mounds of northern exposure, covered with melkozem [fine-grained soil] and overgrown with moss and forest, long-eared pikas do not colonize readily. In forest-meadow-steppe and alpine belts, colonies of pikas are found on extensive (more than 5,000 m^2) large-stony and weakly turfaceous mounds. Such mounds are characterized by poor vegetation, and pikas live here on the edges of mounds. In the subalpine belt and the lower part of the alpine belt, often there are small (200-800 m^2) non-turfaceous mounds situated amidst herbage meadows with relatively low shrubs of juniper, honeysuckle and pea shrub. Here long-eared pikas live among stones and forage beyond the limits of mounds. Above the stone-fall zone, pikas can colonize broken rocks with cracks and niches. In this belt sporadic nature of habitation of pikas is manifested most strongly (Bernshtein, 1970).

In the Cis-Issyk Kul'syrts, four main types of habitat of long-eared pikas are identified: 1) small areas of mounds usually at the feet of rocks, surrounded with rich vegetation; 2) rock outcrops with a large number of cracks and niches with rich vegetation around them; 3) extensive accumulation of fragmented material, non-turfaceous or weakly turfaceous, often found in valleys of rivers and in the lower part of slopes with steppe or dry-steppe vegetation; and 4) mounds of varying areas in the zone of "cold deserts" at an altitude of 3,500 m above sea level, with low and sparse vegetation (Bernshtein, 1970).

In the Aksai valley long-eared pikas are found on large stony mounds of granite outliers, rising above the steppe. Quite often, pikas can be met with in destroyed sheep pens and residential structures made of stones and thatch (Dzhoiloev, 1959).

35   *Population.* The number of long-eared pikas depends on landscape belts and also on the availability of shelters and food. In the Pamirs and Badakhshan, the maximum population density of these pikas is observed on coarse rubbly mounds surrounded by relatively rich herbaceous vegetation. Thus, in July-August, 1971, on the northern slopes of the Shakhdarinsk Range (3,700-3,800 m above sea level) on the coarse rubbly mound, with an area of 2.2 ha and relatively rich vegetation, the population density of pikas was about 23 individuals per ha, or roughly 8 families per ha. On the southern slopes of the Yuzhno-Alichursk Range and on the Kokbai Pass in 1965, the density of pikas was somewhat lower—about 11 and 10 individuals per ha, respectively (Davydov, 1974).

The population density of long-eared pikas in the forest meadow-steppe belt of the Terskei Alatau [Range] at some places may reach 20 families per ha. As also in the Pamirs, the number of long-eared pikas in this part of the range depends on the quality of shelters and abundance of food (Table 2.1).

On the Cis-Issyk Kul'syrts, the maximum population density of long-eared pikas (7 families per ha) is observed on mounds and rocks with a large number of cracks and with rich vegetation. On the weakly turfaceous clusters of stones with steppe vegetation, the population density of pikas constitutes, on average, 4 families per ha. Very low population density is observed in the zone of "cold deserts" at altitudes above 3,500 m above sea level-on average, 1 family per ha (Bernshtein, 1970).

Shelters. Long-eared pikas do not dig burrows (Bernshtein, 1970). In the forest belt of mountains, as shelters they use niches and voids under rocks. On the syrts, long-eared pikas were sometimes caught in traps placed near the burrows of marmots, 1-2 km from the nearest mounds (Sludskii et

Table 2.1. Population density of long-eared pikas on mounds of different types in the forest meadow-steppe belt of northern slopes of the Terskei Alatau [Range] (May-June: 1962-1963) (Bernshtein, 1970)

Regions of investigations, altitude above sea level	Type of habitat	Area, m²	No. of adult individuals		No. of families per ha
			Females	Males	
B[olshaya]. Kokpa River	Small mounds on overgrown slopes of northern exposure	12,000	14	14	10
M[alaya]. Kokpak River (2,400 m)	Extensive, partially turfaceous mounds on slopes of southern exposure	400	4	4	10
Same (8,000 m)	Extensive, weakly turfaceous mounds on meadows and steppefied slopes of various exposure	10,000	4	4	4
Chon-Kyzylsu River (2,500-3,200 m)	Extensive partially turfaceous mounds on slopes of eastern exposure	18,000	33	40	18
Karabatkan River (3,100 m)	Small nonturfaceous mounds on meadow slopes of various exposures	5,000	3	4	6

al., 1980). However, long-eared pikas apparently use marmot burrows only as temporary shelters during the period of dispersal of young ones.

As far as known to us, there is only one description of the nest of a long-eared pika found in stony mounds in the Aksai valley (Dzhoiloev, 1959). The tunnel to this nest, 14-16 cm wide and 135 cm long, passed along a winding crack ascending to 45° (Fig. 10). The nest was situated in a small chamber; the nest walls, 5-6 cm thick, were made of dry twigs and grass; inside the nest, there was bedding of bird feathers and fur.

The tunnels, through which long-eared pikas move in stone mounds, sometimes extend from one end of the mound to the other over many tens of meters. Long-eared pikas build lavatories either on the mound under stones and in niches or under shrubs.

Long-eared pikas in the Pamirs use niches under the over-hanging ledges of rock outcrops as winter shelters (Abdusalyamov, 1962).

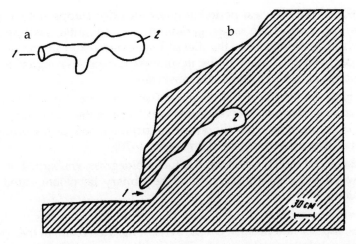

Fig. 10. Schematic drawing of the burrow of a long-eared pika. *a*—plan, *b*—vertical section. *1*—entrance, *2*—nest chamber (from Dzhoiloev, 1959).

36

Feeding. Long-eared pikas feed only on plants. The composition of their food depends on the plants growing near their colonies, or on their availability for the animals. In Badakhshan, long-eared pikas use as food 50 species of plants (Davydov, 1974); on the Terskei Alatau Range, 53 species; on Cis-Issyk Kul'syrts, 47 species (Bernshtein, 1970); in Aksai valley of Kirgizia, 16 species (Dzhoiloev, 1959).

The species composition and specific contribution of vegetative parts of plants consumed [by pikas] change according to seasons. Until the beginning period of vegetation (April to the first half of May), the previous year's vegetative growth predominates in the diet of long-eared pikas— the previous year's shoots of winter fat, wormwood, blue grass, fescue, brome grass, bulbs, as well as twigs and bark of currant, rose, honeysuckle, false tamarisk, and occasionally—fruits and bark of ephedra and juniper.

Until August, long-eared pikas feed predominantly on green parts of plants, among which grasses predominate. Legumes, mosses, lichens, and also berries and other fruits play a great role in the summer diet of pikas of Badakhshan. Preference is shown for buckwheat, sedges, geraniums, milk vetches, woody betony, corydalis, Jerusalem sage, and cinquefoil. At the end of summer and in the fall the diet of long-eared pikas is supplemented with berries, bark and twigs of shrubs. In humid habitats— by springs and near glaciers, these pikas feed on sedges and kobresias (Davydov, 1974).

The species composition of plants eaten by long-eared pikas in the mountains of Terskei Alatau, as well as the seasonal variability of foods, are available in the works of A.D. Bernshtein (1970) and A.A. Sludskii et al. (1980). In the snowless period in forest-meadow-steppe and subalpine belts in this part of the range, geranium, lady's mantle, saxifrage, and buckwheat predominate in the diet of long-eared pikas. In spring, pikas feed mainly on grasses and bulbs; in summer, the species composition and plant parts consumed by them are most varied.

37   On high-mountain syrts of the inner Tien Shan, because of the absence of large shrub thickets, browse plants are absent in the diet of long-eared pikas. Here pikas feed mostly on barley, richteria, larkspur, wormwood, as well as mosses and lichens (Bernshtein, 1970).

In the Aksai valley, *Leucopoa*, Kentucky bluegrass, rhubarb, Tien Shan wormwood and giant fennel constitute the primary diet of long-eared pikas (Dzhailoev, 1959).

In winter, long-eared pikas feed mainly on the branches of shrubs, needles, mosses and lichens. On the syrts in the Tien Shan, in the absence of shrub vegetation and large snow cover, the composition of winter diet of long-eared pikas almost does not change from that in the summer (Sludskii et al., 1980).

Unlike other pika species long-eared pikas do not stock food for the winter (Bernshtein, 1970; Sludskii et al., 1980). Data indicating location of piles of hay belonging to long-eared pikas (Rozanov, 1935; Zimina, 1962; Abdusalyamov, 1962) are apparently erroneous—these stocks probably belonged to Pamir high mountain voles (*Alticola argentatus*), which almost everywhere live alongside long-eared pikas. However, it is not ruled out that, in some parts of the range, long-eared pikas can make food stocks. Thus, on the southern slope of Kungei Alatau in the upper reaches of the Srednyaya Uryukta River, in niches under stones, large quantities of grass were found, cured most likely by long-eared pikas (Sludskii et al., 1980).

Activity and behavior. For long-eared pikas round-the-clock activity is characteristic (Bernshtein, 1970; Sludskii et al., 1980). Other investigators (Dzhoiloev, 1959; Abdlusalyamov, 1962; Zimina, 1962) suggest that long-eared pikas are active mainly in the morning and evening.

In spring, long-eared pikas come to the surface as the sun shines on the stone mound; in summer they surface earlier—before the mound is exposed to light. During the day the pikas hide in shelters for not more than an hour, and the remainder of the time they either feed or hop from place to place, or rest sitting on the stones. Long-eared pikas move with small hops, sometimes without a halt, covering a distance of 40-50 m (Bernshtein, 1970), and the height of their hops may reach 3 m (Smirnov, 1987). During foraging pikas may move away from the edge of the mound to 5-10 m (Bernshtein, 1970), or even to 110 m (Smirnov, 1987). Adult individuals are less timid and let humans close to them. Young animals are more cautious.

In winter, long-eared pikas often come to the snow surface. They nimbly move even on loose snow, sinking not more than 2-3 cm. In long-eared pikas, the load on the supporting surface of the paw is not great—15 g/cm² (Zimina, 1962). In the snow layer, pikas can make small tunnels leading to shelters in the mound. In the forest-meadow-steppe belt, where the snow is not deep, pikas are active on the snow surface throughout the winter. In the alpine belt, in the second half of winter, when snow becomes deep and dense, they apparently completely change over to life under the snow (Zimina, 1962).

Vocal signaling in long-eared pikas is weak compared with other species of pika. In the spring-summer period, their calls may be heared just occasionally (Shnitnikov, 1936; Dzhoiloev, 1959; Zimina, 1962). When scared, long-eared pikas produce a low interrupted hiss. Closer to fall, especially in places with high numbers, pikas produce sounds more frequently. In this case, their whistle (quite sonorous and uninterrupted) differs from the sounds produced when scared (Sludskii et al., 1980).

Area of habitation. Long-eared pikas apparently do not have individual or familial territories. Even in May-June, at the peak of reproduction, the same territory is visited by several adult pikas, both males as well as females. Sometimes several individuals stay close to each other, but encounters between them do not occur. Animals with traces of bites are a rare occurrence. In fall, on termination of reproduction, adult pikas occasionally chase young ones; however, fights between them do not take place (Bernshtein, 1970).

38 *Reproduction.* The period of reproduction in long-eared pikas extends for 4 months—from April to the middle of August. In the Pamirs, females of long-eared pikas bear two litters during the summer. The first litter happens in the end of April to the beginning of May (start of reproduction, as a rule, is not synchroneus); the second litter, at the end of July to the first half of August (Abdusalyamov, 1962). Pregnant females caught in the middle of July were simultaneously nursing, that is, fertilization occurred during confinement (Davydov, 1974). In long-eared pikas pregnancy extends about a month.

In Kazakhstan the major part of females, during the period of reproduction, bear not less than three litters. In the valley of Chon-Kyzylsu River at an altitude of 2,600 m above sea, long-eared pikas begin to reproduce in the first half of April, and large-scale pregnancies occur in the second half of this month. Here, the difference in the periods of fertilization in different individuals is apparently two weeks. In August pregnant females were noticed just in the first decade, and by the end of the month females stopped nursing young of the last brood (Bernshtein, 1970). Individual females that gave birth not so long ago can also be found

48

in September. Spermatogenesis in males already terminates by the end of August (Zimina, 1962).

In the Pamirs and Badakhshan, the number of embryos in long-eared pikas varies from 2 to 5 (Abdusalyamov, 1962; Davydov, 1974); in Terskei Alatau, from 2 to 6 (Bernshtein, 1970). From Table 2.2 it is seen that the number of embryos in the first brood is significantly lower than in the subsequent ones. According to the data of R.P. Zimina (1962), in the eastern part of the Terskei Alatau [Range] the number of embryos in long-eared pikas varies from 3 to 7 (average 5). Pikas inhabiting syrts have higher fecundity than pikas living in the forest belt. There, the number of young ones in a litter of over-wintered females varies from 4 to 8 (average 6.2). Moreover, on syrts the current year's females from first litter take part in reproduction; they begin reproducing in June without having attained the weight of adult individuals; moreover some of them succeed in bearing two broods during summer. Fecundity of the current year's females is somewhat lower than over-wintered females—the number of embryos in them does not exceed 5 averages 4.5 (Bernshtein, 1970). Males of long-eared pikas on syrts, as also in the forest belt, attain sexual maturity only in the following year after birth.

39   *Growth and development.* Prenatal embryos of long-eared pikas have a length of 65-70 mm and weigh 11-12 g. Their back is covered with dense, short, black hairs and the belly with sparse white hairs. The eyes and auditory meatuses are closed, the ear flaps [pinnae] are free, the digits have dark claws. Lower incisors begin to cut by the time of birth.

On reaching a weight of 48-50 g, young long-eared pikas begin to come out of shelters. In the Terskei Alatau large-scale exodus of young ones from the first brood begins in the end of May to the first days of June; and of the second brood, in July and August. By August young of the first brood almost attain the size of adults. Long-eared pikas become sexually mature at the age of 7-10 months and, except for females living on the syrts, the current years' individuals do not take part in reproduction until the following summer (Zimina, 1962; Bernshtein, 1970).

Molt. In long-eared pikas there are two molts during a year—spring and fall. Spring molt begins at the end of March to the beginning of April; in females it starts earlier. In the majority of animals only the head and the anterior part of the body molts completely, while winter fur may be retained on the rump and partly on the back until the fall (Abdusalyamov, 1962; Bernshtein, 1970).

In some animals the fall molt begins in the middle of August, and adult males are the first to molt. By the beginning of September the change of hair coat is observed in all animals. In the beginning, winter coat appears

Table 2.2. Number of embryos in long-eared pikas in the forest-meadow-steppe belt of Terskei Alatau (2,400-2,700 m above sea level) during different periods (Bernshtein, 1970)

Type of capture	Region of capture	Females with number of embryos					Average number of embryos	Total of females investigated
		2	3	4	5	6		
April-1st decade* of May (1st brood)	Chon-Kyzylsu	5	18	–	–	–	2.8	23
Second and third decade of May-1st decade of June, 1963 (second brood)	Chon-Kyzylsu	1	3	19	3	3	4.2	29
Second decade of June 1963 (3rd brood)	Chon-Kyzylsu	–	–	2	2	2	5.0	6
June 1962	M[alaya]. and B[olshaya]. Kokpak rivers	–	1	7	5	3	4.6	16
July 1962	Kuilyu River	–	–	8	2	5	4.8	16

*Ten-day period—General Editor.

50

on the back, followed by the head and breast, and last on the belly. Fall molt proceeds more intensively than the spring molt.

Young animals change their juvenile fur to the adult one in the second month of their life, upon attaining a body weight of 80-120 g; the juvenile molt does not follow a definite sequence (Bernshtein, 1970).

Sex and age composition of population. Two year observations on populations of long-eared pikas in the forest-meadow steppe belt of the Terskei Alatau confirms the predominance of females (57%) in embryos; while in growing animals, males (53%). Among adults, in April-June the male to female ratio is roughly equal, while in August and September a predominance of females (60%) has been noted (Table 2.3).

Table 2.3. Sex composition of long-eared pikas in the forest-meadow-steppe belt of the Terskei Alatau in different age groups (Bernshtein, 1970)

Age groups	Of them		Total of individuals examined
	% females	% males	
Embryos	57.0	43.0	86
Growing	47.0	53.0	160
Adults (April-June)	49.5	50.5	233
Adults (August-September)	60.0	40.0	64

Differences in the sex ratios in different age groups can be explained by the dissimilar activity of males and females in different seasons. As regards the predominance of females in the prenatal period, it may be attributed to a methodological error, or that the sex ratio in long-eared pikas is unusual for mammals, at least at birth.

In spring (before the appearance of the first brood) the population of long-eared pikas consists of animals between the ages of 8 months to 2-3 years. Thus, of 43 pikas caught in April-May in the forest meadow-steppe belt of the Terskei Alatau, 14 (more than 30%) survived two winters, 4 (almost 10%) 3 winters, that is, the latter are not less than two-years and ten-months old (Bernshtein and Klevezal', 1965). Possibly, individual animals may live longer.

On completing the period of reproduction, adults continue to form a considerable part of the population. Thus, of 190 pikas caught from the end of August to October (1962-1963) in the Barskaun and Chon-Kyzylsu* ravines, 33.7% were adult (Bernshtein, 1970). Thus, with relatively low fecundity of long-eared pikas, mortality of young in the summer period is

---

*In Russian original, Kazylsu—General Editor.

quite high. For example, of 12 young born to one female in the forest meadow-steppe belt, on average, 3.3 survived to the fall, that is not less than 70% of young died in their first four months (Bernshtein, 1970).

On the Issyk Kul'syrts the age composition of populations of long-eared pikas was somewhat different: here, of 166 pikas caught in August 1962-1963, adults constituted 33%, that is, slightly less than in the forest-meadow-steppe belt. On syrts each female, including the current year's reproducing ones, during the season bore, on average, 24 offspring. By fall, on average, 6.1 young survived for every female, that is, almost twice than that in the forest-meadow-steppe belt, which is explained by the higher fecundity of pikas living on the syrts. Here in this belt, the mortality of young was slightly higher (75%) than in the forest-meadow-steppe belt (Bernshtein, 1970).

Population dynamics. The population of long-eared pikas is apparently not subject to considerable change. In any case, in the Chon-Kyzylsu River basin, during two years (1962-1963) the population density of pikas was roughly the same (Bernshtein, 1970). The absence of significant change in the population of long-eared pikas in different years was also noticed in Badakhshan and in the Pamirs (Davydov, 1974).

Enemies. Apparently ermines are the main enemies of long-eared pikas. They are especially dangerous for young pikas. In some regions, Polecats could be the enemies of long-eared pikas. Stone martens are usually so rare, that they cannot significantly affect populations of long-eared pikas. On syrts, where pikas are far more mobile than in the forest-meadow-steppe belt and often move away from their shelters in search of food, foxes pose considerable danger to them (Sludskii et al., 1980). In the Tien Shan, avian predators are very few in number in the habitats of long-eared pikas and cannot inflict significant harm. In the Pamirs, remnants of fur and bones of pikas were found in regurgitates of falcons and kestrels. Here the Himalayan eagle-owls also prey on long-eared pikas (Andusalyamov, 1962).

Competitors. High-mountain vole (*Alticola argentatus*) lives throughout the range of long-eared pikas and in its neighborhood. The gray hamster (*Cricetulus migratorius*), and very rarely the forest mouse (*Apodemus sylvaticus*), is found in Badakhshan, at the border of the lower limit of the vertical distribution of the long-eared pika. The gray hamster lives alongside pikas also in the high-mountain zone of the Pamirs, but it lives there not in the rocks, but on wet meadow patches at the base of mounds. In the Tien Shan, the Tien Shan vole (*Cleithrionomys frater*) and the gray marmot (*Marmota baibacina*) live alongside long-eared pikas Apparently, there is no great competition between pikas and rodents. In some cases, long-eared pikas use marmot burrows as temporary shelters (Sludskii et al., 1980).

Diseases and parasites. Fleas, ixodid and gamasid mites parasitize long-eared pikas. In the forest-meadow-steppe belt of the Tien Shan, 5 species of fleas are known to parasitize long-eared pikas: *Amphalius clarus, Ceratophyllus caspius, Paraneopsylla ioffi, Amphypsylla primaris,* and *Frontopsylla ornata* (Sludskii et al., 1980). Parasitic among the gamasid ticks on long-eared pikas are *Haemogamasus dauricus, Laelaps* sp., and *Dermanissus* sp. Among ixodid ticks *Haemophysalum warburtoni* (Sludskii et al., 1980), *Ixodes persulcatus* (Zimina, 1962), *Dermacentor marginatus* (Davydov, 1974) have been reported. In the Pamirs, the ectoparasites of long-eared pikas are the fleas *Hoplopleura ochotonae* (Davydov, 1974).

Among endoparasites 3 species of flatworms have been found in long-eared pikas: *Schizorchis altaica, Dicrocoelium lanceatum,* and *Hasstilesia ochotonaae;* also 4 species of roundworms: *Cephaluris andrejevi, Dermatoxis schumakovitshi, Labiostomum vesicularis* and *Murielis tjanschaniensis* (Tokobaev, 1976).

Pathogenic infection in pika has not been studied further.

Practical Significance

In the 1930-50s skins of long-eared pikas were processed as second grade fur raw material. Starting from 1930, processing reached several hundred thousand skins (total number of all species of pikas and gerbils). The share of long-eared pikas is not known. In recent years, the skins of long-eared pikas were not processed.

Long-eared pikas can serve as the food base of fur animals. In some places with high population, they may inflict some damage to woody and shrub species.

The epizoological and epidemiological significance of long-eared pikas is not understood to date.

Red Pika
Ochotona (Conothoa) rutila Severtzov, 1873

1873. *Lagomys rutilus.* Severtzov, N.A. Vertikal'noe i gorizontal'noe raspredelenie turkestanskikh zhivotnykh [Vertical and horizontal distribution of Turkestan animals]. Izv. Ob-va Lvubitelei Estestvoznaniya, Vol. 8, Vyp. 2, p. 83, Kazakhstan, Trans-Ili Alatau, mountains south of Alma-Ata.

53

Diagnosis

Dimensions relatively large—body length surpassing 200 mm. Skull weakly bulged, quite large—condylobasal length surpasses 41 mm. Interorbital distance broad (5.2-6.3 mm), quite flat. Edges of intermaxillary bones closely converging between themselves, but not partitioning incisoril fossa. Formina on frontal bones, as a rule, absent. Hair coat on upper side of body, particularly on head, bright reddish-ruffous; in winter—ruffous on head, sides and in hind part of body, and on back to gray, often with ochreous tone; hairs on sides of paw lighter in color. Vibrissae long—to 94 mm.

Description

Body length of adult animals from the Trans-Ili Alatau—215-230 mm (average 220 mm) (n = 285) (Sludskii et al., 1980), from Gissar and Turkestan ranges—180-260 mm (Davydov, 1974). Length of foot—36-39 mm (average 38 mm); length of ear 27-29 mm (average 28.5 mm). Body weight varies with the season. Thus, the weight of adult females from the Trans-Ili Alatau during the reproductive period (May-August) varies from 220 to 320 g (average 280 g; without embryos); in fall and winter—from 225 to 295 g (average 270 g). The weight of males during the reproductive period is somewhat less (average 265 g) than in fall and winter (average 280 g) (Sludskii et al., 1980).

Summer fur on back and upper part of head is bright reddish-ruffous, on sides of neck darker reddish-brown-ruffous spots; behind ears—an inconspicuous light spot. Sides of the body and paws are considerably lighter than the back, yellowish-ruffous. Belly is light gray, with yellowish tone. Chin and throat white.

Winter fur is thick and dense, ash-gray on back, with yellowish-ruffous tone in the hind part of the body. The upper part of the head is ruffous. Sides of the body and feet are roughly of the same color as in summer, and the belly is lighter, without a yellowish tone. Soles of feet are covered with short stiff gray hairs.

Vibrissae long—80-95 mm.

Skull is weakly bulged, elongated; its condylobasal length from 42 to 49.6 mm. The interorbital space is broad—5.2-6.3 mm. Frontal bones lack crests. Zygomatic width—22.8-25 mm. Incisor foramina partitioned by protrusions of intermaxillary bone, but not completely; the width of hiatus between the prominences of intermaxillary bones varies widely. Foramina in the anterior part of the frontal bones are present in 20% of individuals (n = 30); the size of foramina varies greatly. The preorbital foramen is small

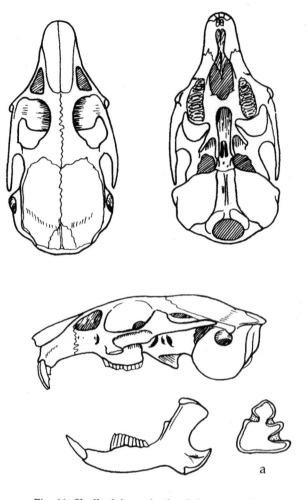

42

Fig. 11. Skull of the red pika *Ochotona rutila*.
a—third lower premolar tooth of the red pika.

and triangular in form. Nasal bones are broadened in the anterior section. Zygomatic arches are parallel to each other. (Fig. 11).

The mandible is quite massive, relatively high; the incisor section is absent. The angular process is not broad. The angular notch is shallow (Fig. 11).

The anterior segment of the third lower premolar tooth is quite small, variable in form, the inner and outer folds, separating it from the posterior segment are of the same depth. The posterior segment is broad, often without the inner fold (Fig. 11a).

The os penis is relatively long and thin; its length is more than 8.5 times the width (Aksenova and Smirnov, 1986).

There are 62 chromosomes in the diploid complement; 7 pairs of metasubmetacentrics; 5 pairs of subtelocentrics, 18 pairs of acrocentrics; X-chromosome—submetacentric, Y-chromosome—small acrocentric (Vorontsov and Ivanitskaya, 1973).

Systematic Position

Morphologically, ecologically and cytologically O. rutila is close to O. macrotis. M.A. Erbaeva (1988) separates O. rutila with O. erythrotis, O. kamensis and O. ladacensis in an independent subgenus Buchneria, the diagnostic character of which is the form of the incisor foramen, partially partitioned by the protuberances of the inner maxillary bones. Howeves it should be noted that such a form of the incisor foramen is characteristic of O. (Lagotona) pusilla. Given the absence of sufficiently complete information on pikas of Central Asia, apparently, it is more appropriate to leave O. rutila and O. marotis in the circumscription of the same subgenus.

Geographic Distribution*

Red pikas are found in the mountains of Tadzhikistan, Uzbekistan, Kazakhstan and Kirgizstan. Their distribution, as also of long-eared pikas, is extremely sporadic (Fig. 12).

44 The valley of the Pshart River (collections of 2M MGU) should apparently be considered as the most southeastern localities of red pikas. From the valley of the Pshart River, the southern boundary of the range of red pikas passes to the southwest. In Badakhshan, the southernmost populations of red pikas have been recorded on the right bank of the Gunt River (E. Yu. Ivanitskaya). From here the range boundary of red pikas turns northwest; pikas were reported on the left bank of the Dzhizevdara River (northern slope of the Rushansk Range) and the right bank of the Murgab and Bartang rivers (southern slopes of the Yazgulemsk Range) and in the environs of the village of Rushan (collections of IZIP, Tadzh.). The extreme western localities of red pikas in Badakhshan—vicinity of Shidz Kishlak on the Pyandzha River (Rozanov, 1935; Davydov, 1974). North of Badkhshan pikas were caught in the upper reaches of the Vanch River and in the region of the Geographic Society glacier on the Akademiya Nauk [Academy of Sciences] Range (collections of IZIP, Tadzhikistan). Red pikas have not been found so far on the Peter the Great and Darvaz ranges but, according to the suggestion of G.S. Davydov (1974), they ought to live here and, consequently, these [ranges] should be considered as the western

*In Russian original, relegated to a side heading—General Editor.

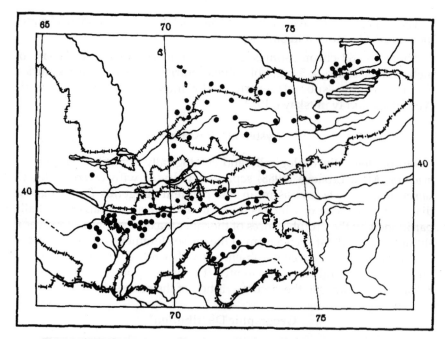

Fig. 12. Locations of red pikas in the territory of Russian Central Asia
and Kazakhstan.

boundary of the range. Further, the range boundary of red pikas passes to
the north and northwest—to the convergence of the Turkestan, Zeravshan,
Gissar, and Alai ranges.

The southern distribution boundary of red pikas, in the western part of
its range passes from the Zeravshan glacier through the mouth of the
Dekhisor gorge, Shakhi-Safed gorge (Dikhauza region) and farther—along
the left tributaries of the Zeravshan [River] and valley of the Yagnob River
(Davydov, 1974). Along the southern slopes of the Gissar Range the
boundary passes to the south up to the upper reaches of Takob and
Maikhura rivers and environs of the Takob mines (Davydov, 1974). On the
southern slopes of the Gissar Range red pikas were found during 1952-
1964 on Anzob Pass, in 1952-1953—in the area of Zidda Kishlak, in many
places along the Dushanbe-Leningrad motorway, and in the Kizil-Tash
tract on the border with Uzbekistan (Minin, 1938; Davydov, 1974). In the
Lake Iskander-Kul' area, red pikas were reported from the end of the last
[19th] century up to the present time (collections of ZM MGU, ZIN; Ognev,
1940; Davydov, 1974). Further, the southern distribution boundary of the
red pika passes west along the southern slopes of the Zeravshan Range
through Archmaidon and Lake Maryuzar-Kul' (Ognev, 1940; Davydov,
1974). On the northwestern spurs of the Gissar Range, red pikas were

caught in the upper drainage of the Shink River, in the upper reaches of Kyzyldar'ya and Urudar'ya rivers, as well as in the Aksu tract (Tret'yakov et al., 1989). According to the data of S.I. Ognev (1940), the westernmost habitations of red pikas are found in southeast Uzbekistan in the region of Shaar (Shakhrisyabz) and northward—in the Nura-Tau mountains (A.D. Bernshtein, personal communication).

Later the range boundary of the red pika turns to the east and proceeds along the northern slopes of the Turkestan Range through the upper reaches of the Kshemyasha, Rovchilik and Devrov (Devchilik) gorges and the upper reaches of the Khodzhametk [River] (Davydov, 1974). On the northeastern slopes of the Turkestan Range, red pikas were reported in the upper drainage of the Karavshin River in Kirgizstan and in the environs of Vorukh Kishlak in Tadzhikistan (Davydov, 1974). Along the northern slopes of the Alai Range, the northern boundary of this part of the range of red pika follows farther east bending into the Fergana valley from the south. Red pikas were found in the Gandakush tract (64 km south of Fergana) and on the northern slopes of the Alai Range on the Pengalbai (Tengizbai) Pass, near the village of Shakhimardan along the Ak-su River and in the upper reaches of the Karaganda [River] (Ognev, 1940; Kuznetsov, 1948a; Yanushevich et al., 1972; collections of ZIN and IZIP, Tashkent); in 1877-1879, A.N Severtsov caught red pikas in Fergana down from the Shart Pass (collections of ZIN). The southernmost colonies of red pikas in Kirgizstan were observed on the northern slopes of the Trans-Alai Range in the Alai valley—on the Kara-Su River and in the Sary-Tash tract (Kuznetsov, 1948b; Sludskii et al., 1980; collections of Saratov University). On the northern slopes of the Alai Range, red pikas were found along the valleys of the Gul'cha and Kichibulalu rivers (Yanushevich et al., 1972; collections of ZIN). Bending the Fergana valley from the east, the boundary of distribution of the red pika passes along the western spurs of the Fergana Range, proceeds along the Baibash-Ata (Bubash-Ata) Range 6 km north of Arslanbob and north of the Fergana valley—in the environs of Lake Sary-Chilek and higher—between the spurs of the Chatkal Range (Minin, 1938; Ognev, 1940; Yanushevich et al., 1972). In the northwestern part of the Fergana valleys, droppings and haypiles of red pikas were found on the northeastern spurs of the Kuraminsk Range in the upper reaches of the Aktash River, but to the west of this range red pikas were not found (Davydov, 1974). In the northeastern part of the range, red pikas were caught along the northeastern slopes of the Ugamsk Range in the upper reaches of the Shungul'duk and Ular rivers (east of Chimkent region), as well as in the midstream of the Antas River on the territory of the Aksu-Dzhabagly preserve (Sludskii et al., 1980).

From the Antas River the distributional boundary of red pikas turns

45

58

east and later passes along the southwestern and northern spurs of the Kirgiz Range and northern slopes of the Trans-Ili Alatau [Range]. On the southwestern spurs of the Kirgiz Range, red pikas were caught in the Kara-Bura gorge south of Dzhambul (Minin, 1938; Ognev, 1940). Along the northern spurs of the Kirgiz Range, red pikas were found practically everywhere.

The northeastern distributional boundary of red pikas passes along the northern slopes of the Trans-Ili Alatau through Kaskelen, Aksai gorge, environs of Alma-Ata, the valley of Talgar River and the Talgar pass (Ognev, 1940; Sludskii et al., 1980). In the northeast the boundary ascends to the western bank of Lake Issyk and the Issyk forest guest-house—roughly 45 km east of Alma-Ata, where pikas are found quite rarely (Ognev, 1940). Southward, red pikas were reported on the northern slope of the Kungei-Alatoo Range—in the valleys of the Kutarga (Kuturga) and Kurmekty (Kurmenty) rivers, as well as in the upper reaches of the Chilik River (Shnitnikov, 1936; Minin, 1938; Kuznetsov, 1948; Sludskii et al., 1980).

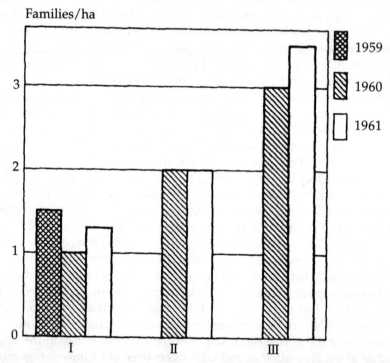

Fig. 13. Population density of red pikas in 1959-1961 on mounds of different types at the upper limit of forest in the Trans-Ili Alatau. *I*—large stony weakly turfaceous mounds; *II*—edge of mound; *III*—partially turfaceous and overgrown mounds (from Bernshtein, 1974).

According to some data (Shnitnikov, 1936), red pikas are not found south of Issyk Kul; however, there are reports of finding this species in the central Tien Shan in the valley of the Kuilyu River (Keolyuu) (Minin, 1938).

Thus, the range of red pika has an insular character and includes mainly the peripheral ranges of Russian Central Asia (Fig. 13). In Tadzhikistan, this species is found on the Turkestan, Zeravshan and Gissar ranges, occasionally—on the Rushansk, Yzgulemsk, and Pshartsk, Vanchsk ranges and Akademiya Nauk Range in Badakhshan and, possibly, on Peter the First and Darvaz ranges. In Kirgizstan, the red pika is distributed north of the Trans-Alai and Alai ranges, and also along the western slopes of the Fergana Range. In the central Tien Shan, this species lives in the northwestern part along the Moldo-Too Range. On the Terskei Alatau Range, and south of it, red pikas are practically not found. North of the Kirgizstan part of the range, red pikas are found on the Talass and Kirgiz ranges and along the southern slopes of the Kungei Alatau Range, and in Kazakhstan—along the northern slopes of the Kirgiz Range and Trans-Ili Alatau to the western spurs of the Ugamsk Range. In Uzbekistan, red pikas are found along the northeastern slopes of the Kuraminsk Range on the border with Tadzhikistan.

The red pika is unknown beyond the limits of Russian Central Asia and Kazakhstan. D. Ellerman and T. Morrison-Scott (1951) include the Central Asian species *O. erythrotis* in *O. rutila*, because of which eastern Tibet finds a mention in the area of distribution of red pika in the monograph of these authors.

Geographic Variation

Ochotona rutila is a monotypic species. The morphometric characters of red pikas inhabiting the Tien Shan and Gissar-Alai are overlapping; there are no clear-cut differences between them in the color of fur.

Biology

Habitat. The red pika is a typical rock dwelling species. Throughout the range its colonies are sporadic, and only in select places more or less extensive. Sporadic distribution of red pikas is associated not only with the availability of stony mounds suitable for shelters, but also with the climatic conditions, altitude above sea level, and abundance of vegetation.

In the Gissar-Alai, populations of red pika are most frequently found at the junction of two vertical [altitudinal] belts—woody-shrubby and high-montane (Davydov, 1974). Vertical distribution of red pikas in this region is restricted by altitudes of 2,000-3,600 m above sea level. In the Trans-Ili

Alatau, populations of red pikas are reported at altitudes of 1,800-3,700 m above sea level, but red pikas reach the maximum population density in the upper part of the forest-steppe belt at an altitude of 2,300-2,900 m. Above the limit of woody-shrubby vegetation, populations of red pikas are found rarely in the Trans-Ili Alatau because of the quite sparse vegetation in summer and a thick snow cover in winter (Bernshtein, 1963).

47 Red pikas inhabit only large-stony, well turfaceous mounds or piles of boulders of moraine origin. In the forest belt of central Tadzhikistan, mounds, as a rule, are surrounded by juniper and shrubs of rose, barberry, ephedra or fruit trees and maple (Davydov, 1974). In the Trans-Ili Alatau, stony mounds with colonies of red pikas are found in sparse forests of Schrenk spruce. On the southern slopes usually grow leafy shrubs—many species of honeysuckle, rose, cottoneaster and others; often here one finds willows and mountain ash. Juniper of the trailing type predominates at the upper limit of forest. Stony mounds are rare in the Talass Alatau, their area is small, and populations of red pika are less numerous and removed from each other by distances of up to 1.5 km (Sludskii et al., 1980).

Population. The population level of red pikas depends on the altitudinal belt, availability of shelters and food. In red pikas high population density, apparently, does not occur and their numbers are fairly stable in different years (Table 3.1; cf. Fig. 13).

Table 3.1. Population density of adult red pikas on mounds of different types at the upper limit of forest in Tadzhikistan (Davydov, 1974)

Habitats	Number of families/ha
Gissar Range, large-stony, nonturfaceous mounds	0.3-0.7
Gissar Range, southern slopes, large-stony, weakly turfaceous mounds with dense shrubs	3.3
Zeravshan Range, rocks and quite turfaceous large-stony mounds	2.0
Gissar Range, rocks and large-stony mounds with abundant herbs and shrubs	2.7
Turkestan Range, southern slopes, large-stony, weakly turfaceous mounds	2.1

46

Shelters. Apparently, red pikas do not dig burrows. For them hollows in rock piles, deep cracks and niches serve as shelters. In red pikas claws are not adapted to digging burrows—they are quite short and blunt. In captivity, red pikas only occasionally made small (not more than 10 cm) hollows under stones, but did not dig burrows (Bernshtein, 1963). In recesses between stories, red pikas make reserves of food and prepare numerous lavatories. These places are used by pikas for several years. They make nest chambers apparently also under stones, but at greater depths than food reserves. In captivity, only pregnant females built true nests. These nests, built exclusively from dry grass, were open above (Bernshetin, 1963).

Feeding. Red pikas feed only on plants. The food of pikas usually includes plants that grow near their shelters. As a rule, the background plant species are most frequently found in the food reserves of red pikas. The species composition of the diet of red pika is exceptionally large throughout its range. In Tadzhikistan, in the stomach contents and winter reserves of pikas, 100 species of plants belonging to 19 families have been recorded: Compositae (18 species), Poaceae (17 species), Apiaceae (8 species), Scrophulariaceae (6 species), Labiatae and Polygoniaceae (5 species each), Boraginaceae, Geraniaceae, Fabaceae, Rosaceae (4 species each), Cupressaceae, Caprifoliaceae, Liliaceae, Cyperaceae (3 species each), Ephedraceae, Ranunculaceae, Brassicaceae (2 species each), Gentianaceae and Amaryllidaceae (1 species each). The list of diet plants also includes 10 species of trees and shrubs (Davydov, 1974).

On the Kirgiz Range 63 plant species have been recorded that are used as food by red pikas (Yanushevich et al., 1972). In the Trans-Ili Alatau, the species composition of plants consumed by red pikas is very rich—89 species from 37 families. Besides herbaceous plants, in this part of the range, the diet of red pikas includes 10 species of trees and shrubs, 2 species of green mosses and 2 species of encrusted lichens (Bernshtein, 1963).

48 The composition of diet preferred by red pikas changes according to the season and month. In spring, grasses predominate in the diet of red pikas and, by the end of summer, their diet becomes more diverse (Table 3.2).

Table 3.2. Change in composition of the diet of red pikas in the spring-summer period in the Trans-Ili Alatau (from Bernshtein, 1963)

Month	Number of stomachs examined	Frequency of occurence, %				
		Grasses	Herbs	Spruce	Mountain ash	Shrubs
May	50	82	76	6	–	–
June	30	70	93	7	3	–
July	40	50	75	10	5	7
August	30	23	50	7	17	27

47

In Tadzhikistan, in spring, the diet of red pikas showed a predominance of grasses; in April in juniper forests, the diet included bluegrasses, roegenerias, *Leucopoa*; in May-June on the southern slopes of the Turkestan

Range, wheat grass, brome grass, orchard grass, rice grass, bluegrass and timothy (Davydov, 1974). In spring and summer, red pikas prefer the most succulent plant parts—leaves, flowers, and less often stems and shoots. In fall they feed on herbs as well as on shoots, branches and leaves of trees and shrubs.

Red pikas begin curing food for winter in the first half of summer—from the middle of June or beginning of July, depending on the weather conditions and state of vegetation. For example, in the Trans-Ili Alatau in 1961, when spring was very warm, red pikas started accumulating their first stocks of food from June 14-15, from early plant growth stages; and in the preceding two years with a cold and protracted spring, storage of food began on July 2-4. The most intensive food storage occurs in July-August. In September, curing of food gradually ceases. Apparently, this is linked with a drying of plants and loss of their nutritional quality. At the end of the curing season, pikas mainly store twigs of spruce or juniper. For one day pikas may carry 35 to 150 g of green plants (Bernshtein, 1963).

The food reserves of red pikas often show, and sometimes are predominated by, species of plants that are absent in their summer diet, for example, golden ray, Jerusalem sage, and rhubarb—pikas do not eat these plants even in captivity. At the same time, grasses used by red pikas in their summer diet are found in their reserves only in small quantity. Thus 3 piles of red pikas from the Gissar Range contained exclusively the leaves and inflorescence of golden ray; 4 other piles contained 15% wild buckwheat and 5% desert candle, while the main mass of these piles consisted of golden ray; grasses were present as isolated plants (Tret'yakov et al., 1989). In winter, even browse plants are cured—mostly honeysuckle and spruce and, in some places, juniper (Sludskii et al., 1980).

Analysis of the contents of winter reserves of red pikas from different parts of their range (Bernshtein, 1963; Yanushevich et al., 1972; Davydov, 1974) shows that, of 60-100 plant species stored, 15-20 species predominate, which in 20% of cases are present in all reserves of the given region. However, the composition of plants in haypiles of different individuals is extremely varied and may differ even for individuals living in the neighborhood of each other. The composition of stored plant species also changes in different years (Bernshtein, 1963).

Red pikas gather their haypiles in niches under stones, under overhanging ledges, and in crevices or cracks of rocks. The distance between haypiles depends on the availability and safety of places for storing hay. Often, the very same place is used for storing haypiles year after year. Plants are gathered without their preliminary curing. Whole plants or large leaves predominate in haypiles, but twigs of shrubs, spruce and juniper are 20-30 cm long. Apparently, red pikas do not make open piles (Bernshtein, 1963).

The size of haypiles varies widely—from very small to large. For example, one of the observed haypiles of red pika occupied an area of 4-5 m^2 and the height of the layer of dry plants reached 50 cm (Sludskii et al., 1980). Most often, the mass of individual haypiles varies in the range from 100-300 g to 1-2 kg, but in the presence of extensive and convenient recesses under stones, the mass of haypiles may reach 3-5, and sometimes also 8 kg (Davydov, 1974). According to the observations of A.P. Strelkov (1989), in the Trans-Ili Alatau, in the storage period the haypiles of red pikas are of three types: small reserves (50 × 20 cm) cured in the current season; the preceding year's partly used reserves, which are replenished in the current year (the sizes of such haypiles are usually large—1-2 × 1 m), and finally the discarded many-year-old haypiles.

In winter, red pikas usually completely use up their reserves. At this time, they feed also on the needles of spruce and juniper, bark of mountain ash and shrubs. In the mountains of the Trans-Ili Alatau, over a 3 km stretch not a single undamaged tree could be found on the stone mounds with a large number of large mountain ash trees; the bark of trunks and branches of some trees were entirely stripped to 0.5-1 m (Bernshtein, 1963). Red pikas can also damage spruce to some extent.

Activity and behavior. The nature of daily activity of red pikas depends on the time of the year. In summer, pikas are active throughout the day; but the peak activity is found in the early morning and evening. The activity of adult pikas increases in the second half of summer, when they begin to store food. Males are the first to begin storing food, and females join them roughly after a month, when young of the last brood change over to independent feeding. Young pikas, in the Trans-Ili Alatau, practically do not participate in storing food, while in the Talass Alatau, Turkestan and the Gissar ranges, young pikas store food on par with adults (Bernshtein, 1963; Davydov, 1974). Storage occurs throughout the brighter part of the day. Most often, pikas collect food in the direct proximity of shelters and do not venture farther than 2-3 m from the edge of the mound, but sometimes they may collect plants 30-40 m away from shelters (Bernshtein, 1963).

In spring and fall, the activity of red pikas depends, to a considerable extent, on the weather conditions. In calm warm weather, pikas are active throughout the brighter part of the day. In the morning, with sunrise, pikas usually sit for long periods on stones or make short runs and rarely hide in shelters. By 16.00-17.00 hr the activity of pikas increases, attains the maximum, and then gradually decreases after 20.00 hr. In the period of maximum activity, pikas run about the mound with great speed and to a considerable distance but, often and for long period, hide in shelters.

In winter pikas surface very rarely, and only in the brighter part of the day. There are no long runways (more than 1-2 m) on the surface of snow. Most often these runways connect neighboring shelters. Under the snow they make short tunnels, which lead to voids between stones. Even in spring in the period of the rut, red pikas rarely venture on the snow surface. The load on the supporting surface of the paws in red pikas is 15 g/cm^2 (Sludskii et al., 1980).

Red pikas move in light hops between stones as also on their surface; the length of their leap reaches 118 cm. Pikas also climb even overhanging rocks to a height of 3-4 m, but rarely climb trees and shrubs. The average speed of movement of red pikas on the mounds is 160-170 cm/sec, but pikas can perform individual leaps at a speed of 310-315 cm/sec (Strelkov, 1989).

In red pikas vocal signals forewarning danger are absent. At the approach of the enemy, pikas most often quietly hide in shelters. But at times their low calls can be heard, which resemble humming. During the rut males chasing females issue a low "urrr," and females squeek quite shrilly (Bernshtein, 1963, 1964; Davydov, 1974).

Area of habitation. Red pikas apprently live in families. Males and females live in pairs in one territory throughout their life, and do not undertake large migrations. According to observations in the Trans-Ili Alatau, on the small mounds of 1,250 m^2, where only one pair of pikas live, the male had a territory 350 m^2; the female—750 m^2. These territories overlapped only at the entrance to the nesting shelter. In September, when storing of food was over, the female not only had a larger territory, but used it more actively. The areas for collection of grasses, used by the male and female may not overlap each other (Strelkov, 1989). The distance between neighboring family territories depends on the population density of pikas and varies from 20-30 to 50-100 m. The family territories may intersect in regions of foraging grounds. A family pair uses a common food stock and in the course of time young pikas join them. After the appearance of young of the second brood, residents of the first brood begin to avoid contact with parents.

As a rule, there are no encounters between neighbors. In the Trans-Ili Alatau, for example, in the period of reproduction, individuals with traces of bites were not observed. The elements of aggressive behavior were also not observed in pikas held in captivity (Bernshtein, 1963, 1964). However, observations on red pikas on the Kirgiz Range confirm individual fights between pikas of different family groups (Yanushevich et al., 1972).

Reproduction. Red pikas are apparently monogamous since adult individuals usually live in pairs. Each adult male, as a rule, guards his family territory.

In the Trans-Ili Alatau, spermatogenesis in males begins probably in March (precise data are not available). In the beginning of April males already have strongly enlarged testes, and on April 21, 1961, a recently pregnant female was caught, that was evidently mated with in the middle of March (Bernshtein, 1963). Thus, the period of reproduction in red pikas begins long before the snow begins to recede and the first greens appear. In May, females were caught that were pregnant for the first as well as a second time, which confirms the lengthiness of the first reproduction cycle. From April to June the majority of adult females participate in reproduction (Fig. 14). Mating apparently takes place in niches and voids under stones.

In July in the Trans-Ili Alatau, there is a sharp drop in sexual activity. In males the testes became reduced in size, lose elasticity and turn dark (Bernshtein, 1964). It is at this time that males begin storing food. The time of termination of reproduction changes depends on the weather conditions. Thus in 1960, in the Trans-Ili Alatau, spring was cold and protracted,

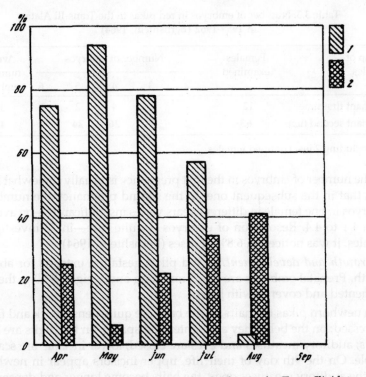

Fig. 14. Dynamics of reproduction of red pikas in the Trans-Ili Alatau. 1—pregnant females (%); 2—lactating females (after Bernshtein, 1974).

66

reproduction was delayed, but spermatogenesis in males continued 15 days longer than in the next year, when spring was early and stormy. Similar dependence of the change of period of reproduction is characteristic also for females (Sludskii et al., 1980). In conditions of the Gissar Range, the reproductive period of red pikas continued somewhat longer, because by July females from the first generation participated in reproduction; they became sexually mature in 3-3.5 months. Moreover, here over-wintering females bear a second, and sometimes a third brood at the end of July to the beginning of August (Tret'yakov et al., 1989).

With the reproductive season continuing roughly four months (April to July), adult females apparently bear not less than two-three broods. In the majority of females, a second pregnancy follows directly after the first, that is, the female is fertilized immediately after giving birth. Thus, according to the data of A.D. Bernshtein (1964), embryos in early stages of development are found in the just confined females.

In red pikas the number of embryos varies from two to six, most commonly 4-5 (Table 3.3).

Table 3.3. Number of embryos in red pikas in the Trans-Ili Alatau in 1959-1962 (Bernshtein, 1964)

Group of females	Females examined	Number of embryos					Average number of embryos
		2	3	4	5	6	
Pregnant first time	12*	–	6	4	2	–	3.8
Pregnant second time	63	2	7	26	24	4	4.3

*Including 2 just confined females.

The number of embryos in the first pregnancy is usually somewhat lower than that in the subsequent one. In the second pregnancy the number of embryos in one female in different years varies insignificantly—on average, from 4.1 to 4.4. Resorption of embryos is quite rare—in 73 investigated females, it was noticed in 6.8% of cases (Bernshtein, 1964).

Growth and development. In red pikas gestation extends for about a month. Prenatal embryos are 73-75 mm long and weigh 18-20 g; they are pigmented and covered with hairs.

In newborn pikas the hairs on the back are quite dense, dark and firmly appressed; on the belly they are lighter and sparse; on the digits are black claws; auditory meatuses and eyes are closed; lower incisors are scarcely visible. On the 9th day of their life, upper incisors appear in newborns and the auditory meatuses open, the hairs become longer and denser, and the body weight increases. On the 13-14th day eyes open, molar teeth begin to cut, the body weight reaches 40-43 g, and young ones begin to venture out of the nest and nibble at grass. Roughly on the 20th day, when the body weight reaches 60-70 g, young pikas change over to independent life. The growth of hairs by this time ceases, and the fresh pelage becomes lighter (Sludskii et al., 1980).

Up to the first molt young red pikas have a furry hair coat, brownish-gray with rusty tone on head and sides. The change of juvenile fur begins on the 25-27th day, when the animals weigh 90-110 g, and terminates after 1-1.5 months. Some young pikas, by the age of two months, already have the characteristic rusty color of the upper part of their body. By the age of three months, young pikas attain adult sizes. In the year of their birth red pikas, as a rule, do not participate in reproduction; they become sexualy mature only by the spring of the following year, that is, at the age of 8-11 months (Bernshtein and Klevezal', 1965).

Molt. In red pikas the spring molt begins in April. First of all changes occur in the hair coat of the upper part of head, and in May heads of the majority of pikas have a rusty summer fur. Molting continues throughout the summer, but its character varies in different individuals. Thus, in August in the fully molted males, there are still areas with the winter fur on hind part of back. In females, at this time, the winter fur is retained not only on the rump, but also on back and sides. During winter the molt involves not only the summer pelage, but also the previous year's remaining winter hairs are replaced. Thus, the spring molt not only extends longer, but is also incomplete (Bernshtein, 1963).

The fall molt begins in September and proceeds most intensively. In a month the animal is almost fully clad in its winter coat. Molting starts from the middle of the back and terminates in the area of the head and belly. It has been reported that, in pikas living in the Pamir-Alai high mountains, the fall molt occurs in a more condensed period (Davydov, 1974).

In young red pikas the change of the juvenile coat to the adult summer coat occurs at the age of 25-27days (Sludskii et al., 1980) or 40 days (Davydov, 1974). This molt is intensive. To begin with, hairs in the midde of the back and in the area of the paws begin to molt, followed by hairs on the head and sides. The last area to molt is the belly. Young molted pikas differ from the over-wintered adults by their brighter color (Bernshtein, 1964).

Sex and age-wise composition of population. In red pikas, judged from the analysis of embryos in the Trans-Ili Alatau, a similar number of males and females are born, although in later samples males are slightly predominant (Table 3.4), which is rather related to their higher activity. On the whole, the sex ratio in red pikas is close to 1:1 (Bernshtein, 1964).

Table 3.4. Sex composition (in %) of a population of red pikas
in the Trans-Ili Alatau in 1959-1962 (from Bernshtein, 1964)

Age of animals	n	Females	Males
Adults	301	46.8	53.2
Young	415	47.8	52.2
Total	716	47.3	52.7

52

A quite long life span (to 3 years) is characteristic of red pikas, which is attributed to their low mobility, better protective conditions and a small number of enemies, low fecundity and high level (about 50%) of mortality of the young. Because of this, replenishment of a population of red pikas occurs quite slowly—slower than, for example, in such species as Pallas' pika, Daurian pika and little pika. According to the data of A.D. Bernshtein (1964), in the Trans-Ili Alatau, by the end of the reproductive season, the population of red pikas consists of the resident and overwintering individuals, the precise age of which is difficult to determine. In August, of 207 pikas caught, adult reproducing individuals constituted 28%, and in fall and winter their contribution in the population decreased slightly— to 22%.

Population dynamics. Stable fecundity and relatively long life span of red pikas cause a quite stable level of population of this species over a large part of their range (Bernshtein, 1964; Davydov, 1974; Sludskii et al., 1980).

However, quite sharp fluctuations are possible in populations of red pikas. Thus, in the upper reaches of the Urudar'ya River (Gissar Range) above the limits of juniper forest, on an area of 25 ha, in 1979, 5 families of red pikas were recorded with the total number of individuals at 60-70. In 1984, in the same area, only one pika was found, and in 1985—two. In 1985, a lower number of red pikas was also recorded in the upper reaches of the Kyzyldar'ya River (Tret'yakov et al., 1989).

Enemies. The main enemies of red pikas are ermines, which readily enter the shelters of pikas between rocks, where they destroy, first of all, the young ones. In the Trans-Ili Alatau, red pikas constituted 7.3 to 20% in the diet of ermines (Sludskii et al., 1980). Despite the fact that ermines are quite numerous in stony habitats, they hardly pose a serious threat to the population of red pikas. Thus in 1959-1960, in the region of the Great Alma-Ata Lake with very high number of ermines, the number of red pikas did not decrease (Sludskii et al., 1980). In central Tadzhikistan and the northern Tien Shan, stone martens hunted red pikas (Davydov, 1974). In the Trans-Ili Alatau, red pikas constituted 1.2 to 12.5% in the excrements and stomachs of stone martens (Sludskii et al., 1980).

53

Avian predators, in principle, can hunt red pikas, but do not pose a serious threat to their population.

Competitors. Throughout the range of the red pika, high mountain voles (*Alticola argentatus*), gray hamsters (*Cricetulus migratorius*), forest mice (*Apodemus sylvaticus*), and in the Tien Shan—Tien Shan voles (*Clethrionomys frater*) live together. However, a sufficient quantity of food and shelter make it possible for these species to coexist without appreciable competitive interactions. On the edges of stone mounds, where red pikas live, marmots also live; but they usually forage in grass plots and also do not compete for food with red pikas.

Diseases and parasites. In the Trans-Ili Alatau, 8 species of fleas have been detected on red pikas. *Paraneopsylla jatti* and *Amphalius clarus* are the dominant species, while *Chaetopsylla homoesus, Rhadinopsylla dahurica, Ceratophyllus rectangulatus, C. penicilliger, Frontopsylla ornata* and *Amphipsylla primaris* are found on pikas as individual specimens (Busalaeva and Fedesenko, 1964; Sludskii et al., 1980). In Tadzhikistan, two species of fleas have been found on red pikas: *Ceratophyllus lagomys* and *Frontopsylla ambigua* (Davydov, 1974).

Ixodid ticks were found on red pikas only in Tadzhikistan. Gamassid mites *Eulaelaps stabularis* and *Hirstonisus eusoricus* were found in pikas in the Trans-Ili Alatau, while *Allodermanyssus sanguineus* was found in the area of Lake Iskanderkul' (Davydov, 1974; Sludskii et al., 1980).

Among the endoparasites in red pikas, 8 species of helminths have been detected: *Dicrocoelium lanceatum, Hastilesia ochotonae; Schizorchis altaica, Cephaluris andrejevi, Dermatoxys schumakovitschi, Labiostomum vesicularis, Trichostrongylus colubriformis* and *Murelius tjanschaniensis* (Gvozdev, 1962; Tokobaev, 1976).

In Tadzhikistan (Gissar Range), in August 1953, one red pika showed 12 larvae of the warble fly *Oestromia laporina* (Davydov, 1974).

Practical Significance

In 1936, experimental processing of hides of red pikas was initiated, which continued until the 1950s. In recent times processing of hides of red pikas has not been done.

As already said, the red pika is one of the food objects of some species of mustelid—valuable fur animals.

Red pikas can apparently inflict some damage to woody and shrubby species and, at places, they can delay replenishment of spruce saplings by nibbling at the growing parts under the root (Bernshtein, 1963).

Infectious diseases, epizoological and epidemiological importance of red pikas have not been studied.

54

Afghan Pika
Ochotona (Ochotona) rufescens Gray, 1842

1842.   *Lagomys rufescens* Gray. Annals Magazine of Natural History, Vol. 10, p. 266, Afghanistan, Kabul.

1911.   *Ochotona rufescens regina* Thomas. Annals Magazine of Natural History. Ser. 8, Vol. 8, p. 760. Turkmenia, Kopetdag Range, west of Ashkhabad.

1911.   *Ochotona rufescens vizer* Thomas. Ibid. Iran, northern end of Kohrud Range, north of Isfahan.

1961.   *Ochotona rufescens shukurovi** Heptner. Zoologicheskii Zhurnal, Vol. 40, Issue 4, p. 621, Turkmenia, Krasnovodsk region, Greater Balkhans, Kendyrli.

Diagnosis

Skull large, condylobasal length more than 43 mm. Interorbital distance narrow—less than 4.8 mm. Frontal bones with crests above the orbits; lacking foramina. Incisor fossa not partitioned by processes of inter-maxillary bones. Summer fur differs little from winter, quite light in color, brownish-gray with ochreous bloom on back, gray on belly. Length of vibrissae not exceeding 58 mm.

Description

Body length of adult animals—150-220 mm (average 190 mm), length of foot—30.5-35 mm (average 33.5 mm), length of ear 20-25 mm (average 22 mm). Body weight of adults varies considerably and depends on the sex, age and geographical disposition (155-236 g).

Summer fur on back with predominance of gray tones with brownish-ochreous tinge in the middle of the back: the top of head rusty-ochreous; the upper part of neck, behind the ears and cheeks, has a collar of dull gray hairs; below the ear there is a reddish-brown rusty spot that is variable in size and intensity of coloration: sides of the body are lighter than the back, with a yellowish tone: the belly and feet are light gray with a dull yellow tone; ears are gray with an admixture of rusty tones. Rusty tones are absent in the coloration of young; brown color predominates.

Winter fur is dull, with a predominance of gray tones; the color of the head is the same as that of the back or slightly more rusty; sides of the body and head do not differ in coloration from the back or are lighter (Fig. 15).

---

shucurovi in the synonymy—General Editor.

55

Fig. 15. The Afghan pika, *Ochotona rufescens*, in winter coat
(photo by E.Yu. Ivanitskaya).

Vibrissae of moderate length—52-58 mm.

The skull is relatively narrow and the frontal section is raised in an arc and well sculptured. The condylobasal length of the skull is 32-49.9 mm; zygomatic width, 23-25.1 mm; and interorbital width, 3-4.9 mm. Frontal bones lack foramina and have supraorbital crests of various degrees of manifestation. Incisor foramen is not partitioned by the processes of intermaxillary bones. The preorbital foramen is triangular in form and relatively small. The nasal bones are fused with each other. The orbits are large and oval in form. The zygomatic arch is relatively massive, and the zygomatic processes are short and curved in the region of the orbit. The mandible is relatively long and low. The ascending ramus of the mandible is quite high. The angular process is broad. The angular notch is deep and semicircular (Fig. 16). The third lower premolar tooth is triangular; its anterior segment is very small, located closer to the outer margin of the posterior segment; the posterior segment is large, trapezoidal, and often without inner folds (cf. Fig. 16a).

The os penis is relatively long and thick—the ratio of its length to thickness is, on average, 5.6 (Aksenova and Smirnov, 1986).

The diploid complement has 60 chromosomes: 9 pairs of meta-submetacentrics, 5 pairs subtelocentrics and 15 pairs of acrocentrics; X chromosome—submetacentric, Y-chromosome—smallest, acrocentric (Vorontsov and Ivanitskaya, 1973).

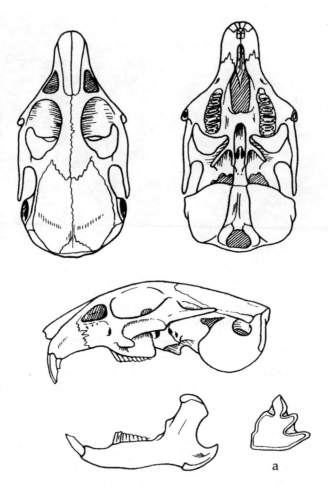

56

Fig. 16. Skull of the Afghan pika *Ochotona rufescens*.
a—third lower premolar tooth of the Afghan pika.

55 ## Systematic Position

Phylogenetic links of Afghan pikas have been interpreted variously and are based on different characters—morphological or ecological. In morphological and cytogenetic characters, *O. rufescens* is undoubtedly close to *O. daurica* and *O. curzoniae*.

Geographic Distribution

Afghan pikas are widely distributed in the mountains of Iran, Afghanistan and the northwestern part of Pakistan, and only the extreme northeastern end of the range of this species enters southwest

Turkmenistan, where Afghan pikas live on the bordering ranges from the Greater Balkhans in the west to the Kopetdag in the east (Fig. 17a).

The Greater Balkhans are the extreme northwestern limit of the range of the Afghan pika. Here, Afghan pikas were caught in the environs of Kendyrli and Kosh-Goi, at the Sakka, Bash-Mugur and other springs (Sapargel'dyev, 1987; collections of ZM MGU and ZIN). On the Lesser Balkhan Range, the Afghan pika was reported near the Chal-Su springs and in the environs of Torangla (Sapargel'dyev, 1987). Reports are available about their presence in the environs of Kazandzhik and Danat Kishlak— at the junction of the Lesser Balkhans with the Kyurendag Range (Ognev, 1940; collections of KGU). On the Kyurendag Range itself, the Afghan pika was caught by M. Sapargel'dyev (1987).

In the western Kopetdag, Afghan pikas have been reported in the environs of Nukhur, Kara-Yantam gorge, on the Chandyr Pass, in Chokrok tract (between Naarli and Kizil-Imam), and in the environs of Kara-Kala (Bondar' and Zhernovov, 1960). Afghan pikas were not found south of the Chandyr River. In the collections of ZIN there are specimens of Afghan pikas caught by Zarudnyi in 1884 in the environs of Kizil-Arvat; here, two Afghan pikas were caught in 1965, but probably they were imported here with the hay cured in the mountains (Fomushkin et al., 1967). In 1964, two Afghan pikas were caught in the environs of Bakharden (collections of KGU). However, according to the scheme of E.P. Bondar' and I.V. Zhernovov (1960), in western Kopetdag, the northern boundary of the range of the Afghan pika passes along the northern spurs somewhat south of Bakharden and Kizil-Arvat (Fig. 18).

According to the data of S.I. Ognev (1940), in the central Kopetdag Afghan pikas are occasionally found in the mountains near the village of Verkhneskobolevskoe. They were more numerous in the vicinity of the village of Mikhailovskoe (Germob), at the mouth of the Sakis-Yaba River, along the gorge of the Mergen-Uli River, and in Kurkulabsk gorge. Afghan pikas are common in the Firyuzinsk gorge 15 km south of Geok-Tepe and in the Chul' and Tutla gorges 8 km south of Ashkhabad (Minin, 1938; collections of ZM MGU and ZIN). Farther to the east Afghan pikas were met with in the valley of the Firyuzinsk stream near Gaudan, in the vicinity of the village of Yablonovskii in the ruins of stone structures in Khiviabad, in the vicinity of the city of Markou (Minin, 1938; Ognev, 1940; Formushkin et al., 1969; collections of the "Mikrob" Institute, Saratov). N.V. Minin (1938) considered Dushak and Kel'techinaru as the extreme eastern distribution of the Afghan pika. At the present time the eastern boundary of distribution of the Afghan pika on the territory of Turkmenistan passes along the latitude of Tedzhen (Fomushkin et al., 1967; Sapargel'dyev, 1987).

Beyond the borders of Turkmenistan, Afghan pikas live in Iran (northern

74

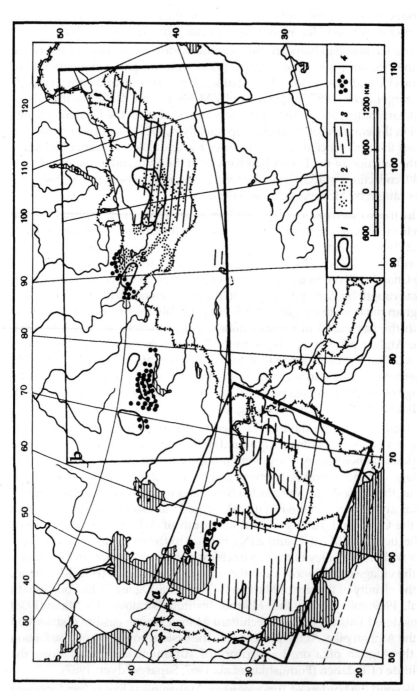

Fig. 17. See next page for caption.

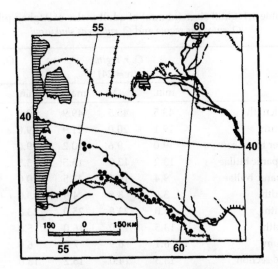

Fig. 18. Locations of Afghan pikas on the territory of Turkmenistan.

part of the Kohrud Range), north-western Pakistan (central Baraguch Range), and in eastern Afghanistan (the Baba, Pagman and Khingil ranges).

Geographic Variation

There are two subspecies of Afghan pika in Turkmenistan.

1. *Ochotona rufescens regina* Thomas, 1911. The Kopetdag Afghan pika. This form differs from the type form by having a larger body (175-220 mm) and skull (condylobasal length 43-49.9 mm) sizes, a larger tympanic bullae (length—12.1-15.5 mm, width 9.1-11.5 mm), and in having lighter coloration. Distributed in the southern part of Turkmenistan—the Kopetdag Range.

2. *Ochotona rufescens shukurovi* Heptner, 1961. Shukurov's Afghan pika. This form differs from the Kopetdag subspecies by having a shorter skull (condylobasal length—43-45.1 mm) with relatively identical zygomatic width (Table 4.1), relatively larger tympanic bulla, and duller coloration. Distributed in southwest Turkmenistan—in the Greater Balkhans.

Outside of the borders of Turkmenistan two more subspecies of Afghan pikas are known—the type subspecies, distributed in eastern Afghanistan,

Fig. 17. The range of the Afghan pika (a) and Pallas's pika (b). *1*—schematic range according to S.I. Ognev (1940); *2*—range of Pallas's pika in the territory of Mongolia (Sokolov and Orlov, 1980); *3*—schematic range according to Corbet (1978); *4*—our data.

Table 4.1. Skull sizes (in mm) of adult males of Afghan pikas
O. rufescens regina and *O. rufescens shukurovi*

Indices	O. r. regina (n = 24)			O. r. shukurovi (n = 10)		
	min.	M	max.	min.	M.	max.
Condylobasal length	43.5	45.5	49.9	43.5	43.4	45.2
Length of rostrum	19.1	20.0	21.9	19.0	19.5	20.0
Length of upper tooth row	9.0	9.6	10.2	9.0	9.4	9.9
Length of tympanic bullae	13.2	13.9	15.5	13.5	14.5	15.0
Width of tympanic bullae	9.4	10.4	11.5	10.0	11.0	12.0
Interorbital width	3.0	3.8	4.9	3.0	3.6	4.5
Zygomatic width	23.0	23.8	25.0	23.3	24.2	25.1
Postorbital width	13.5	15.0	16.0	15.0	15.6	16.5
Width of rostrum	6.1	6.9	7.8	6.0	6.2	7.0
Hight in frontal section	13.3	14.0	15.2	13.0	13.7	14.1
Height of rostral section	7.8	8.4	8.9	7.8	7.9	8.1

and *O. rufescens vizeri* Thomas, 1911, which differs from the type subspecies by having a smaller body size, shorter ears, smaller tympanic bullae and brighter coloration, it lives in Iran—in the northern part of the Kohrud Range. The subspecies *O. rufescens vulturna* Thomas, 1911, was described from a lone aberrant specimen from Baluchistan (Kharboi near Kelat), hence cannot be regarded a separate taxon.

Biology

Habitat. Afghan pikas are inhabitants of arid montane slopes. They live in areas differing in altitude—from the foothills of mountains and bottoms of stony gorges at sea level to the upper xerophytic mountain belt at an altitude of 1,500-2,000 m above sea level. Afghan pikas do not enter the region of foothill plains, nor do they occupy level montane slopes and their summits. Typical habitats of Afghan pikas are the montane valleys, gorges and ravines with stone mounds, outliers and large stone barriers, as also montane slopes of different gradients and exposures with rubbly ground and individual stone mounds. Vegetation in the habitats of the Afghan pika is sparse, and predominated with cereal-wormwood associations, ephedra and pea shrub (Sapargel'dyev, 1987).

The area occupied by colonies of Afghan pikas depend primarily on the geomorphological peculiarity of the locality and availability of suitable habitats and then on the number of individuals. Afghan pikas most often colonize only local areas of dissected relief, not subjected to spring floods. The area of such colonies varies from 0.25 to 2-3 ha (Sapargel'dyev, 1987).

Usually the colonies of Afghan pikas extend along slopes, and sometimes one colony closely touches the other. Subtropical regions of mountains of the western Kopetdag are the most favorable for Afghan pikas, where an abundance of sites for shelters are associated with rich vegetation, whose growth period is longer than, for example, in dry valleys of central Kopetdag. However, very wet places with dense thickets and water sources are avoided by Afghan pikas throughout.

At places where pika colonies are in the neighborhood of human dwellings, pikas dig burrows in old ruins, and they live in recesses during temporary storms.

Population. The maximum population density of Afghan pikas is observed in isolated and small areas. This regularity is manifest in all parts of the range of the Afghan pika. For regions of western Kopetdag a higher density of colonization of Afghan pikas has been observed than in other regions (Table 4.2). Western Kopetdag with its subtropical climate is the most favorable region for habitation of Afghan pikas.

Table 4.2. Population density of Afghan pikas in some regions of Turkmenistan (Sapargel'dyev, 1987)

Region	Year	Time of observation	Area, ha	Total number of individuals	Number of individuals/ha
Greater Balkhans					
	1967	May-June	0.5	14	28
	1967	May-June	1.2	15	12
	1967	May-June	1.5	22	15
	1968	April-May	3.1	13	4.3
	1968	April-May	0.9	9	10.5
Western Kopetdag					
	1967	June	1.8	37	21.1
	1968	May	0.2	9	57.1
	1968	May	0.7	23	36.8
Central Kopetdag					
	1967	July	6.1	43	7.0
	1967	July	1.9	14	7.0
	1967	July	5.2	25	5.1

Shelters. Reliable hideouts for Afghan pikas are the large stone mounds and outliers with a large number of niches and cracks. In such places, the ground is usually firm, and Afghan pikas almost do not dig burrows, using natural hollows as temporary and permanent shelters. In places with a loose ground, Afghan pikas sometimes broaden and deepen the tunnel, one wall of which is formed by the stone and the other by the soil layer. The depth of burrow tunnels in such habitats does not exceed 10-15 cm and is limited by the length of natural niches and pliability of the ground (Sapargel'dyev, 1987). In habitats with stone barriers but a softer ground, often found in montane valleys and small depressions, Afghan pikas dig burrows with rather complex underground tunnels that have many outlets. Often the outlets are masked by natural heaps of stones.

60   During the reproductive period digging activity of Afghan pikas increases. Deeper burrows appear with complex tunnels and nest chambers that are located at a depth of 80-110 cm. Sometimes the nest chamber is situated closer to the land surface—at a depth of 30-40 cm (Fig. 19). The burrow of Afghan pikas is a rather simple system of tunnels 8-10 cm in

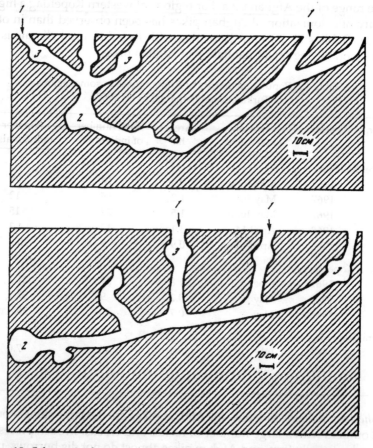

62

Fig. 19. Schematic drawing of burrows (vertical section) of Afghan pikas. 1—entrance opening; 2—nest chamber; 3—broadened tunnel (from Sapargel'dyev, 1987).

diameter with several chambers for various purposes. Usually 25-30 cm from the entrance there are two small chambers 15-20 cm in diameter, which are sometimes used by animals for underground storage of cured food. There is one nest chamber, usually spherical in form; its diameters about 25 cm. Before the nest chamber, 10-15 cm from it, there is a branch burrow that serves as a lavatory for the pikas. The total length of burrow tunnels in the reproductive period reaches 1.4-4.0 m (Sapargel'dyev, 1987). Afghan pikas pack the burrow with a bed of soft grasses, usually bluegrass. Near human dwellings, rags, animal wool and bird feathers are used as bedding in burrows. A female begins to dig her burrow 2-3 days before the first confinement, the same burrow is used also for feeding the second brood.

Besides the main burrow, both adult and young Afghan pikas use temporary shelters—cracks and recesses between rocks, where usually grass reserves are stored. These temporary shelters are located, most often, not more than 20 m from the main burrow (Zaletaev and Sapargel'dyev, 1975).

Feeding. Afghan pikas feed exclusively on plants. As a rule, Afghan pikas use those species of herbaceous plants and shrubs that grow near their habitations. The species composition of plants eaten and stored by Afghan pikas changes over their range. The composition of food also changes according to the season. In cold winters and dry summer periods Afghan pikas feed mainly on plants from their haypiles. In Kopetdag, in warm winters, winter vegetation of different ephimers is observed, and in such periods Afghan pikas almost do not touch their haypiles but feed on green food.

Afghan pikas store food twice a year. Winter haypiles are made only by adult animals and used during the dry summer period. Fall storage of food starts from the middle of September and continues until the onset of frosts; both adult and young individuals take part in this act.

Almost all species of plants surrounding the habitations of Afghan pikas are stored. The list of plants eaten and stored by Afghan pikas in central and western Kopetdag is available in the monograph of M. Sapargel'dyev (1987). The species composition of plants in haypiles of these pikas in central Kopetdag is twice as large as in haypiles of these pikas in western Kopetdag.

61   The weight of stored food varies from 0.5 to 5 kg (Sapargel'dyev, 1987). The haypiles of Afghan pikas living on the southern slopes of level areas, as a rule, have lesser mass than haypiles of pikas from the northern slopes and stony areas. The fall haypiles are usually smaller than spring haypiles and brouse plants predominate in them.

Afghan pikas do not make open piles. They store hay under stones and in cracks and niches without preliminary drying. The hay is pushed in the burrow where it is stored in small chambers. Most often each individual makes an individual haypiles, but occasionally collective haypiles have been observed. These pikas collect plants for storage often near the shelters, but sometimes gather plants from 40-50 m away from them (Sapargel'dyev, 1987).

In the course of a year pikas store much more food than they can use in unfavorable seasons. In captivity an adult Afghan pika may eat 180-200 g of green food in a day (Sapargel'dyev, 1987).

Activity and behavior. Afghan pikas lead a diurnal mode of life. In summer, they are active from dawn to the onset of darkness. In the morning or evening hours, when pikas remain on the land surface for long, they are either feeding or sitting immobile or lying on rocks not far from their shelters. During the period of food storage, the maximum activity is noticed in the morning hours. On particularly hot days, pikas are active also at dusk, and in midday do not venture out of their burrows. Information on nocturnal activity of Afghan pikas is not available (Sapargel'dyev, 1987).

In winter, the nature of daily activity of Afghan pikas changes somewhat—the maximum activity is observed in midday, while in the morning after sunrise and at dusk they rarely leave their shelters. In January-February (before the start of reproduction), Afghan pikas are more active in their burrowing activity than in other seasons.

Afghan pikas are aggressive only in the period of mass dispersal of young—in the middle of August. At this time, fights are common, during which pikas can injure each other with varying degree of severity (Sapargel'dyev, 1987). In the beginning of September, when limits of individual territories and population hierarchy are established, aggressiveness of behavior of Afghan pikas disappears—they forage in the same areas, use the same temporary shelters, and make collective food reserves. A male most often helps two-three nearby females in digging burrows and curing food and also mates with these very females. During the entire first reproduction cycle until birth of young of the first brood and subsequent copulation, a male lives in burrows of females and uses their temporary shelters. After the second mating males lead a solitary life, though not far from females and offspring.

62   After the appearance of the second brood, young of the first litter remain in one nest chamber. Hostile relationship of female toward young of the first litter was not observed after the appearance of the second brood. Usually the nursing female remains outside the nest burrow. In the brighter time of the day, females feed their young 3-4 times, 6-12 min each time. Young suckle milk stretching their head upward and, after they become strong, they rise high on front feet; in the more advanced stage they feed standing on their hind feet (Sapargel'dyev, 1987).

Thus, for Afghan pikas a complex family is characteristic, in which there is one male, many independently living females, and young, sometimes of two generations. One or several such family colonies form habitations of Afghan pikas (Zeletaev and Sapargel'dyev, 1975).

Despite their colonial mode of life, vocal signals are weakly developed in Afghan pikas. It is possible to hear the call of Afghan pikas only from a close distance. In the absence of danger, pikas produce a weak melodious squeek. Their frightened squeek is prolonged and louder. A vocal signal from an Afghan pika begins with a short, scarcely audible whistle, which turns into a trill resembling a suppressed crackle of an alarm clock (Flerov and Gromov, 1934). According to the observations of M. Sapargel'dyev (1987) the vocal signal in Afghan pikas is developed weakly and even in the event of danger, they quietly hide in the burrow, only occasionally producing a low squeek.

Area of habitation. Data on the individual territory or family territory of the Afghan pika are not available.

Reproduction. Afghan pikas bear two litters in a year; the reproductive cycles are proximate and repeated estrus in females starts a day or two after the first births. The reproductive period extends 3-3.5 months. The most complete information on reproduction of Afghan pikas of Turkmenistan is available in the monograph of M. Sapargel'dyev (1987). It has been reported that the start of reproduction of Afghan pikas depends on the conditions in the preceding winter, on weather in spring, food reserves and the age structure of the population. In different years the time of beginning and end of the reproductive period may shift by 7-10 days. Much greater differences in the period of reproduction are observed in different subspecies of Afghan pikas. Reproduction in Shukurov's pika starts 3-4 weeks later than in Kopetdag pikas, which is associated with a later onset of spring and less rich plant cover in the Greater Balkhans.

In the central Kopetdag the rut in Afghan pikas begins in the middle of February. In 1967, large-scale mating was observed here on March 1-5, and in 1968—starting from February 15. In the end of February (1968) and beginning of March (1967), in the vicinity of Firyuza, females were found already with embryos. In the end of March-April, in central Kopetdag, 90.3% of females participated in reproduction. In May the number of pregnant females drops—in 1967 they were only 31%, and in June-July pregnant Afghan pikas were not found in Kopetdag (Table 4.3).

In the western part of the range (Greater Balkhans Range), pregnant females of Afghan pikas are found from March to the middle of June. The nature of dynamics of reproductive processes in the two Turkmenian subspecies of Afghan pikas is similar, and differs only in the timing of the beginning and end of reproduction (Sapargel'dyev, 1987).

Table 4.3. Reproductive characteristics of female Afghan pikas in 1967-1968 in different regions of Turkmenistan (from Sapargel'dyev, 1987)

Region	Month	Total number of females investi-gated	Pregnant	Nursing	Barren	Average number of embryos per female	Average number of placental scars
Central	Feb.	4	2	–	2	6.2	–
Kopetdag	Mar.	20	15	–	5	6.7	–
	Apr.	18	12	6	–	7.6	7.3
	May	4	–	4	–	–	9.3
	Jun.	18	–	18	–	–	7.5
Western	May	21	11	8	2	6.3	7.1
Kopetdag	Jun.	7	–	7	–	–	7.0
Greater	Apr.	15	9	1	5	6.3	8.0
Balkhans	May	35	11	13	11	6.0	6.5
	Jun.	7	–	7	–	–	7.1

In years with an increasing population, some resident females from the first brood also participate in reproduction. They attain sexual maturity at the age of 5-6 weeks. Young males of the year do not participate in reproduction.

According to data from 1966-1968, in the Kopetdag the number of embryos in one female did not exceed 11; while in the Greater Balkhans, 9 (Sapargel'dyev, 1987). The average number of young in first and second litters from the Greater Balkhans, as well as the Kopetdag Afghan pika, is somewhat different. The mean value of litters in the Kopetdag is somewhat higher than in the Greater Balkhans (Table 4.4). The number of young in litters of Afghan pikas of the Kopetdag varies from 3 to 11; but in the Greater Balkhans (according to data from April-May, 1967-1968), from 3 to 9. During summer one female Afghan pika bears 1 to 16 young; in the Kopetdag populations the maximum is 16; in the Greater Balkans, 11.

Growth and development. In Afghan pikas gestation lasts apparently for 26 days.

In Afghan pikas young ones are born absolutely naked, with weakly pigmented skin. In newborns eyes are closed, incisors appear like protruding white tubercles, and labial vibrissae are developed. The weight of newborns of the first brood is 9.1-9.7 g (average 9.5 g); weight of newborns of the second brood, 5.7 to 7.5 g (average 5.9 g). Observations on newborn Afghan pikas in captivity (Sapargel'dyev, 1987) showed that on the 3rd day young are entirely covered with sparse short fur. Cutting of

Table 4.4. Litter size of Afghan pikas in different regions
of Turkmenistan (from Sapargel'dyev, 1987)

Region	n	$M \pm n$
Firyuza	35	7.1 ± 0.23
Karak-Kala	25	6.8 ± 0.29
Greater Balkhans	35	6.3 ± 0.31

teeth occurs on the 8-9th day, and eyes also open at that time. On the 11th day young begin to eat grass, which is brought to them in the nest by the female. On the 12th day young begin to leave the nest. Independent life begins in Afghan pikas eighteen days after birth. Young Afghan pikas grow very fast. Their body weight triples by the ninth day. Body length, which at birth, on average, is 48.0 mm, doubles in 15-20 days. By the 53rd day of life young Afghan pikas, on average, weigh 145 g, and their average body length is 159.5 mm. Length of hind feet and ears, which at birth is 9 and 3 mm, respectively, increases particularly intensively during the first three weeks of life of these animals.

Young Afghan pikas completely change over to feeding on plant food 20 days after birth. Initially they eat only tender parts of plants, later they change over to coarse food.

Molt. Afghan pikas molt twice a year—in spring and fall. The spring molt begins in April and extends to the end of July. The upper part of head molts first, followed by the anterior part of body; change of hairs in each section of the body bears a mosaic character. By the end of May molting reaches the sides of the body and limbs. As a rule, in Afghan pikas there is no complete change of winter coat (Sapargel'dyev, 1987). For Afghan pikas living in the territory of Pakistan, it has been reported that the higher the altitude of the locality of their habitation above sea level, the longer the winter fur stays on them and the earlier the fall molt (Fulk and Kokhar, 1980).

The fall molt in the Turkmenistan subspecies of Afghan pikas begins in September and continues, unlike the spring-summer molt, intensively and for a relatively short period; by November almost all animals have a winter hair coat.

Young animals change their juvenile coat to that of an adult at the age of 25-30 days. This molt proceeds faster and simultaneously covers almost the entire body surface of animals. Juvenile molting in Afghan pikas coincides in time with the period of dispersal of young (Sapargel'dyev, 1987).

Sex and age composition of the population. Most often females predominate in populations of Afghan pikas. During the year, depending on

Table 4.5. Sex ratio (%) in different age groups of Afghan pikas in Turkmenistan (from Sapargel'dyev, 1987)

Region and time of observation	Adult			Young			Residents		
	n	Females	Males	n	Females	Males	n	Females	Males
Central Kopetdag									
March-April, 1967	31	64.5	35.5	17	64.7	35.3	9	55.5	44.5
July, 1967	59	59.3	40.7	–	–	–	67	68.7	31.3
September-December, 1968	12	50.0	50.0	19	89.5	11.5	–	–	–
February-April, 1968	11	63.6	36.4	9	44.4	55.6	–	–	–
Western Kopetdag									
June, 1967	38	34.2	65.8	–	–	–	67	58.2	41.8
May, 1968	10	70.0	30.0	15	86.7	13.3	25	72.0	28.0
Greater Balkhans									
May, 1967	33	72.7	27.3	18	66.7	33.3	7	28.5	71.5
May, 1968	11	81.8	19.2	19	57.9	42.1	21	57.1	42.9
May, 1973	16	62.5	37.5	21	59.2	40.8	8	75.0	25.0
May, 1974	14	71.4	28.6	17	61.3	38.7	–	–	–

the season and age composition of population, the sex ratio of Afghan pikas changes somewhat (Table 4.5). In the reproductive season, when a population comprises sexually mature individuals of different ages, females predominate, which determines the social structure of populations of Afghan pikas. In the reproductive period family groups are formed, which include 2-3 females and 1 male. Colonies of Afghan pikas are usually comprised of 8-15 such groups.

In spring the bulk of the population of Afghan pikas, both in the Kopetdag and the Greater Balkhans, comprise individuals that have survived one to two winters. Three-year-old Afghan pikas are found rarely in natural populations (Table 4.6). Although possible, individuals may live even to five years of age (Zaletaev and Sapargel'dyev, 1985).

Table 4.6. Population age structure of Afghan pikas (from Sapargel'dyev, 1987)

Region and period collection		n	Pikas of different ages, %			
			Less than 1 year	1 year	2 years	3 years
Firyuza	(April)	28	–	50	36	14
	(July)	71	69	17	11	3
Greater Balkhans	(May)	22	–	54	41	4
Karakala	(June)	47	60	30	8	2

In summer the bulk of the population is comprised of resident individuals. In spring overwintered Afghan pikas are comprised of two groups, one of which consists of the preceding year's individuals of the first brood, which in size and weight do not differ from older individuals; the second group consists of smaller individuals from the second brood. Both groups of Afghan pikas actively participate in reproduction.

By fall the population of Afghan pikas consists of resident individuals. Total replenishment of the population takes place in three years (Sapargel'dyev, 1987).

Population dynamics. The population of Afghan pikas is not uniform in different parts of the range, in different habitats, and in different seasons of the year. For this species a periodic fluctuation of population is also characteristic, when years of high population alternate with its sharp or gradual decrease. Thus, on the Greater Balkhans Range in 1933, a very high population of Afghan pikas was reported (Laptev, 1934), and after a year their population fell sharply in this region (Vinogradov, 1952). A high population of Afghan pikas in the Greater Balkhans was reported in 1946, 1948, 1954, 1959, 1964-1966 (Laptev, 1934; Bondar' and Zernovov, 1960; Shukurov, 1962; Fomushkin et al., 1967; Sapargel'dyev, 1987). In the central

Kopetdag in a typical habitat of Afghan pikas, according to the observations of M. Sapargel'dyev (1987) in 1965, over a 1 km-route, on average, 21.5 individuals were found; in 1956, 13.7; in 1967, just 3.6; and in 1969 not even one animal was recorded in this area. In the western Kopetdag 11 Afghan pikas were found over a 1-km route, but one year later only 4.

Drought, severe winter combined with starvation, and sometimes predators adversely affect the population of Afghan pikas. In high-population years Afghan pikas may suffer from an insufficiency of available habitat. In such periods Afghan pikas are forced to live in habitats that are not typical for them, which leads to accentuated competitive interactions as well as to their death due to predators. As a result, the population of Afghan pikas falls.

Enemies. Mustelids, in particular weasels, are serious enemies of Afghan pikas; weasels may readily enter shelters of pikas. Stone martens also feed on Afghan pikas, but nowhere is their number high. Weasels are widely distributed in Turkmenistan; they very frequently hunt Afghan pikas. Predatory birds also feed on Afghan pikas: kestrel, little owl, vulture, black kite, eagle owl, and goshwak, as well as crows (Sapargel'dyev, 1987). One of the main enemies of Afghan pikas is the blunt-nosed viper. In high-population years Afghan pikas are attacked by monitors, and young Afghan pikas are also preyed on by common sheltopusik [*Ophisaurus apus*—Translator].

Competitors. Persian jird, gray and mouse-like hamsters, house and field mice, and mouse-like dormouse live in the habitats of Afghan pikas. However, these species of rodent lead an exclusively nocturnal mode of life and are not significant competitors of Afghan pikas. Moreover, the set of plants used as food by rodents and Afghan pikas differ somewhat. Thus, hamsters are predominantly granivorous, while Persian jird, and forest mice in unfavorable periods of the year, may satiate themselves with dry remnants of grasses. Competitive interactions of Afghan pikas with birds and mice are more probable in spring and fall when their diets match partially. Competition from other herbivores (hares, crested porcupines, ungulates) is also insignificant, since these species feed usually beyond the habitations of Afghan pikas, only occasionally entering stony mounds.

Diseases and parasites. Among ectoparasites 4 species of fleas have been reported in Afghan pikas: *Ceratophyllus* tiflovi, C. turkmenicus, Ctenophyllus rufesetus*** and *Wagneria schelkovnikovi*; 5 species of ixodid ticks: *Ixodes redikorzovi, Haemaphysalis punctata, H. octophila, Dermacentor daghestanicus,* and *Rhipicephalus rossicus*; 4 species of red mites and their larvae:

---

*In Russian original, *Cetotophyllus*—General Editor.
**In Russian original, *rufesetns*—General Editor.

Eriocotronmbidium coropinius and *Leptotrombidiiium** sp. (Dubinin and Bregetova, 1952; Paramonova et al., 1958; Zagniborodova, 1960; Bakhaeva, 1960; Amanguliev and Sapargel'dyev' 1970).

The helminth fauna of the Afghan pika is unique and species specific. In the territory of Turkmenistan, 16 species of helminths belonging to 13 genera have been recorded in Afghan pikas. The species composition of parasitic worms found in Afghan pikas is characterized by geographic variation (Table 4.7). Of 19 species of helminths found in Afghan pikas, 2 species from the genera *Pikaeuris* and *Fastigiuris* are found only in *O. rufescens*; members of the genera *Cephalurus* and *Labiostomum* are also found in other species of pikas. For 6 species of parasitic worms, Afghan pikas are not the usual hosts (cf. Table 4.7). For *H. krepkogorski* Afghan pikas are an intermediate host; for *S. lupi*, a reservoir host (Babaev and Sapargel'dyev, 1970).

Table 4.7. Species composition of helminths of Afghan pikas in different regions of the Kopetdag (from Babaev and Sapargel'dyev, 1970)

Species of helminths	Regions		
	Kiryuzak and Chuli	Garmab and 3i-Birleshikh	Kara-Kala
Schizorchis* altaica	–	–	+
Hydatigora krepkogorsk**	–	–	+
Dermatoxys schumakovitschii	–	+	+
Cephaluris andrejevi	+	+	+
C. chabaudi	–	+	+
Fastigiuris devexus	–	+	–
Labiostomum naimi	+	+	–
L. vesicularis	+	+	–
L. akhtari	+	+	+
Syphacia sp.**	+	–	+
Oxyyurata gen. sp.**	–	+	–
Trochocephalus sp. larvae**	–	+	–
Pseudophyaloptera sp. larvae**	–	+	–
Physalopteriata sp. larvae**	+	–	–
Spirocera lupi**	–	+	–
Nematodirus aspinosus	–	+	–

"+"—species present, "–"—species absent.
*In Russian original, *Schisorchis*—General Editor.
**Species of helminths, for which Afghan pikas are not the usual hosts.

---

*In Russian original, *Leptotronmbidium*—General Editor.

Practical Significance

In high population years Afghan pikas can inflict some damage to cucurbitaceous crops and fruit trees, but such damage is of a local nature. Hence, Afghan pikas should not be referred to as pests of agricultural crops. Afghan pikas do not inflict significant damage to pastures and meadows, since their habitations are usually situated away from such farmsteds and are not easily accessible for cattle (Sapargel'dyev, 1987).

Skins of Afghan pikas, because they easily tear, are valued low, though until 1940 they were processed for fur. At the present time curing of hides of Afghan pikas has been discontinued.

In some countries Afghan pikas are used as laboratory animal, since they live and breed in captivity better than other pika species.

Epidemiological significance of Afghan pikas has been little studied, but the abundance of ectoparasites and endoparasites found in this species testify to a definite role of Afghan pikas in the conservation and dispersal of certain transmissive diseases (Sapargel'dyev, 1987).

Daurian Pika
Ochotona (*Ochotona*) *daurica** Pallas, 1776

1776. *Lepus dauuricus** Pallas. Reise durch verschiedenen Provinzen des Russischen Reiches. Bd. III, S. 692. Mountains along Selenga River (Ognev, 1940) or; Onon River, Kulusugai (Ellerman and Morrison-Scott, 1951).

1773. *Lepus ogotona* Pallas. Novae species quadrupedum e glirium ordinae, p. 59-70. Mountains along Selenga, Chikoi and Dzhida rivers.

1795. *Ochotona minor* Link. Beitrage zur Naturgeschichte. Bd. 1, Abr. II, S. 52-74. Mountains south of Siberian and Mongolia, east of Baikal.

1890. *Lagomys daurica* Büchner. Nauchnye rezul'taty puteshestviya Przhevalskogo po Tsentral'noi Azii [Scientific Results of Przhevalsky's Expedition in Central Asia]. Vol. 1, p. 172.

1904. *Ochotona daurica* Bonhote. Proc. Zool. Soc. London, Vol. 2, p. 205-220.

1908. *Ochotona bedfordi* Thomas. Proc. Zool. Soc. London, p. 45, China, Shansi.

1908. *Ochotona huangensis* Matschie. Wiss. Ergebn. Exped. Filchner, Bd. 10, s. 214-217.

---

*Incorrect in Russian original; *daurica* retained throughout text, although properly it should be *dauurica* (ATS)—Editor.

1911. *Ochotona annectens* Miller. Proc. Biol. Soc. Washington, Vol. 24, p. 54, China, Gansu.

1911. *Ochotona daurica altaina* Thomas. Ann. Mag. Natur. Hist., Vol. 8, p. 760, Mongolia, Lake Achit-Nur.

1951. *Ochotona daurica murzaevi* Bannikov. Mongolia, Khangai, Dakhtindaba Pass.

Diagnosis

Skull of moderate size; condylobasal length less than 44 mm; length of upper tooth row not less than 9 mm. Frontal bones of adults with crests above the orbits, lacking foramina. Incisor fossa not partitioned by processes of the intermaxillary bones. Summer fur light yellowish-pale gray to darker grayish-pale with ochreous tinge. Sides lighter and more yellow. Belly whitish. Ochreous spots absent on sides of neck. Paws whitish or grayish-brown. White border on margin of ears. Winter fur lighter, gray with yellowish-pale tinge. Majority of vibrissae white, 45-50 mm long.

Description

Body length of adults from 170-220 mm (average—180 mm); length of hind foot 25-31 mm (average—29 mm); length of ears 18-24 mm (average—21 mm).

The color of summer fur on the back varies from light ochreous-sandy to darker ochreous-gray. Sides are lighter with a yellowish tinge. The belly is white (with base of hairs gray); on the throat and along the central line of the stomach sometimes with a yellowish tinge. The upper surface of the paws is whitish or light brown; lower side of the paws is gray or brownish; the digital pads are covered with hairs. Claws are long and thin. The length of guard hairs on the back reach 11 mm. The ears are covered with sparse hairs, but have a distinct white border on the margin.

Winter fur is longer and denser than summer fur (length of guard hairs on the back reach 21 mm). The color of the back and belly is a monotone light gray with a sandy tinge. Paws and ears are more strongly fured than in summer. The light border on the edge of the ears is almost indistinct.

The skull is of a moderate size, narrow and less bulged. Condylobasal length of skull is 37-44 mm; zygomatic width, 19.5-24.2 mm; interorbital width, 2-4.5 mm; and alveolar length of the upper tooth row, 7.6-9 mm. Frontal bones lack foramina; in adults, with supraorbital crests. Incisor fossa are not partitioned by protuberances of the intermaxillary bones. The preorbital foramen is triangular in form and elongated. Nasal bones are broad, their outer margins parallel to each other in the posterior half. Orbits are large and round in form. Zygomatic arches are parallel to each other;

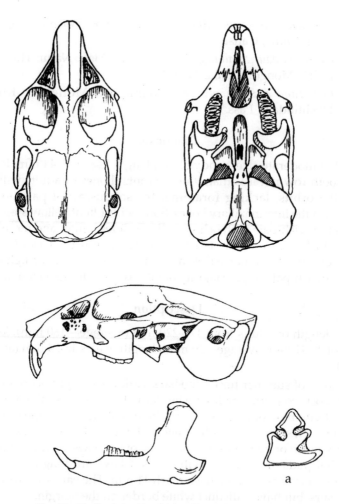

68

Fig. 20. Skull of the Daurian pika *Ochotona daurica*.
a—third lower premolar tooth of Daurian pika.

posterior zygomatic processes are thin and long. The tympanic bullae are relatively large. The mandible is short and low. The angular process is broad. The angular notch is shallow (Fig. 20).

The anterior segment of the third lower premolar tooth is large, rhombic in form, and the entrant folds separating the anterior segment from the posterior are deep and of equal length. The posterior segment is not large, and is rectangular in form (Fig. 20a).

The os penis is relatively short and thick, the ratio of its length to thickness, as a rule, does not exceed 8 (Aksenova and Smirnov, 1986).

The diploid complement has 50 chromosomes: 10 pairs of meta-submetacentrics and 14 pairs of acrocentrics; the X-chromosome is submetacentric and the Y-chromosome (the smallest) is acrocentric (Vorontsov and Ivanitskaya, 1973).

Systematic Position

The daurian pika is close to *O. curzoniae* and *O. rufescens*.

Geographic Distribution

In Russia the Daurian pika enters just the northern part of its range, where it is distributed in three areas: western (southeastern Altai and southern Tuva); central (southeastern Buryatia); and, eastern (southeast of the Chita region) (Figs. 21, 22b).

The western boundary of the range of the Daurian pika began in the past from the Alakha (Aklakha) River in Gornyi Altai, from where two specimens of Daurian pikas are available in the collections of ZIN, caught in 1899 during Kozolov's expedition. At the present time the western boundary of the range of the Daurian pika passes along the eastern edge of the Ukok table mountains to the east along the western part of the Sailyugem Range through the middle fork of the Kalguty River (Yudin et al., 1979). Along the southern edge of the Chui steppe, colonies of Daurian pikas are found along the northern spurs of the Yuzhno [southern] Chuisk and Sailyugem ranges, in the upper reaches of the Irbistu and Tarakhty rivers, along the Chagan-Burgazy River, along the valleys of the Malaya and Bolshaya Shibeta rivers, and along the upper and middle forks of the Ulandrik River (Ognev, 1940; Kir'yanova, 1974; collections of ZM MGU). To the north of the Chui steppe, the distributional boundary of the Daurian pika passes along the Chui River (vicinity of the villages of Ortalyk, Kosh-Agach, Aktal and Tashanta), and later it proceeds to the east along the Chui tract and enters Mongolia (Demin, 1960; Lazarev, 1971; collections of BIN).

Later the range boundary of the Daurian pika in the Gornyi Altai passes from Durbet-Daba in the north along the spurs of the Chikhichev Range along the Bogutsk lakes, along the right bank of the Yustyd River to the upper reaches of the Bar-Burgazy and Buguzun rivers (Vassil'ev and Lazareva, 1968; Kir'yanov, 1974; Yudin et al., 1979). In the Ulagansk region colonies of Daurian pikas are reported only in its southeastern part, in the vicinity of Lake Kyndykty-Kul' (collections of BIN). These habitations may be considered the northeastern boundary of the distribution of the Daurian pika within the bounds of the Altai territory.

In Tuva, the water-divide line of the mountain chain formed by the

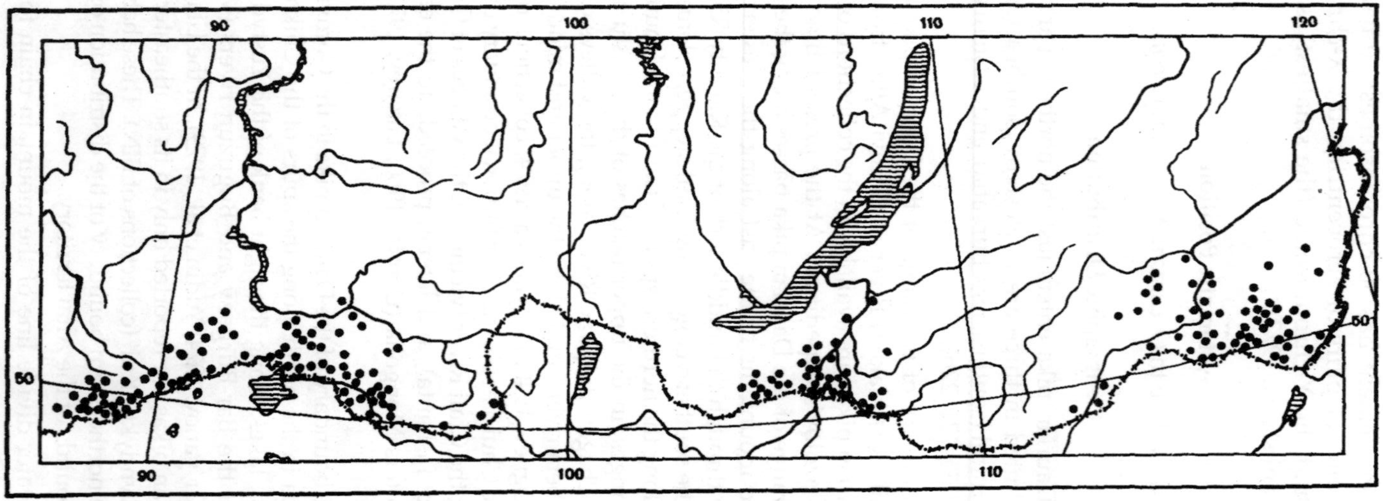

Fig. 21. Localities of Daurian pikas in Russia.

69

93

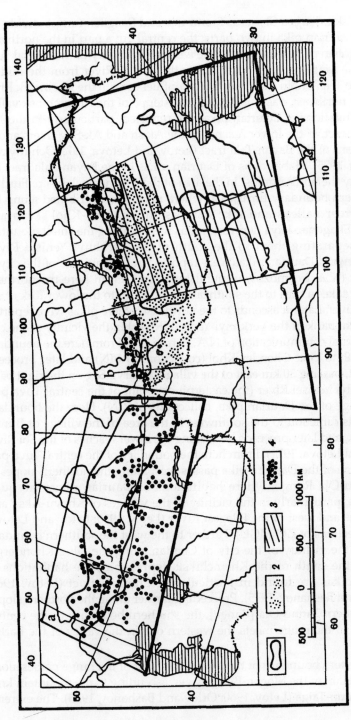

Fig. 22. Range of (a) little and (b) Daurian pikas. 1—schematic range according to S.I. Ognev (1940); 2—range of Daurian pikas in Mongolia (Sokolov and Orlov, 1980); 3—schematic range of the Daurian pika according to Corbet (1978); 4—our data.

70

Shapshal'skii, Tsagan-Shibetu, Tannu-Ola and Sengilen ranges divide the range of Daurian pika in two parts: the central Tuva part in the north and the Mongolian part in the south. The western boundary of the central Tuva part of the range ascends north along the Barlyk River from the middle fork to the Teeli-Sholu tract west of its confluence with the Khemchik River and turns northeast, passing through the steppe of the Khemchinsk valley, where habitations of Daurian pikas are reported, particularly in the Ak-Dovurak tract, in the lower reaches of the Alash and Ak-Sug rivers and in the environs of the Bayan-Tala tract (Letov and Letova, 1971; Ochirov and Bashanov, 1975). Habitations of Daurian pikas in the Bayan-Tala tract are apparently the northernmost in the area of the Khemchinsk basin. Further, the northern boundary turns east, where pikas are distributed along the ancient river valleys and lake depressions in Chaa-Khol'skaya and Torgalyg-Shagonarskaya depressions, as well as along the small mountain massifs separating them to the south of the Verkhnyi Yenisei River. Northeastward Daurian pikas were reported in the environs of the city of Shagonar (N.A. Formozov, unpublished data). From here the northern boundary takes a turn to the southeast and east up to the lower fork of the E'legest River. Later it ascends to the northeast, where pikas were reported on the right bank of the Verkhnyi Yenisei [River] in the vicinity of the city of Kyzyl (oral communication of N.A. Formozov). From here the boundary turns south to Lake Chender-Khol (collections of BIN) and later proceeds to the east, passing 30 km east of the village of Saryg-Seg on the right bank of the Malyi Yenisei River (the easternmost point of the central Tuva part of the range of the Daurian pika; collections of ZIN). Here the boundary changes its direction to the southwest in the area of the village of Buren-Bai-Khaak, and later proceeds west along the lower reaches of the Soi River (Letov and Letova, 1971). From here the boundary of the central Tuva part of the range of the Daurian pika passes west along the northern slopes of the Tannu-Ola Range, where habitations of Daurian pikas are quite numerous, particularly in the vicinity of the villages of Khovu-Aksy and Khenderge and west of the E'legest River (Letov, 1960; Letov and Letova, 1971: collections of ZM MGU). Farther to the west the southern boundary passes in the vicinity of the city of Chadan along the Ulug-Khondergei River; to the south of the Khemchiksk depressions, the habitations of Daurian pikas are found in particular in Chirgaki tract (Letov, 1961: Ochirov and Bashanov, 1975). Proceeding farther along the northern slopes of the western Tannu-Ola [Range], the southern boundary of the central Tuva part of the range meets the western one in the valley of the Barlyk River.

73   The western boundary of the Mongolian part of the range passes along the flood plain of the Mogen-Buren River, ascending to the mountain knot of the Mugun-Taiga (Letov, 1960: Orlov and Bashanov, 1975). The extreme

western habitations of Daurian pikas are reported on the southern spurs of the Tsagan-Shibetu Range in the area of Lake Khindiktig-Khol'. To the east, habitations of Daurian pikas in this part of the range are associated with the latitudinal span of the Tannu-Ola Range. Along the state border the habitations of Daurian pikas are noticed in the flood plains of the Bukhé-Muren and Khuv-Usny-Gol rivers in the regions of the Khopshu ravine, along the northern spurs of the Tsagan-Shibétu, in the valley of the Kargy River, in the upper reaches of the Barlyk River (Letov, 1960; Ochirov and Bashanov, 1975; Nikol'skii, 1984). Eastward, Daurian pikas are found along the valleys of the Terekhtyg-Khem, Khandagaity and Torgalyg rivers, near the Dus-Tag and Chaa-Suur "somons" (Vlasenko, 1954; Ochirov and Bashanov, 1975). Along the southern foothills of the eastern Tannu-Ola in the area of the Ubsunursk depression, habitations of Daurian pikas are reported in the interfluve of the Irbiteya and Kholu rivers, as well as in the valley of Orkhin-Gol (Ochirov and Bashanov, 1975; Yudin et al., 1979). Eastward, Daurian pikas are found in the flood plain of the Tes-Khem River, in the vicinity of the town of Oo-Shynaa, west of Ushino "somon" and near Khol'-Oozhu (Vlasenko, 1954; Ochirov and Bashanov, 1975; collections of ZM MGU, ZIN, and BIN). From here the northern boundary passes along the interfluve of the Uzharlug-Khem and Terektyg-Khem, along the foothills of the Samagaltai and 15 km east of the village of Samagaltai, ascending later in the northeast to the village of Shuurmak. Later the northern boundary turns south and passes along the flood plain steppes of the Tes-Khem River and its tributaries, in particular the Uzharlug-Khem and E'rzin, and then to the east—along the upper and middle forks of the Naryn River to Kachik "somon" (Vlasenko, 1954; Letov, 1960; Ochirov and Bashanov, 1975; collections of ZM MGU and BIN). From the east the Mongolian part of the range is limited by the Ikh-Tarisiin-Gol River. The southern habitations of Daurian pikas within the limits of Tuva have been found in the vicinity of Sangan-Tologoi "somon" (valley of the Tes-Khem River) and in the area of Lake Tere-Khol' (Vlasenko, 1954: Yudin et al., 1979).

The central part of the range of the Daurian pika stretches west to east along the southern spurs of the Malyi [Lesser] Khamar-Daban and Khamar-Daban ranges—from the Ulekchin River to the vicinity of the village of Mukhorshibir'. In the north, this part of the range of the Daurian pika borders with the Ulan-Udinsk region and in the south, with the northern spurs of the Dzhidinsk Range south of the Kyakhtinsk district. The western boundary of the central part of the range of the Daurian pika ascends in the northeast along the Ulekchin River up to the middle stream of the Armak River (Ognev, 1940). Farther to the east the boundary passes again along the left tributaries of the Dzhida River from the mouth of Armak River through to the Toreiskie lakes, valley of the Tsagantui River and the

Bargaiskaya steppe (Leont'ev, 1968: collections of IGU, Institute of Zoology of the Turkmenian Academy of Sciences and "Mikrob" Institute, Saratov). From here the western boundary meets the northern and passes in the northeastern direction along the left bank area of the Selenga River through the edges of the Burankhan (Zund-Burin-Khan) mountains, along the lower and middle streams of the Iro River (collections of IGU, and Kyakhtinsk division of Chita Antiplague station). Along the left bank of the Temnik River Daurian pikas are uncommon, and in the area of Lake Gusinoe, along the valley of Selenga and in the vicinity of Novoselenginsk, the habitations of Daurian pikas are found frequently (Ognev, 1940: collections of ZIN and IGU). Later the boundary of the central part of the range proceeds northeast along the Selenga River. On the southern spurs of the Khamar-Daban Range, Daurian pikas were caught at the mouth of the Orongoi River and in the vicinity of the village of Kokorino (collections of BIN and IGU). The northernmost place where Daurian pikas are encountered is in Buryatia—in the vicinity of Ulan-Ude (collections of ZIN). From Ulan-Ude the boundary turns southwest, meets the eastern, and passes along the right bank of the Selenga River through the town of Barykino-Klyuchi [springs] (Barykino), where it turns southeast toward the village of Mukhorshibir' (Ognev, 1940; Fetisov, 1942). Later the eastern boundary distributional of the Daurian pika again turns southwest and proceeds through the village of Bilyutai in the valley of the Khilok River and along the Chikoi River (Shvetsov and Moskovskii, 1961; collections of Borzinsk and Kyakhtinsk divisions of the Chita Antiplague station). Here the boundary turns east, passing along the southern spurs of the Malakhansk Range, where Daurian pikas were caught along the valleys and dry islands of the Kiret' and Kudara rivers and southward in the vicinity of Kudara-Somon (Shvetsov and Moskovskii, 1961). In the extreme southeast of the central part of the range, Daurian pikas were caught in the vicinity of the town of Shara-Gol (collections of IGU). From the towns of Shara-Gol and Narin-Kundui, the distributional boundary of the Daurian pika turns northwest and proceeds along the valley of the Chikoi River along the southern Kyakhtinsk district at the border with Mongolia (Shvetsov and Moskovskii, 1961). In the collections of ZIN there is one specimen of the Daurian pika, caught by Mikhno in 1925, in the vicinity of the village of Naushki. On the border with Mongolia, the habitations of Daurian pikas are found in the vicinity of the towns of Tsagah-Usun and Enkhory (Ognev, 1940; collections of IGU and BIN). Westward, along the southern spurs of the Dzhidinsk Range, Daurian pikas are found rarely, and here the distributional boundary does not touch Mongolia (Ognev, 1940). Daurian pikas were reported along the right bank of the Dzhida River, at the mouth of the Zheltura River and in the middle fork of the Khuldat River (collections of ZM MGU, IGU, and Geography Faculty of

MGU). The Ulekchin River is apparently the western limit of distribution of Daurian pikas in this part of its range (Fetisov, 1942).

The eastern part of the range of Daurian pikas includes the Trans-Baikal steppes south of the lower forks of the Ingoda and Shilka rivers to the border with Mongolia and China. The northwestern boundary of this part of the range passes along the eastern slopes of the Stanovik, Mogoituisk and other ranges, while the eastern passes along the state border with China southeast of the Chita region. The extreme southwestern habitations of Daurian pikas, in this part of the range, are reported in the vicinity of the town of Kyra (collections of IGU). Along the valley of the Onon River, Daurian pikas were encountered west of Kyrinsk and east of the Ashkinsk districts of the Chita region (Fetisov, 1940; Nekipelov, 1954). Later the boundary distributional of Daurian pikas turns northeast and follows along the southern spurs of the Mogotuisk Range along the valley of the Agi River—through the villages of Ilya and Aginskoe, passes south of the Buryat station and the town of Baishin, through the village of Chiron in the lower reaches of the Onon River and to the northeast up to the lower reaches of the Shilka River (Nekipelov, 1961). Then along the western spurs of the Borshchovochnyi, Ononskii and other ranges the boundary turns south and along the flood plain of the Onon River and passes through the town of Karaksar and Ust'-Ulyatui; then it turns east toward the village of Arenda and descends south to the Khada-Bulak station (Nekipelov, 1954; Nikol'skii, 1984). Proceeding upward along the course of the Borzya River, the boundary turns east and along the northern spurs of the Nerchinsk Range and proceeds to the villages of Ust'-Ozernoe and Belektui to the Kadminskoi stony mounds in the vicinity of the Aleksandrovsk plant (Nekipelov, 1954; Nikol'skii, 1984; collections of ZM MGU and IGU). Northwards, Daurian pikas were reported in the former Verkhneudinsk district (Ognev, 1940). In the Nerchinskozavodsk district, Daurian pikas are absent. The eastern habitations of Daurian pikas are found in the valley of the Kalga River and south and southwest of it, in the Urulyungui ravine, Klichkinsk Range and in the depression southeast of it, in the lowlands between the Argunsk and Klichkinsk ranges and along the steppe areas of the Argunsk Range (Nekipelov, 1954; Leont'ev, 1968). The habitations of Daurian pikas are found farther along the flood plain of the Argun River and in the vicinity of the town of Kailastui (Smolina, 1958; collections of IGU). The extreme southeastern localities of Daurian pikas in the Chita region is at the Abagaitui mines (collections of Borzinsk division of Chita antiplague station). From the Abagaitui mines the eastern distributional boundary of the Daurian pika turns west and northwest. The habitations of Daurian pikas are reported in the vicinity of the village of Zabaikal'sk (the former Otpor station), near Lake Zerdé-Azakhara, along the valley of the Sharasun River, in the southeast of the Nerchinsk Range

98

in the area of the Kharanorsk depression, in the vicinity of the Kharanor station and on the northern slopes of the western end of the Nerchinsk Range, in the depression of Lake Bolshoi Chindant (Nekipelov, 1954; collections of IGU, Astrakhan antiplague station, Borzinsk section of the Chita antiplague station and "Mikrob" Institute, Saratov). Habitations of Daurian pikas were repeatedly encountered southeast of the Ononsk district, in the depressions Lake Zun-Torei; on the northern bank of Lake Tarun-Torei, the southernmost habitations in the depressions of these lakes lie in the vicinity of Solov'evsk and near the mouth of Ul'dza River on the border with Mongolia (Nekipelov, 1954; Guzhevnikov and Tarasov, 1968; collections of ZIN). In this part of the range, the spread of Daurian pikas along the Mongolian border is apparently hampered by the Érman Range, which stretches along the southeastern border of the Akshinsk district.

75

The Daurian pika is widely distributed in the territory of Mongolia, except the montane forests and deserts south of the Mongolian and Gobi Altai. In the east, the range of the Daurian pika passes to northern China (Greater Khingan) in the south—to central China (Shansi Province) (Fig. 21b).

Geographic Variation

Two subspecies of Daurian pika are found in Russia.

1. *Ochotona daurica daurica* Pallas, 1776 (syn. *O. daurica altaina* Thomas, 1911, and *O. daurica mursaevi* Bannikov, 1951)—eastern Daurian pika.

The type subspecies is characterized by a narrow tympanic bullae flattened on the sides, their width usually exceeding 10 mm. The color of summer fur is highly variable, especially in Mongolian populations. Daurian pikas of the type subspecies are distributed in the upper reaches of the Argun' River and in the basin of the Onon River (Chita region), in the basin of Selenga (Buryatia) and in the Gornyi Altai—from the upper reaches of the Argut River in the west to the upper reaches of the Chulyshman River in the east, as well as almost throughout the entire territory of Mongolia, except the basin of Lake Ubsu-Nur.

2. *Ochotona daurica latibullata* ssp. nov.—Ubsu-Nur Daurian pika.

Differs from the type subspecies by having round, broader tympanic bullae, their width usually exceeding 10 mm (Table 5.1). In the color of summer fur, ochreous tone is weakly manifested; on the sides of the body there is a sharp border between a brownish-gray color on the back and a white color on the belly; the inner surface of the ears is dark, almost black; a white border along the margin of the ear is well-developed. Altai populations of the Daurian pika in their metric characters belong to the type subspecies, although in some parameters they are intermediate

Table 5.1. Dimensions of skull (in mm) of adult Daurian pikas
(*O. d. daurica* and *O. d. latibullata*)

Indices	*O. d. daurica* (n = 179)			*O. d. latibullata* (n = 39)		
	min.	M	max.	min.	M	max.
Condylobasal length	37.0	40.0	44.0	38.7	40.9	43.9
Length of rostrum	15.2	16.6	18.6	15.5	16.6	17.3
Lenght of upper tooth row	7.9	8.5	9.0	7.9	8.7	9.0
Length of orbit	12.0	12.9	15.0	12.0	12.6	13.2
Length of tympanic bulla	12.0	13.5	15.0	13.0	13.6	15.1
Width of tympanic bulla	8.5	10.8	11.8	9.5	13.8	12.0
Interorbital width	2.5	3.9	5.0	3.1	3.7	4.1
Zygomatic width	19.5	21.4	24.2	20.0	21.2	22.3
Width of cranium	15.1	16.7	19.2	16.0	16.8	18.0
Width in the region of tympanic bulla	18.7	20.5	23.8	20.0	21.0	22.0
Postorbital width	12.5	13.8	15.6	12.5	13.6	16.3
Width of rostrum	5.0	6.5	7.5	5.3	6.2	7.0
Occipital height	10.9	11.9	13.1	11.0	11.9	13.0
Height in frontal region	11.0	11.9	13.1	11.2	11.9	12.9
Height of rostrum	6.0	6.9	7.0	5.6	6.9	7.5
Height of cranial part in the region of tympanic bulla	12.5	14.8	16.1	13.5	14.7	16.0

between the Chita and Tuva populations (Table 18). The Ubsu-Nur Daurian pika is distributed in Tuva, where its northern boundary coincides with the range boundary of the species; the western boundary passes along the Shapshalsk and Tsagan-Shibetu ranges; the eastern boundary, along the basin of the Balyktyg-Khem River. This subspecies of Daurian pika lives also in Mongolia along the edge of the Ubsu-Nur basin.

Two more subspecies of Daurian pika have been described from the territory of China: *O. daurica bedfordi* Thomas, 1908 (Shansi) and *O. daurica annectens* Miller, 1911 (Gansu).

Biology

Habitat. In the southeastern Altai the Daurian Pika lives mainly on the northern and eastern slopes of mountains from the foothills to the summits. They occupy relief depressions, valleys of small rivers and streams with a well-developed grass stand, and shrub thickets (Lazarev, 1971). Daurian pikas are found also along the dry chee grass and pea shrub steppes. In tundra habitat they are more numerous at the junction of tundra with mountain steppe areas, while in grassy tundra they are rarely found. In the Chui Alps [high mountains] Daurian pikas are found along the edge of the Chui steppe, at the foothills and above—predominantly along the gentle steppe slopes of the range ascending in the east to 2,000-2,500 m above sea level (Kolosov, 1939). Here Daurian pikas most often remain in small depressions [lowlands] and ravines with green vegetation. On the gently sloping areas, Daurian pikas remain in the valley of small rivers: sometimes they can be found in wet and even swampy bottomlands. In spring, when the depressions are inundated with meltwaters, Daurian pikas move to higher dry places, sometimes occupying abandoned burrows of Pallas's pikas.

In Tuva Daurian pikas live in grassy-wormwood, wormwood-grass, cinquefoil-wormwood, grassy-herbage hillocks, lyme grass-chee grass associations and in river flood plains, in poplar-hilly thickets (Ochirov and Bashanov, 1975). They are also found on fixed sands with grass-herb-leguminous vegetation and rare pea shrubs (Ognev, 1940; Vlasenko, 1954). Loose sands and rubbly semidesert ridges of the Tannu-Ola and Agar-Dag ranges are clearly avoided by Daurian pikas. In southeastern Tuva along the rubbly montane steppes, Daurian pikas ascend to 2,000 m above sea level, where they live on slopes of northeastern exposures on the edges of deciduous forest groves (Berman et al., 1966). Still rarely, but regularly, Daurian pikas are found in valley forests; along the Tes-Khem River—in poplar forests with dense bushes of pea shrub; along the Khandagaita River—in birch willow wet forests with occasional larch trees. They are also found in damp hummocky flood plains with rare groves of stunted birches and patches of dense willow bushes in valleys of the Orgin-gol and Naryn-gol rivers, where pikas construct burrows in high sedge hummocks.

In foothills of the Tannu-Ola Range, Daurian pikas live on the bottom and southern part of the depressions and valleys in the lower parts of slopes. Here the grass cover is dense and has a larger number of islands of wormwoods, where Daurian pikas make their burrows. In the southern regions of Tuva, habitations of Daurian pikas may be found in cereal and other agricultural crops (Mokeeva and Meier, 1969). Here they build their burrows predominantly on the edges of fields and on the boundary, but sometimes also all over the field.

South of Tuva and northwest of Mongolia, valleys of rivers and temporary water streams with high and dense pea shrub bushes are the refugia of Daurian pikas. In these bushes the burrows are situated usually at the base of shrubs, which provides these pikas protection against predators and temperature fluctuation. In these places in summer, the air temperature in the day changes in the range of 1° (from 12 to 13°), whereas in the open, the temperature fluctuates in the range of 17° (from 20 to 37°). In winter the snow in pea shrubs is deep and the temperature in burrows under snow is higher than in the open (Letov, 1960).

According to the data of A.N. Formozov (1929), east of Trans-Baikal (along the road from Ulyastui to Khubsugul) Daurian pikas are found in ravines and lowlands, and southward (in the zone of semideserts and deserts)—along river valleys, preferring places with bushes of iris and derris. In northeast Mongolia Daurian pikas are most numerous in dry and meadow steppes—finely turfaceous grass cattail associations with the presence of wild rye, that is, even in dry steppes. Daurian pikas prefer to live in more damp places with fertile soils (Kucheruk et al., 1980). In the Khingan and Mongolian Altai, Daurian pikas inhabit sheep's fescue steppe, particularly in lowlands and depressions, diverse meadows, thickets of pea shrub along river valleys, grass-wormwood-steppes and thickets of iris in lake depressions. They also live in bordering areas of sheep's fescue mountain steppes with feather-grass steppes, although here they are not so numerous (Bannikov, 1954).

Thus within their range habitats of Daurian pikas are very diverse. Mainly they are related to the zone of dry steppes and semideserts, where pikas prefer to live in more damp places with well-developed grassland. Daurian pikas are also found in mountain meadows. One of the factors limiting habitation of Daurian pikas is the compact ground and large number of stones in it.

Population. The population of Daurian pikas depends on the quality of soil, humidity, and nature of vegetation. As already mentioned the turfaceous mountain slopes with thickets of pea shrubs are the most favorable places of habitations of Daurian pikas. Here Daurian pikas occupy both lower and upper parts of the slope. In the dry loamy plain, with rare and low thickets of pea shrubs, Daurian pikas are quite common and their habitations are evenly distributed, but here their population is lower than in the optimal habitats. In such places pikas usually poorly survive the winter, and during summer a rise in their population occurs mainly on account of migrants from the surviving colonies.

According to the data of G.S. Letov (1961), in Tuva Daurian pikas are most numerous in pea shrub-wormwood steppes, where their density may reach 172 individuals/ha (Table 5.2). According to other data (Boyarkin, 1984), the highest density of habitation of the Daurian pika is in grass-wormwood steppes in the foothills of mountain massifs (on average, 36 individuals/ha), while in wormwood-grass steppes with rubbly and insufficiently moist soil the habitation density of pikas is much lower—on average, 8.3 individuals/ha. The grass-herb steppes south of Tuva

Table 5.2. Population of Daurian pikas in 1957-1958 in Tuva (Letov, 1961)

Place	Date	Habitat	Average per ha		
			colonies	burrows	animals
Khopsu	July	Grass-herbage steppe along mountain river	12	207	60
E'legest	August	Pea shrub-herbage steppe in flood plain	4	48	28
Town of Te'li	August	Grass-wormwood steppe	8	152	24
Lake Khak	August	Pea shrub-wormwood steppe	12	252	172
Krasnaya Gorka	May	Shrubs in flood plain	10	200	40

usually serve as dispersal colonies for pikas. Here the number of burrows of Daurian pikas is quite high (27/ha), but in the beginning of summer the majority of burrows are empty and the remaining ones are occupied sparsely. In August the population of Daurian pikas in such places increases considerably (D.G. Derviz, in print).

In the lake basins of Achit-Nur and Tsètsèg-Nur, Daurian pikas are particularly numerous in wormwood-cinquefoil steppe in association with saltwort, characteristic for the depression of Lake Ubsu-Nur. In such habitats 3-4 colonies/ha are encountered. In dry steppes with thickets of iris and derris, there are, on average, 3 colonies/ha, in sands overgrown with thermopsis 2-3 colonies/ha, and in sheep's-fescue mountain steppes 2 colonies/ha (Bannikov, 1959).

In the central part of Mongolia, Daurian pikas live most densely in low-hill ravines and lower parts of mountain slopes. In Khingan in the most depressed areas, the number of their burrows may reach 28/ha; moving up along the slope, the density of distribution of burrows and the area of colonies decreases (Dmitriev, 1991). In southern Mongolia Daurian pikas are found in islands of mesophytic vegetation near springs or in sandy places among thermopsis thickets. According to the data of D. Tsybegmit (1950), along the Ul'dzya River the maximum number of colonies of Daurian pikas (4-5/ha) is found in ravines in wormwood-feathergrass steppe, 2-3 colonies/ha in tansy-feather grass steppes on ridges, in feather grass-tansy steppe on plains and in iris thickets in feathergrass steppes.

In winter the population of the Daurian pika decreases in all habitats, and under favorable conditions usually does not exceed 14 animal/ha (Zonov and Okunev, 1991).

Shelters. Daurian pikas dig quite complex burrows. The tunnels lie usually at a depth of 30-40 cm. However, in the Trans-Baikal, individual burrows of Daurian pikas were found at a depth of 1.5 m (Fetisov, 1936). The area of tunnels of one permanent burrow varies from 4 to 700 m², and the number of entrances from 4 to 40. According to the data of N.V. Ol'kova (1954), in Trans-Baikal the area of a burrow of Daurian pika was, on average, 190 m² (n = 61). Here in August-September, 1 to 10 Daurian pikas live in each burrow. The entrances are connected to each other with beaten paths, which at the entrance to the burrow become deeper (4-5 cm). The diameter of entrances is usually 5-8 cm. Depending on the age of the colony and the intensity of use of the burrow, more or less large heaps of earth are located before the entrances; their volume varies from 0.01 to 0.02 m³ (Berman et al., 1966). In some parts of the range, pikas periodically clear the already existing tunnels and then the volume of the thrown out mass increases to 0.4 m³ (Dmitriev, 1991).

As a rule, a burrow has one nest chamber with dimensions roughly 30 × 40 cm, the bottom of which is packed with dry grass, usually the same as in the food reserves. Most often, two tunnels lead to the nest chamber. A large number of branches are characteristic, which the animal uses as lavatory. The hay is usually stored not in the branches, but in the expanded areas of the tunnels, where apparently it is better aerated.

In eastern Mongolia (Rashan-Buèrè), a burrow of Daurian pika situated in feathergrass-wormwood steppe occupied an area of about 1,000 m²* (20 × 50 m) and had 40 entrances, the majority of which had small hay piles. The distance between entrances often was 1-1.5 m; in some cases to 3 m. Numerous entrances were interconnected and some terminated in blind alleys to 20 cm long. The blind alleys are used by these animals mainly as lavatories. Tunnels most often were at a depth of 15-20 cm; their diameter was 10-12 cm. In the center of this system of tunnels, somewhat deeper (to 30 cm), there was one nest chamber 60 cm in diameter the bottom of which was packed with dry grass (Tsybegmit, 1950).

In eastern Khingan the total area of burrows in one colony of the Daurian pika varied from 10 to 200 m² (average 50 m²); the length of all tunnels may reach 90 m, and the number of entrances, 80. Here, the depth of tunnels does not exceed 30 cm, and the chamber diameter, 40 cm. The surface of soil in the area of a colony is not used uniformly by animals. The maximum activity is attributed to the ecological center occupying even in large burrows an area not more than 10 m². The major underground galleries pass from here and have a large number of exits and a denser network of runways (Dmitriev and Guricheva, 1978; Dmitriev, 1991). Temporary

---

*In Russian original, 100 m²—General Editor.

burrows situated usually at the periphery of the colony have one or two exits, and their length varies from 10 cm to 8 m, and the depth does not exceed 30 cm. Such burrows are used as temporary shelters or for storing reserves of grasses (Dmitriev, 1991).

The burrows inhabited by young ones are quite simple in construction. The area of such burrows depends on the number of Daurian pikas and the time of the year. Thus, in the vicinity of the Shamar somon (northern Mongolia), at the edge of pine forest, permanent burrows which accommodated an adult male and female were located under a spirea shrub. This burrow had 6 exits and was practically inaccessible for investigation. Around it, on the dry meadow with an overgrowth of cinquefoil, wormwood and grasses, there were four unconnected underground tunnels of burrows, of young ones. The total area of the system of these burrows was 200 m². At the end of July, when observations were conducted, young of the first brood had already attained adult size. Young did not approach the burrow of adults and spent all their time in their burrows. One such burrow had nine exits and, judged from the large quantity of old grass and excreta in special chambers, was not used for a year (Fig. 23). The maximum depth of tunnels of this burrow was 30 cm and, from all accounts, it was not only a temporary shelter for young, but also served as a permanent burrow (E.Yu. Ivanitskaya, unpublished data).

Thus, in each inhabited area of Daurian pikas there may be one to three complex burrows with a large number of entrances; moreover only one complex burrow or part of entrances to it was actively used (Proskurina et

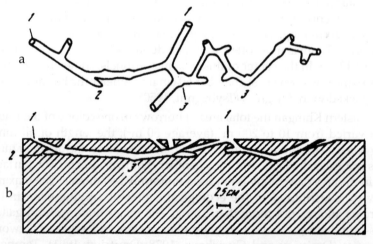

78

Fig. 23. Schematic drawing of a burrow of the Daurian pika in north Mongolia. a—plan; b—vertical section. 1—entrance opening; 2—branched burrow; 3—nest chamber (E.Yu. Ivanitskaya).

al., 1985). Moreover, a large number of protective burrows with one to two outlets were scattered over the entire area of habitation. Numerous walkways on the surface connected these protective burrows with each other and with the main burrow.

Often Daurian pikas use burrows of rodents (voles, marmots, and sousliks) modifying them for themselves. In winter they can make trenches in the snow by building a network of burrows under the snow (Letov, 1961).

In summer, as a rule, adult males and females with their progeny live in one burrow; in winter, only the male and female. Investigations on 63 burrows of Daurian pikas in Tuva showed (Zonov, 1983) that by February only 1-2 animals remained in each of the burrows.

Feeding. Feeding in Daurian pikas has been studied predominantly on the basis of analysis of the haypiles stored by them in winter. The main diet of Daurian pikas consists of plants present directly on the sweepings from the burrows. In pika colonies plants usually rapidly complete all stages of vegetative growth and attain larger sizes than plants in the background community (Dmitriev and Guricheva, 1978).

Forty species of plants were recorded in the diet of Daurian pikas living in the territory of Tuva. In spring and in the beginning of summer, up to 20% of their stomach contents may contain underground plant parts (Boyarkin, 1984). The principal spring diet of Daurian pikas in southeast Tuva consisted of stemless cinquefoil, Turchaninov anemone [*Pulsatilla*], and squrrose zmeevka* [*Diplachne squarrosa*]. In summer, their diet predominately consists of prostrate summer cypress**, old-world winter fat, Amman's glorybind [*Convolvulus ammani*], and Lena alyssum [*Alyssum lenense*]. In fall, they prefer Altai aster, pygmy pea shrubs and roots of stemless cinquefoil. In winter, besides the stored plants, pikas often feed on subaerial and often aerial parts of plants growing beyond the limits of their shelters. In all seasons the diet of pikas in Tuva includes fringed sage brush, which is a background species throughout (Ondar, 1989).

For winter, Daurian pikas make haypiles of grasses mainly at the base of different shrubs and sometimes trees. The height of such piles may reach 40 cm, and weight 6 kg. In north Khingan, the analysis of 8 piles showed that the weight of haypiles varies from 136 to 828 g (Tsybegmit, 1950). In some places in Tuva, south-eastern Altai, Trans-Baikal and Mongolia, Daurian pikas do not collect hay in piles, but store their food in the broadened tunnels of burrows or in niches. The weight of such reserves varies from 300 to 700 g (Boyarkin, 1984).

In the steppes of southeastern Tuva, the size of haypiles depends on

*Common name could not be confirmed—Translator.
**Kochia prostrata*—General Editor.

the productivity of grasses in each specific habitat. In the steppes of the northern macroslope, where the haypiles of aerial green mass are always higher, storage of food progresses more intensively. Here, in each colony of Daurian pika, 8-10 piles are to be found, each of which weighs, on average, 0.5 kg. On the southern macroslope, there are 2-3 haypiles for each colony, the weight of which does not exceed 0.3-0.4 kg, and the species composition of haypiles here is considerably leaner (Ondar,1989).

The beginning and duration of food storage by Daurian pikas changes year-to-year and depends probably on the weather conditions and the state of vegetation. In eastern Khingan, for example, there were years with a complete absence of haypiles (1980) and years with intensive reserves, which continued for three months (Dmitriev, 1985). Possibly there is a correlation between the degree of productivity of plant cover and the size of haypiles: the higher the productivity (in different years or habitats differing in productivity), the higher and more compact are the haypiles made by Daurian pikas. In dry steppes large haypiles are found only in years with late summer intensive vegetation.

In the southern part of Tuva, Daurian pikas begin to make haypiles earliest of all in places with poor and early fading vegetation (end of July and, sometimes, also in June). Here, curing of food continues until late fall—at this time long-growing wormwood, saltwort and goosefoot are stored (Prokop'ev, 1957; Boyarkin, 1984). In summer, Daurian pikas leave uprooted plants for drying for a day and only then place them in piles and push them in to their burrow. Plants collected in fall, as a rule, are not cured.

The species composition of plants is different in haypiles made by Daurian pikas in different parts of the range and sometimes even among neighboring colonies, and depends mainly on the species composition of plants surrounding their burrows. In eastern Trans-Baikal haypiles have been found which predominantly consisted of wormwood, and less often, cinquefoil. In other areas of the Trans-Baikal, haypiles may predominantly contain fringed sagebrush, meadow rue, anemone, sawwort, thermopsis, tanacetum*, hemp nettle, cinquefoil, and furcate saussuria* (Nekipelov, 1959). In the vicinity of Kyakhta, some stacks consisted of stems of licorice, and others stems of myshyak* and silver sagebrush or only myshyak* (Moskovskii, 1936). According to the data of A.I. Argiropulo (1935), in steppe areas the haypiles made by Daurian pikas consisted mainly of wormwood and leaves of iris, but in river flood plains, usually horsetail. In the Borzinsk steppes the haypiles made by Daurian pikas showed 60 species of plants (Fetisov, 1936). In the number of species grasses (11 species)

---

*Common name could not be confirmed—Translator.

and composits (9 species, including 5 of wormwood), legumes (5 species), buttercups and roses (4 species) predominated.

On the northern macroslope in southeastern Tuva, the food reserves includes 21 species of plants with a predominance of pygmy pea shrub, Turchaninov anemone, woolly speedwell, fringed sagebrush, old-world winter fat, and prostrate summer cypress. Grasses, constituting the bulk of herbaceous vegetation in the steppe, are found rarely in food reserves of Daurian pikas. On the southern macroslope summer cypress predominates in haypiles with an insignificant admixture of fringed sagebrush; sagebrush appears in reserves only in fall after flowering (Ondar, 1989). In the steppes of south Tuva, Asian goosefoot, fringed sagebrush and summer cypress are most common in haypiles of Daurian pikas (Prokop'ev, 1957).

In north Khingan, haypiles of Daurian pikas include 31 species of plants belonging to 11 families. On average, each haypile contained 9 species of plants, and sometimes only one or two species (Tsybegmit, 1950).

In east Khangai, in the vast majority of cases, haypiles of Daurian pikas consist of plants not relished by domestic animals, particularly horses. Some of these plants, for example, thermopsis, may be used as food of cattle in the form of hay. Other plant species stored by pikas are poorly relished by cattle even as hay (Nikol'skii et al., 1984). Thus, Daurian pikas can use for their haypiles plants that are left behind after grazing by cattle; moreover, a large part of their food reserves is not destroyed by cattle.

The observations of D. Tsybegmit (1950), made on the right bank of the Dal'gen-Muren River, confirm the selectivity of Daurian pikas during the curing of food. Geobotanical studies in areas with colonies of Daurian pikas has revealed 29 species of plants, among which dominants were fescue, koeleria, aster and anemone. However, in the food reserves of pikas 84.6% were made of anemone, 3.6% of wormwood, and there were absolutely no grasses in the piles. In eastern Mongolia, of the six background species only fringed sagebrush was present in haypiles of pikas.

Activity and behavior. The nature of activity of Daurian pikas depends on the time of the year. In spring in northeastern Mongolia, they are active from sunrise to sunset; in summer and fall the maximum activity is seen in the morning and evening hours. In the hot time of the day, pikas hide in burrows. On overcast or calm days the activity of Daurian pikas increases, but on windy days they almost do not venture out of burrows (Tsybegmit, 1950).

According to the observations of some researchers (Formozov, 1929; Okunev, 1975), Daurian pikas are active throughout the day over the large part of their range. For instance, in rainy or overcast weather, squeaks of pikas can be heard even at night.

In winter, Daurian pikas leave their burrows only on warm sunny days. At this time, they warm up themselves under the sun or run about on the snow. Tunnels under the snow apparently are not found everywhere. Thus in the valley of the Sel'ba River (north of Ulan-Bator) in January, 1944, tunnels under the snow were not found, although there were many Daurian pikas in this area (Bannikov, 1954).

In southeast Trans-Baikal and southwest of Tuva, in relatively rich-snow winters with loose snow, Daurian pikas appear on the surface almost at any time of the day with peak activity from 12.00 to 16.00 hr and low from 20.00 to 22.00 hr. The distance to which Daurian pikas can run from their burrows varies from 10 to 200 m, and depends on the depth of the snow cover. In 1981, in the Trans-Baikal, Daurian pikas did not cure food in winter and, it was found that in January they were active at lower temperatures (from -22°C to -35°C) and with less deep snow cover than in February (-15° ... -24°C) (Zonov et al., 1983; Zonov and Okunev, 1991). In winter Daurian pikas make lavatories, and there are no excreta near air holes leading to places of feeding.

In the southeastern Altai and southwest of Tuva, in mid-January with thick snow cover, Daurian pikas generally do not come out at night. They are rare in the day (mostly from 13.00 to 15.30 hr) and only in good weather do they come out from under the snow. At such time they feed under the snow, leaving the burrows to a distance of 3.5 m. Thus the nature of winter activity of Daurian pikas is mostly determined by the density of snow cover, and to a lesser extent on its depth and the air temperature (Zonov et al., 1983; Zonov and Okunev, 1991).

82   At no time of the year do Daurian pikas move away from the burrows. Depending on the distance from the burrows, pikas become more careful and, in the event of danger, attempt to run for cover in the burrows. Having reached the entrance, a pika always halts, inspects and listens. If danger approaches, it hides in the burrow, but 10-15 min later surfaces again; if the danger is not noticed, the animal creeps to the surface and raises itself on hind feet and inspects. Apparently, hearing is better developed in Daurian pikas than other sensory organs.

Vocal signals in Daurian pikas are well developed. Their sound signal can be divided into three types: song, trill, and alarm. Trills, as a rule, are never solitary, issued at small intervals and usually simultaneously by many individuals (calling). During calling, sound signals are produced by an individual near the exit of a burrow or not far from it. After a signal by one pika, a response trill or song is heard after 1-2 min. Most often the response signal is of the same type as in the starting one, that is, in response to a trill, a trill is sounded, and a song is uttered in response to a song. Males usually respond to females and vice versa. A song differs from a trill both in the nature of sonorrousness and duration; it is 3-4 times as long as the trill. Moreover, songs are used considerably rarely, practically never repeated a second time, and exchanged among only 2-3 individuals. Apparently, males sing more often than females (Smirnov, 1988; Proskurina, 1991). Observations on Daurian pikas in southern Tuva in July-August 1977 showed that the peak activity of calls is in the evening hours (from 17.00 to 21.00 hr); calls are least heard from 10:00 to 16.00 hr. Starting from July 25, the frequency of both calls and songs drops sharply. Despite the presence of seasonal variability in the acoustic variation, the function of trills and songs of Daurian pikas should not related only to the territorial or sexual behavior. Probably, vocal signalization of Daurian pikas is multifunctional (Proskurina* 1991).

In Daurian pikas the alarm signal is not a very loud whistle/hiss. On giving such a signal all pikas in a radius of 100 m start repeating it. The frequency of giving signals depends on the degree of danger. If the source of danger is near, the number of alarm signals increases.

The Daurian pika is a less aggressive animal. Only during dispersal of young, adult individuals may chase them away. On meeting a long-tailed Siberian souslik, Daurian pikas do not make aggressive contact with them, most often they avoid any encounters with them (D.G. Derviz, personal communication).

In a study of contacts of Daurian pikas with each other, elements of nonaggressive forms of behavior were revealed most often. Friendly contacts (72%), manifest in flocking and mutual cleaning, occur between females and males from family couples. Aggressive behavior (28%) on running into alien animal on the territory defended by it is manifested as a short chase without pushes and bites (Prosrurina* et al., 1985).

Observations on Daurian pikas in captive conditions and in natural habitations in Tuva revealed six forms of intraspecific interactions (cleaning, surrender to cleaning, chasing, crowding, creeping and boxing), which under different motivating situations (friendly, aggressive, surrender, sexual, playful) constitute practically the entire gamut of behavioral repertoire of Daurian pikas (Proskurina and Smirin, 1987).

Area of habitation. Sizes of individual territories of Daurian pikas in south Tuva vary from 480 to 900 m^2; on average 610 m^2 (n = 9) (Proskurina et al., 1985). Territories of males and females differ little in size: in males, on average, 675 m^2 (n = 4); in females, 555 m^2 (n = 4). The territory is divided in two zones: the inner, confined to the main burrow, and the peripheral. In the inner zone are situated practically all haypiles, favorite

*In Russian original, Proskurin—General Editor.

places of rest, permanant places in which the animals sing, and sometimes where they feed. The host defends the inner part of the area from its
83   neighbors, actively chasing them out from its territory. The size of the defended territory is 20-30% of the total area of the habitation area (on average 150 m²). In the peripheral part of an individual territory, Daurian pikas forage and collect grass for their haypiles. These areas may overlap significantly. On encountering neighbors in the area of a personal territory, aggression between them is not noticed; here, as a rule, any type of active interaction between them is absent (Proskurina et al., 1985). For a male and a female forming a family couple, the peripheral part as well as the defended territory overlaps.

Daurian pikas mark individual territories with their excreta, urine, secretion of the buccal gland and, in addition, indicate their presence through acoustic signals. Marking with the help of urine and excreta is done at the entrance in specially dug pits. Also marked are large conspicuous objects, for instance, stones. The stones are marked with the secretion of the buccal gland. Such markings are observed only at the entrance to the burrow (D.G. Derviz, personal communication).

Reproduction. Daurian pikas reproduce, apparently, not less than twice a year. Overwintering females have two broods by July. Females of the first brood attain sexual maturity approximately three weeks after birth, and in the end of June-beginning of July also produce litter (Moskovskii, 1936; Gaiskii and Altareva, 1944).

In the Trans-Baikal, reproduction of Daurian pikas begins in the beginning-end of April (Smolina, 1958; Leont'ev, 1968), and in Tuva, in the second half of March (Ochirov and Bashanov, 1975). Start of reproduction depends on weather conditions, mainly on the mean daily temperature (Leont'ev, 1968). In May and June reproduction occurs intensively, in July it slackens somewhat (only 15% of females reproduce), and in August it almost stops (5% of reproducing females). Roughly the same dynamics of reproduction were observed in different years in Tuva pikas (Table 5.3). In individual cases, gestating females are found also in September (Leont'ev, 1968). Depending on meteorological conditions, the number of broods in a female may be two-to-three (Ochirov and Bashanov, 1975), although according to the data of I.V. Boyarkin (1984), in 1972 in west Tuva, only 55.2% of females had two litters, 9.4% adult females did not participate in reproduction, and not a single female had three litters.

Over a large part of Mongolia, reproduction in Daurian pikas begins in the first days of April and terminates by September. In eastern Mongolia
84   in the end of April-beginning of May, gestating females and young pikas with a body length of 11-13 cm and weight of 35-60 g are found simultaneously (Bannikov, 1954). In Khingan and the Mongolian Altai,

Table 5.3. Reproduction of Daurian pikas in Tuva
(from: 1—Ochirov and Bashanov, 1975; 2—Boyarkin, 1984)

Month	Females investigated	Of these			
		number		percent*	
		gestating	confined	gestating	confined
1					
April	10	3	–	30.0	–
May	61	41	11	67.2	18.0
June	216	100	61	46.3	28.2
July	96	14	48	14.6	50.0
August	73	3	43	4.1	58.9
2					
May	20	9	7	45.0	36.0
June	161	76	135	47.2	90.6
July	65	4	57	6.1	87.7
August	18	2	17	11.1	88.8

gestating females and young individuals with sizes up to 15 cm were caught in June and July, and gestating females were found until September 2 (Kolosov, 1939; Bannikov, 1954).

The number of young of the first brood in west of Tuva in 1972 varied from 2 to 8 (on average 6.5); and in the second, from 3 to 12 (on average 6.1). In resident females fertility is much lower, on average 2 babies per female (Boyarkin, 1984). There is information to suggest that Daurian pikas may bear 14 babies each year (Moskovskii, 1936).

The brood size in Daurian pikas varies from year to year. Thus, in 1955 in the eastern Trans-Baikal area, during a peak of population, there were 15 babies for every female, but in 1963-64 during a period of population rise, 30-35 babies (Leont'ev, 1968).

Growth and development. Young Daurian pikas are born naked. Their growth and development has not been studied.

Molt. In south Buryatia, spring molt in Daurian pikas begins in the middle of May and terminates by end of June (Moskovskii, 1936). In eastern Mongolia, the spring molt apparently begins somewhat earlier—first animals with a molted head and anterior part of the back (roughly to one-third) were caught here on May 9. By May 15, animals were caught in which almost half of the back was covered with summer fur. In the first days of June, winter fur is retained only in the posterior part of the back and on the thighs, and by July 18-20 completely molted pikas began to be encountered. In northeastern Mongolia (Khurkhu and Ul'dzya rivers of Khentei "somon") until the last days of June, animals were caught with remnants of winter hairs on the sacrum. In Khangai and in the Mongolian Altai, spring molt begins and ends later—even on July 5-9 in Daurian pikas in this region, winter fur is still retained on the sides and sacrum. In the Gobi Altai (Deun-Saikhan), on July 19 winter fur remained in Daurian pikas on the posterior third of the back and on the sides of the body (Bannikov, 1954).

Fall molt in Daurian pikas in the western Trans-Baikal area terminates by October 10. In northeastern Mongolia, starting from the middle of September, pikas caught were completely covered in their winter coat. Pikas caught on September 8 in the vicinity of Ulan-Bator still had summer fur (Bannikov, 1954).

Sex and age composition of population. In populations of Daurian pikas of western Tuva, there are more males among the same year's individuals. Their predominance is particularly noticed in May (Table 5.4). The sex ratio evens out with age, and among adults, by the end of summer, some predominance of females is noticed (Boyarkin, 1984). In other parts of the range, among adult individuals a small predominance of females has been noticed—from 51 to 59% (Smolina, 1958; Eshelkin et al., 1968).

Table 5.4. Sex- and age-composition of a population of Daurian pikas west of Tuva in 1972 (Boyarkin, 1984)

Month	Total indivi- duals caught	Adults				Current year's			
		Abso- lute	%	Males, %	Fema- les, %	Abso- lute	%	Males, %	Fema- les, %
May	42	27	62.8	51.9	48.1	15	21.2	60.0	40.0
June	330	156	47.3	43.0	57.0	174	52.7	55.2	44.8
July	155	9	5.8	44.5	55.0	146	94.2	55.5	44.5
August	36	1	3.0	–	–	35	97.0	54.3	45.7

In south Tuva in summer, the current year's individuals constitute 82.6% of the population: 14.3% falls to the share of overwintered yearlings; the rest of the population (3.1%) consists of individuals surving two winters (D.G. Derviz, in press). For restoration of the population, the group of yearlings is most important, since they take part in reproduction in the first half of summer. In August the proportion of yearlings drops to 8%, and two-year-olds are already not met with. According to the data of L.L. Smolina (1958), overwintered Daurian pikas constitute 7% of the population by the end of July, and by the beginning of winter they fade away. According to other data (Leont'ev, 1968), overwintered individuals in July (area of Toreisk Lake) constituted 11% of the population.

In southeast Trans-Baikal, in May the under-yearlings constitute about 10% of the population, in June-July—from 30 to 60%, and in August-September—72.5% (Kardash et al., 1961). Thus, data on the age composition of populations of Daurian pikas obtained by different authors are fairly close, and some differences probably are related to the change in the age composition of populations in different years.

The life span of Daurian pikas is apparently two years (D.G. Derviz, personnal communication) and not 15-16 months (Smolina, 1988).

Population dynamics. The population of Daurian pikas is subject to significant fluctuations according to years and seasons. As the census of occupied burrows of Daurian pikas show, their number in the very same year may be different in different places. However, over the large part of the territory, the change of population in years, as a rule, coincides (Nekipelov, 1954). This is linked with the general direction of changes of conditions of the surroundings in the given part of the range. Analysis of such changes showed that early and warm springs favor high fecundity, that leads to an increase in numbers. The duration of the summer period of reproduction depends on the presence of green vegetation; the duration of vegetation, in turn, is linked with the quantity of precipitation. Draughts in the middle of summer, accompanied with a fading out of vegetation, lead to contraction of the reproductive period. In draught years the reproductive period in Daurian pikas terminates in June, whereas in favorable years it continues to September. Abundance of precipitation in fall adversely affects the conservation of food resources of Daurian pikas, which may lead to high mortality in the winter period. Snow-rich winters are favorable for Daurian pikas, since a thick snow cover protects the burrow from freezing and makes hunting difficult for predators. Under favorable conditions, with early and harmonious reproduction, when in May-June 60-80% of females reproduce and each female bears 6 or more embryos, by September the population may increase more than five-fold (Kardash et al., 1991).

In the southeastern Trans-Baikal area in 1956, the number of Daurian pikas was very high and the number of occupied burrows reached 10-20/ha (Leont'ev, 1968). In the end of winter of 1958 and beginning of the 1960s, the number of Daurian pikas here fell sharply; there was only one occupied burrow over tens of hectares. Predators in this territory were few and the reason for disappearance of Daurian pikas, apparently, was some infectious disease. For the period 1962 to 1964, the number of pikas began to be restored, and by 1964 there were 3 occupied burrows/ha (Guzhevnikov and Tarasov, 1968).

114

Enemies. Daurian pikas have many enemies. These are, first of all, corsacs, foxes, wolves, and steppe polecats. Predatory birds, especially buzzards, eagles and horned owls hunt Daurian pikas.

In the Trans-Baikal area, Daurian pikas constitute 24% of the total number of mammals found in the stomachs of steppe polecats, and in the diet of corsac—about 30% in summer and 50% in winter (Brom, 1952, 1954). Daurian pikas constitute 43% of the total number of mammals in the diet of owls; in steppe eagles—from 10 to 80%; in buzzards 10 to 53%; in saker falcons 23 to 65%; and in horned owls, from 45 to 73% (Nasimovich, 1949; Lipaev and Tarasov, 1952; Peshkov, 1957). During the winter time Daurian pikas are the main food of the majority of predators living in this area.

Competitors. In places of combined occurrence of Daurian and Pallas's pikas, competition may develop for habitat and food resources. However, with normal numbers of pikas habitat differentiation of these species is manifested quite well, which weakens their competition.

86

In eastern Mongolia, the oppressive effect of Brandt's vole has been reported on the number of Daurian pikas. For instance, after 1945 when Daurian pikas were quite numerous, there occurred a population explosion of Brandt's vole, which by the end of summer destroyed almost all of the vegetation. By the next spring Daurian pikas in these places had almost disappeared (Tsybegmit, 1950).

87

Diseases and parasites. In Tuva, 31 species, and in the Borzinsk district of Trans-Baikal area, 16 species of fleas were found in Daurian pikas (Table 5.5).

According to the data of N.F. Darskaya (1957), in Borzinsk district, the index of abundance of fleas *Amphalius runatus* and *Ctenophyllus hirticrus**, specific to Daurian pikas, is very high in the cold period of the year. In summer, most numerous, both on Daurian pikas and in their burrows, and common for the steppe zone of the eastern Trans-Baikal area, are *Frontopsylla luculenta* and *Neopsylla bidentatiformis.*

Among ticks, the following have been identified in Daurian pikas: *Dermacentor nuttalii, Haemogamasus kitanoi, Hirstoinyssus ochotonae, Laelaps nilaris, L. cleithronomydis,* and *Eulaelaps cricetuli* (Dubinin and Dubinina, 1951; Koshkin et al., 1978).

Starting from the second half of August in the southeastern Trans-Baikal area, under the skin of Daurian pikas appear larvae of warble flies *Oestromyia dubinini* and *O. prodigiosa*. The most intensive infestation occurs in September-October. In this part of the range, the percentage of infection of Daurian pikas with *Ferrisella ochotonae* is quite high (98.5%) (Dubinin and Dubinina, 1958).

---

*In Russian original, *hurticrus*—General Editor.

Table 5.5. Fleas of Daurian pikas from different parts of the range

Tuva (Letov and Letova, 1971)	Southeastern Trans-Baikal area, Berzhinsk district (Dubinin and Dubinina, 1951; Darskaya, 1952)
Certatophyllus scaloni	–
C. gaiskii	–
C. tesquorum altaicus	*C. tesquorum sungaris* (p, n, g)
C. laeviceps kuzenkovi	–
C. laeviceps ellobii	–
–	*C. avicitelli* (n, g)
–	*C. calcarifer* (p)
–	*C. gallinae* (p)
–	*C. fasciatus* (p)
Echinophaga oschanini	–
Paradoxopsyllus dashidorzhii	–
P. scorodumovi	–
P. integer	–
Rhadinopsylla dahurica sila	*P. dahurica* ssp. (n)
R. li transbaicalica	–
P. altaica	–
R. dahurica dahurica	–
–	*R. rothschildi* (p, n)
Frontopsylla frontalis baicali+	*F. frotalis baikali+* (p, n, g)
p. hetera	–
P. elatoides	–
P. wagneri	*F. wagneri* (n)
–	*F. luculenta* (p, n, g)
Amphipsylla longispina	–
A. primaris	–
A. vinogradovi	–
*Ctenophyllus hirticrus**	*C. hirtiorus** (p, n, g)
*Amphalius runatus**	*A. runatus** (p, n, g)
Neopsylia mana	–
N. bidentatiformis	*N. bidentatiformis* (p, n, g)
N. pleskei orientalis	*N. pleskei orientalis* (n)
–	*N. abagaitui* (p. n, g)
Ctenophthalmus arvalis	–
Pectinoctenus pavlovskii	*P. pavlovskii* (g)
Ophthalmopsylla praefecta ecphora	*O. praefecta* ssp. (p, n, g)
O. kirischenkoi	–
–	*O. kukuschkini* (p, n, g)
Catallagia dacenkoi	–
Wagnerina tecta aemulans	–
–	*Oropsylla silantiewi* (p, n, g)

Note: *—fleas specific to Daurian pikas; p—fleas found on pikas.
n—fleas found at the entrance to burrows; g—fleas from nests.
+ So given in Russian original—General Editor.

The following parasitize intestines of Daurian pikas: *Ctenotaenia citelli, Schizorchis altaica, Diuterionaenia spasskyi, Dermatoxys schumakovitschi, Cephaluris andrejevi, Syphacia obvelata, Graphidiella olsoni, Cephalobus andrejevi, Eugenuris schumakovitschi, Labiostomum vesicularis, Physalopteriata citelli, Tarassospirura* sp. *Moniliformis mooniliformis* (Gvozdev, 1962; Danzan, 1978). Moreover, four species of coccids have been found in the intestines of Daurian pikas: *Eomeria daurica, E. ochotonae. E. metelkini, E. erschovi* (Machul'skii, 1949).

Daurian pikas may suffer from plague. For them it is characteristic to have low susceptibility to this disease. However, with a high population Daurian pikas are involved in the epizootic, which develops primarily in rodents. The role of Daurian pikas in the epidemiology of plague is secondary. In the Trans-Baikal area, in some species of rodents and the Daurian pika, the causal organism of listeriosis, pasterellosis, and erysipelloid was identified; the causal organism of erysipelloid was isolated also in Daurian pikas from the regions of Buryatia bordering Mongolia (Gaiskii and Altarova, 1944; Kucheruk, 1945; Nekipelov, 1959b).

Practical Significance

In southern regions of Tuva, colonies of Daurian pikas are found in fields, which may lead to considerable damage to crops. Daurian pikas particularly damage crops of perennial grasses (Mokeeva and Meier, 1968).

The digging activity of Daurian pikas considerably affects the state of surrounding vegetation—series of low-growing species are covered with earth and die, which leads to thinning out of vegetation for the colony. However, the mean productivity of vegetation in the colonies of Daurian pikas is usually higher or the same as in the uninhabited areas, since the conditions of growth of plants not eaten by them or not covered with earth improves here. The soil in the territory of colony is loosened and better wetted, because of which feathergrass and tall wormwoods grow luxuriantly. In the course of time, wormwood squeezes out feathergrass. In mesophytic habitats of Khangai and Khantai, the digging activity of Daurian pikas leads to the formation of herb and shrub-herb associations (Dmitriev, 1991).

Altai (Alpine) Pika
Ochotona (Pika) alpina Pallas, 1773

1773. *Lepus alpinus* Pallas. Reise durch verschiedene Provinzen des Russischen Reichs. Bd. 2, S. 701. Border of Altai territory and Eastern Kazakhstan region, Tigersk Range, "Tigerskoe".

1842. *Lagomys ater* Eversmann. Addensa ad celeberrimi Pallasii Zoogeograph. Russo-Asiatica, p. 3, Altai tertitory, Gorno-Altaisk autonomousm district, northwestern spurs of Katunsk Range, Uimon.

1858. *Lagomys hyperboreus* var. *cinereo-fusca* Schrenk. Reisen und Forschungen im Amur-Lande. I. S. 150, Amur region ("upper reaches of Amur").

88 1912. *Ochotona nitida* Hollister. New mammals from the highlands of Siberia. Smithson. Miscellan. Collect. V. 60, 14, p. 4. Altai territory, Gorno-Altaisk autonomous district. Upstream of Chui River, Chegan-Burgasy Pass (Cheganburgazy River).

1924. *Ochotona svatoshi* Turov. O faune pozvonochnykh zhivotnykh severo-vostochnogo poberezh'ya oz. Baikal [On the fauna of vertebrate animals of the northeastern banks of Lake Baikal]. Dokl. Ros. Akad. Nauk, p. 110, Buryat ASSR, Barguzinsk district, Sosnovka River, Shumilikha ravine.

1924. *Ochotona sushkini* Thomas. A new pika from the Altai. Annals and Magaz. of Natur. Hist.* V. 8, p. 163-164, Altai territory, Gorno-Altaisk autonomous district], Taltura River.

1928. *Ochotona alpina argentata* Howell. New Asiatic mammals, collected by F.R. Wulsin. Proc. Biol. Soc. Wash., V. 41, p. 1-6, China, Northern Gansu.

1935. *Ochotona alpina scorodumovi* Skalon. Nekotorye zoologicheskie nakhodki v yugo-vostochnom Zabaikal'e. Sb. rabot. protivochumnoi organizatsii Vost.-Sib. kraya [Some zoological finds in southeastern Trans-Baikal area. In: "Collection of Works of Antiplague Organization of East Siberian Territory]. Vol. 1, pp. 85-87. Chita region, Borzinsk district, Kailastui.

Diagnosis

Pika of medium and large sizes. Condylobasal length of skull more than 41 mm, height of skull in occipital part more than 11 mm, length of tympanic bulla more than 13 mm. Interorbital distance broad—more than 4 mm. Incisorial fossa partitioned by protrusions of intermaxillary bones. Length of hind foot more than 28 mm. Frontal bones lacking foramina.

Color of summer fur on the back varies from yellowish-gray-ochreous to saturated rusty-ochreous and black brown-ochreous. Winter fur gray or brownish-gray with dark; almost black, ripples. Vibrissae mostly black or black-brown, their length reaching 60-70 mm. Paw of foot black or black-brown.

*In Russian original, title incomplete—General Editor.

Description

Body length of adult males varies from 175 to 251 mm (average 209 mm); of females from 174 to 230 mm (average 202 mm). Length of foot 27 to 36 mm (average 31.5 mm). Length of ear—from 18 to 26 mm (average 23 mm).

The color of summer fur on the back varies from light yellowish-gray-ochreous to bright ochreous, rusty-ochreous, black-brown-ochreous and reddish-brown. The sides are usually of the same color as the back, but with an admixture of rusty and ochreous tones. The belly is pale yellowish-ochreous or pale reddish-brown. Along the margin of the ears there is an indistinct light border (Fig. 24).

The winter fur is mainly of two types: lighter ash-gray tone with black ripples on account of dark brown and black tips of hairs on the back and dull yellowish tone on sides, and darker—with an admixture of dull reddish-brown tone on the back and ochreous tone on sides and belly.

In Altai pikas from the taiga zone there are total melanists, the proportion of which may reach 20% of the population (Marin, 1984).

The length of vibrissae in Altai pikas reaches 60-70 mm; the majority of them are black or black-brown in color.

Fig. 24. Altai pika *Ochotona alpina* of southeastern Altai
(photo by V.V. Kucheruk).

Frontal bones lack foramina. The incisor fossa is partitioned by the protrusions of the intermaxillary bones. The preorbital foramen is small and triangular in form. Nasal bones are not broad in the anterior part, their lateral margins being almost parallel to each other. Orbits are relatively small and oval in form. The posterior zygomatic processes are long and quite broad. The tympanic bulla is of moderate size. The mandible is long and low. The ascending ramus of the mandible is low and broad. The angular notch is deep and oval in form (Fig. 25).

The form of the third lower premolar tooth varies quite significantly. Its anterior segment is relatively large and of variable configuration. The depth

89

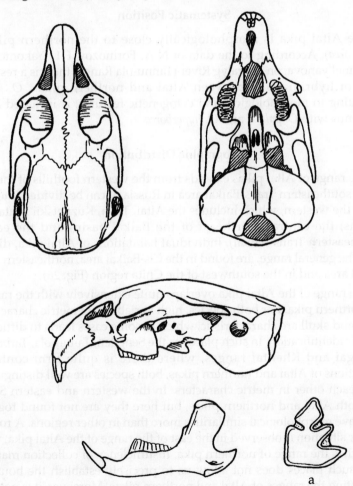

Fig. 25. Skull of the Altai pika *Ochotona alpina*.
a—third lower premolar tooth of the Altai pika.

120

of the inner entrant fold separating the anterior segment from the posterior, as a rule, is more than the outer (Fig. 25a). The posterior segment is variable in form.

The os penis is relatively long and thin; its length more than 8.5 times that of the width (Aksenova and Smirnov, 1986).

The diploid complement has 42 chromosomes: 12 pairs of meta-submetacentrics, 3 pairs of submetacentrics and 5 pairs of acrocentrics; the X-chromosome is submetacentric, and the Y-chromosome is smallest and acrometacentric (Vorontsov and Ivanitskaya, 1973).

Systematic Position

The Altai pika is morphologically close to the northern pika (*O. hyperborea*). According to the data of N.A. Formozov, E.L. Yakontov and L.E. Emel'yanova, on Torgalyg River (Tannu-ola Range) there is a restricted zone of hybridization between Altai and northern pikas. *O. alpina*, according to morphological and cytogenetic features, is included in one subgenus with *O. pallasi* and *O. hyperborea*.

Geographic Distribution

The range of Altai pikas extends from the western foothills of the Altai to the southeastern Trans-Baikal area in Russia. It can be divided into three parts: the western, which includes the Altai, Tuva, Kuznetskii Alatau and Sayans; the central (northeast of the Baikal basin) and the eastern (southeastern Trans-Baikal). Individual habitations of Altai pika, disjunct from the general range, are found in the Cis-Baikal area, northeastern Trans-Baikal area and in the southwest of the Chita region (Fig. 26).

The range of the Altai pika overlaps quite extensively with the range of the northern pika. For both species, high variability of metric characters of body and skull are characteristic, which in some cases leads to difficulties in their identification. In such places as the Sengilen, Tannu-ola, Tarbagatai, Khangai and Khéntai ranges, where there is quite firm contact of habitations of Altai and northern pikas, both species are well distinguished from each other in metric characters. In the western and eastern Sayans live both Altai and northern pikas, but here they are not found together, and have morphological similarities more than in other regions. A roughly similar situation is observed in the east of the range of the Altai pika, where it touches the range of northern pika. Insufficiency of collection materials from such places does not allow us to precisely establish the boundary separating the ranges of Altai and northern pikas. Moreover, it is not ruled out that there also exist other thus far unknown zones of contact of Altai and northern pikas.

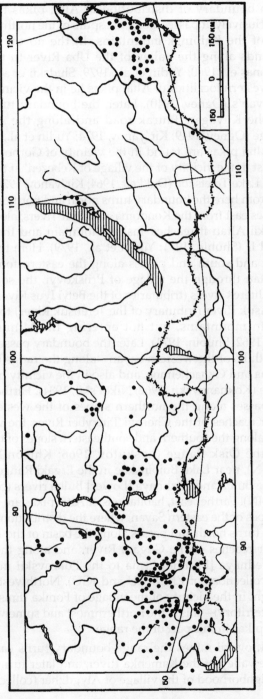

Fig. 26. Locations of Altai pikas in the territory of Kazakhstan and Russia.

The western boundary of distribution of Altai pikas has not been established conclusively. It passes along the spurs of the northwestern Altai in the region of the Kalbinsk mountains. In the north, the western boundary ascends along the valley of the Uba River to the Tigiretsk mountains (Afanas'ev, 1960; Yudin et al., 1979; Sludskii et al., 1980). The extreme northwestern localities of Altai pikas lie in the vicinity of Sinaya Sopka on Kolyvan's (Ognev, 1940). Later, the boundary turns east and passes along the Kuyagan Taurak road and along the spurs of the Cherginsk Range (Ognev, 1949; Kir'yanov, 1974; Yudin et al., 1979). In the eastern Altai, Altai pikas are found in the vicinity of Gorno-Altaisk, and farther to the east in the vicinity of the village of Kebezen' at the exit of the Biya River from Lake Teletskoe (Dul'keit, 1964; Kir'yanov, 1974; collections of ZM MGU). From here the boundary turns north and northeast to Gornoi Shorin and passes east from the Kondoma and Tom' rivers; along the spurs of the Kuznetskii Altau it reaches Taskyl Mountain and the villages of Berikul'skii and Ir (Shubin, 1971; Yudin et al., 1979). Here the boundary turns southeast and south and passes along the eastern foothills of the Kuznetskii Alatau through the village of Priiskovyi, the sources of the Tyurtek and Pikhterek rivers (tributaries of the Belyi Iyus River), the upper reaches of the Asuk River (tributary of the Terensuk River) to the vicinity of the Karlygan mountains, but not entering the Minusinsk basin (Kokhanovskii, 1962; Shubin, 1971). Later the boundary passes east along the western foothills of the western Sayan—along the range between the villages of Maina and Oznachennoe, and also in the vicinity of the village of Ermakovskoe (Kokhanovskii, 1962; Nikol'skii, 1984). Farther southeast the boundary passes along the northern slopes of the western Sayan— along the upper reaches of the Chernyi Tanzybei River, the valley of the Tanzybei River, along the southern and southeastern slopes of the Kulumys Range, along the Oisk Range (Loskutov, 1966; Khlebnikova, 1974; collections of BIN), near Lake Buibinskoe in the Ergaki Range, and along the valleys of the Buiba and Nizhnaya [Lower] Buiba rivers (collections of ZM MGU and BIN). Farther east, habitations of Altai pikas are found along the northern slopes of the eastern Sayan, where the distributional boundary of the Altai pika turns north and passes along the basin of the Tobol' River, along the upper reaches of the Gugara River, and along the Nizhnaya [Lower] and Srednaya [Middle] Erma to the Pokrovskii mines on the Biryusa River (collections of ZM MGU and IGU). Northwestward, Altai pikas were caught in the vicinity of the village of Fomka (Eniseisk district of Krasnoyarsk territory)—this is the northernmost and somewhat disjunct locality of Altai pikas in this part of the range.

From the Pokrovskii mines the range boundary turns eastward and passes along the valley of the Kamenka River, and later turns south and passes in the neighborhood of the village of Alygdzher (collections of ZM

MGU, IGU, and BIN). Southward, Altai pikas are possibly present on the southern slope of the eastern Sayan, in the upper reaches of the Azas River (Yudin et al., 1979).

Southeast of Tuva, Altai pikas were caught in the Sayan-Tuva upland in the middle fork of the Chavach River and in the valley of the Kara-Khem River (collections of BIN). Following a western direction, the southern boundary distributional of the Altai pika passes through the basin of the Karga River, along the upper reaches of the Shurmak River and in the vicinity of the village of Ushino (Yanushevich, 1952; Vlasenko, 1954; collections of ZM MGU). Westward, the boundary passes along the southern and northern slopes of the Tannu-Ola Range, particularly in the basin of the Torgalyg River and in the upper reaches of the Ulug-Khondergei River (collections of BIN). According to the data of N.A. Formozov (oral communication), Altai pikas occur in the upper reaches of the Barlyk River (northern slope of the Tsagan-Shibetu Range), in the upper reaches of the Talaity River (southern slope of the mountain massif of Mogun-Taiga), in the valley of the Orto-Shigetei River and in the valley of the Mogen-Buren River. Farther west along the state border, Altai pikas were noticed near the source of the Yustyd River (spurs of the Chikhachev Range) and in the upper reaches of the Bar-Burgazy River (Vorontsov and Ivanitskaya, 1973; collections of IE'RZh). Maintaining a western direction, the southern boundary in this part of their range passes along the Sailyugem Range, through the Bolshie [Greater] Shibety, the upper reaches of the Dzhaster, Dzhumaly and Kalguty rivers and along the valley of the Karakugoi River (right tributary of Akalakha River) (Potapkina, 1971; Kir'yanov, 1974; collections of BINS). Still westward, pikas were caught in the upper reaches of the Bukhtarma and Sarymskaty river, along the spurs of the Kurchumsk Range, and surrounding Lake Markakol (Potapkina, 1971; Sludskii et al., 1980; collections of BIN). From Markakol the distributional boundary of Altai pikas ascends north, passes along the southern spurs of the Narymsk Range and later passes west to the Kalbinski Alatau (Potapkina, 1971).

In the southwestern Trans-Baikal area, Altai pikas apparently are very rare. It is known that here both Altai and northern pikas are found. Thus, in the collections of Irkutsk University is preserved a skull of a pika, caught in 1938 in the vicinity of the village of Moinda and identified as *O. alpina*. In the very same region N.A. Formozov (oral communication) found northern pikas.

The central part of the range of the Altai pika includes habitations along the eastern bank of Lake Baikal. The southern habitations of Altai pikas are found here in the Svyatoi Nos Peninsula in the valley of the Burtui river, in vicinity of Kurbulik, on the slopes of mountains facing the

Chivyrkuisk Gulf (collections of ZM MGU). Along the western and eastern spurs of the Barguzinsk Range, Altai pikas were caught on the territory of the Barguzinsk Preserve in the valley of the Bolshoi [Greater] Cheremshan River, along the Vorokinsk spring, at the sources of the Nirgili and Shirgili rivers, in the valley of Shumilikha River, and in the upper reaches of the Verkhny [Upper] Levyi [Left] Kurumkan River (Turov, 1936; Ognev, 1940; collections of BIN and ZM MGU).

Southwest of the Chita region (Krasnochinsk district, mouth of the Menza River and in the upper reaches of the Zhergokon River on Chikoisk Golets [Bald peak] apparently Altai pikas occur (collections of IGU). Altai pikas are frequent in the vicinity of the village of Bukun on the Onon-Bal'dzhinsk Range, whereas eastward, in the vicinity of the village of Mangut, they are absent (Nekipelov, 1961).

The eastern part of the range of the Altai pika is restricted from the west by the right bank part of the lower fork of the Onon River. In the southwestern part of this area, Altai pikas were caught in the Ust'-Kopchil and Urtui ravines (Nekipelov, 1961; collections of ZIN). Northward the distributional boundary of the Altai pika passes through the town of Sheronai up to the mouth of the Unda River (Nekipelov, 1961). S.I. Ognev (1940) and S.S. Turov (1936) assumed that northern pikas occur on the Undin Range south of Nerchinsk. In the area of Nerchinsk and the town of Shivka, the boundary turns northeast and proceeds along the right bank of the Shilka (Formozov and Nikol'skii, 1979; collections of ZM MGU and ZIN). According to the data of N.A. Formozov (oral communication), Altai pikas are found 21 km southeast of the mouth of the Kuringa Rier (right tributary of the Shilka) as well as downstream of the Shilka—opposite the village of Ust'-Chernaya (this is the northernmost locality of Altai pikas within the limits of the eastern part of the range). Altai pikas do not extend farther to the northeast. Altai pikas are absent also on the eastern slopes of the Gazimursk and Uryumkansk ranges. Consequently, the eastern boundary of distribution of Altai pikas in the eastern Trans-Baikal area passes from Ust'-Chernoi to Ushmunsk golets* 38-40 km east of Gazimursk plant and later turns east and enters China along the valley of the Urov River (Nelkipelov, 1961; collections of ZM MGU; oral communication of N.A. Formozov). In the territory of the former Soviet Union, colonies of Altai pikas were found in the vicinity of the Nerchinsk plant, towns of Mikhailovskoe, Srednyaya [Middle] Borz'ya, Pokrovka, Chil'gintui, Nizhnyi [Lower] Kalgukan, Byrka and Urlyungui, and along the Argun' River—in the vicinity of the towns of Novyi Tsurukhaitui and Kailastui

---

*Golets—a bald peak (cf. Callaham), or a bare rock (cf. Alexander Knox)—General Editor.

(Nekipelov, 1961; collections of ZM MGU, ZIN, IGU and "Mikrob" Institute, Saratov). Kailastui is the extreme southwestern locality of Altai pikas in Russia.

94   From the town of Novyi Tsurukhaitui, the southern boundary of the eastern part of the range of the Altai pika proceeds west along the left bank of the Urulyungui River through the town of Urulyungui and enters the neighborhood of the town of Mulino. Along the Nerchinsk Range the boundary proceeds to the town of Perednyaya Byrka (Nekipelov, 1961; Formozov and Nikol'skii, 1979).

Thus, the range of the Altai pika stretches in the meridional direction from 82° and 120° E. Long. In the western part of the range its southernmost colonies are confined to the southwestern spurs of the Altai in the area of the Kurchumsk Range approximately at 49° N. Lat., and the northermost at 56.25' N. Lat. (vicinity of the village of Fomka). The central part of the range, which includes colonies northeast of the Baikal basin, is stretched in the meridional direction along the Barguzinsk Range between 53°35' N. Lat. in the south (Svyatoi Nos Peninsula, Burtung River) and 54°20' N. Lat. in the north (upper reaches of the Verkhnyi Levyi Kurumkan River) and, possibly, along the western slopes pushed farther north. In the eastern part of the range, Altai pikas are found from 49°50' N. Lat. in the south (vicinity of Kallastui) to 52°55' N. Lat. in the north (right bank of the Shilka River, vicinity of the village of Ust'-Chernaya). The internal structure of these three parts of the range of the Altai pika is shown in the map (cf. Fig. 26).

Beyond the borders of Russia the Altai pika is widely distributed in the territory of Mongolia—in the Khangai, Mongolian and Gobi Altai mountains and in Khe'ntei, as well as in China—Northern Gansu (cf. Fig. 9a).

Geographic Variation

Seven subspecies of Altai pika occur in the territory of Russia and Kazakhstan.

1. *Ochotona alpina alpina* Pallas, 1773—Middle-Altai pika. The type subspecies is characterized by quite large skull dimensions: condylobasal length varies from 42 to 49.2 mm (average—46.1 mm); length of rostrum—from 17.5 to 21 mm (average—19.3 mm); width of rostrum—5.9-7.8 mm (average—6.6 mm); length of upper tooth row—8.1-9.5 mm (average—8.7 mm) (Table 6.1). The color of summer fur of adults is yellowish-ochreous-gray on back, rusty on the sides and grayish-yellow or the belly. The winter fur is dirty ash-gray with different proportions of black admixture.

95   Distribution: Western and central Altai (excluding the Ivanovskii and Kholzum ranges), in the east—up to the Katun' River, in the south—up to

94

Table 6.1. Skull dimensions (in mm) of adult individuals of seven subspecies of Altai pikas

Indices	O. a. alpina (n = 86)			O. a. nitida (n = 161)			O. a. aura (n = 9)			O. a. nanula (n = 70)		
	min.	M	max.	min.	M	max.	min.	M	max.	min.	M	max.
Condylobasal length	42.0	46.1	49.2	41.0	44.1	48.1	48.0	50.7	52.4	40.0	42.3	44.5
Length of rostrum	17.5	19.3	21.0	17.0	18.5	20.5	19.7	21.4	22.3	16.3	17.8	19.9
Length of upper tooth row	8.1	8.7	9.5	7.7	8.6	9.4	9.1	9.4	10.0	7.8	8.2	9.0
Length of orbits	12.4	13.6	14.9	12.2	13.3	15.3	13.9	14.0	15.5	11.8	13.0	15.7
Length of tympanic bulla	13.0	14.2	15.9	12.7	13.8	15.3	13.7	15.3	16.0	12.5	13.5	14.5
Width of tympanic bulla	8.9	10.7	12.2	9.3	10.5	12.2	10.1	11.4	12.3	9.4	10.2	11.0
Interorbital width	4.5	5.3	6.8	3.6	5.1	6.0	4.5	5.1	5.9	4.0	5.1	5.7
Zygomatic width	21.6	23.3	24.9	21.1	22.7	24.8	24.5	25.3	26.2	20.6	21.9	23.0
Width of cranium	17.3	20.4	22.0	18.0	19.6	21.3	20.6	22.0	22.7	17.8	19.1	21.0
Width in region of tympanic bullae	20.5	23.0	24.6	20.5	22.3	24.9	24.3	25.0	26.2	20.2	21.7	22.9
Postorbital width	13.8	14.8	15.9	13.2	14.4	16.0	14.1	15.0	15.6	13.3	14.5	15.5
Width of rostrum	5.9	6.6	7.8	5.5	6.4	8.0	6.8	7.6	8.2	5.4	6.4	7.0
Occipital height	11.5	12.4	13.2	11.3	12.2	13.5	12.8	13.3	13.8	10.9	12.1	13.2
Height of frontal region	11.5	12.1	13.5	11.0	11.8	12.7	12.3	12.7	13.1	10.0	11.7	12.7
Height of rostrum	6.9	7.6	8.5	6.7	7.4	8.5	7.8	8.4	9.0	6.5	7.2	7.9
Height in region of tympanic bullae	14.6	15.8	17.2	13.8	15.4	17.0	16.2	17.0	17.7	14.0	15.2	16.9

Contd.

Table 6.1 contd.

Indices	O. a. sayanica (n = 43)			O. a. svatoshi (n = 10)			O. a. scorodumovi (n = 13)		
	min.	M	max.	min.	M	max.	min.	M	max.
Condylobasal length	44.0	47.9	52.0	42.1	43.2	45.0	41.5	43.4	45.0
Length of rostrum	18.8	20.6	22.0	16.7	17.6	18.9	17.1	18.3	20.0
Length of upper tooth row	8.9	9.3	10.0	8.1	8.7	8.9	8.0	8.7	9.3
Length of orbits	12.9	14.1	15.8	12.4	12.9	13.5	12.2	13.3	14.0
Length of tympanic bulla	13.4	14.9	16.0	12.9	13.7	14.1	13.0	13.6	15.0
Width of tympanic bulla	9.9	11.1	12.2	9.3	10.2	11.3	9.7	10.1	10.8
Interorbital width	4.4	5.6	6.9	4.0	4.7	5.1	4.6	5.0	5.5
Zygomatic width	22.4	23.9	26.3	21.0	21.6	22.1	21.7	22.3	23.0
Width of cranium	19.0	14.7	23.5	18.0	19.2	20.0	18.0	19.1	21.8
Width in region of tympanic bullae	21.7	23.9	25.7	21.0	22.0	23.0	21.2	22.1	24.0
Postorbital width	13.7	14.8	16.0	13.7	14.0	14.4	13.4	14.1	14.5
Width of rostrum	5.7	7.2	8.1	6.4	6.8	7.3	6.2	7.0	7.8
Occipital height	11.8	12.9	13.8	11.0	11.7	12.1	11.0	12.0	12.6
Height of frontal region	11.4	12.6	14.0	10.9	11.3	11.8	11.0	11.7	12.5
Height of rostrum	7.2	8.3	9.2	6.9	7.1	7.5	6.8	7.4	8.0
Height in region of tympanic bullae	14.9	16.3	18.0	13.9	14.8	15.4	14.0	14.9	16.3

the Narym Range and Kalbinski Altai. The Kuznetskii Alatau populations, apparently, also belong to the type subspecies.

2. *O[chotona] alpina nitida* Hollister, 1912 (syn. *O. sushkini* Thomas, 1924).—Eastern Altai pika. Skull dimensions, on average, smaller than in the type subspecies: condylobasal length varies from 41 to 48.1 mm (average—44.1 mm); length of rostrum—from 17 to 20.5 (average—18.5 mm); width of rostrum from 5.5 to 8 mm (average—6.4 mm); length of upper tooth row—from 7.7 to 9.4 mm (average—8.6 mm) (cf. Table 6.1). The color of summer fur of adults is darker than in the type subspecies: the back is usually dark brown, sometimes dark reddish-brown with brown or black tones. Ochreous or rusty tones are manifest only on the sides. Distribution: Eastern Altai and western Sayan. Beyond the borders of Russia, the eastern Altai pika occurs in the Mongolian and Gobi Altai (Ikh-Bogdo-Ula and Adzh-Bogdo-Ula).

3. *Ochotona alpina atra* (Eversmann, 1842).—Eversmann's Altai pika. This subspecies is characterized by the largest skull dimensions for the species: condylobasal length of skull more than 48 mm; zygomatic width more than 24.5 mm; width of cranium usually more than 21.5 mm; length of rostrum usually more than 21 mm; width of rostrum usually more than 7.5 mm (cf. Table 6.1). Light tones with ochreous-gray tinge predominate on the back; yellowish on sides and grayish-yellow on the belly. Distribution: Eversmann's pikas are known from the Kholzun Range (border of Altai territory and eastern Kazakhstan region) and Ivanovsk (eastern Kazakhstan region) as well as along the upstream of the Kokas River (western Altai).

4. *Ochotona alpina sayanica* Yakhontov et Formozov—eastern Sayan pika. The eastern Sayan subspecies of the Altai pika is characterized by quite large dimensions. The length of the rostrum is usually more than 19.5 mm, length of upper tooth row usually more than 9 mm and length of tympanic bullae usually more than 14 mm; differs from *O. a. atra* by having a lesser zygomatic width, which is usually less than 24.5 mm (cf. Table 6.1). The color of summer fur is dark ochreous with brown speckles on the back, dark rusty on sides and light yellow on the belly. The inner surface of the ears is dark gray with a distinct light border on the margin. The underside of the paws is dark reddish-brown. Distribution: eastern Sayan from the upper reaches of the Gutara and Biryusa [rivers] to the Tunginsk goltsy [bald peaks]. Possibly this subspecies lives in the [Lake] Khubsugul' area.

5. *Ochotona alpina svatoshi* Turov, 1924.—Barguzinsk pika. The Barguzinsk subspecies of the Altai pika is characterized by smaller dimensions: condylobasal length of skull usually less than 44 mm; length of rostrum usually less than 18 mm; length of upper tooth row less than 9 mm; zygomatic width usually less than 22 mm; length of tympanic bullae

usually less thah 14 mm. In skull dimensions it is close to Skorodumov's Altai pika, but has a somewhat greater width of skull in the region of the tympanic bullae, which is usually more than 22 mm (Table 6.1). The color of summer fur is bright, ochreous-rusty on the back. Winter fur is ochreous-rusty with black flecks. Distribution: northeastern coast of Lake Baikal, Barguzinsk Range.

N.A. Forzomov and E.L. Yakhontov, based on the analysis of sound signals, relate pikas from the Barguzinsk Range to *O. hyperborea*. However, specimens available in the collections of the Zoological Museum of MGU, from the Barguzinsk Range, according to the morphometric characters should be referred to *O. alpina*. Unfortunately, there are no data on the chromosome complements of pikas from the Trans-Baikal area and the collection material is absent on the pikas, on which sonographic analysis was done. It may be suggested that in the area of the Barguzinsk Range, both northern as well as Altai pikas live, but definitive conclusions on the taxonomic status of pikas from northeastern Trans-Baikal should be made only based on the study of the complete set of characters—cytogenetic, morphological and acoustic.

6. *Ochotona alpina scorodumovi* Scalon—Skorodumov's Altai pika. A small form of Altai pikas: condylobasal lengtn of skull usually less than 44 mm; length of upper tooth row usually less than 9 mm; height of skull in frontal part usually less than 12 mm. Differs from the Bargusinsk subspecies by having a narrower skull in the region of the tympanic bullae (usually less than 22 mm) (cf. Table 6.1). In the color of summer fur on the back, brown and black tones predominate; sides, with an admixture of ochreous tinge; and the belly paler, rusty gray. Distribution: southeastern Trans-Baikal area, interfluve of the Shilka and Argun' rivers.

97   S.I. Ognev (1940), for Altai pikas living in the Eastern Trans-Baikal area, recognized the existence of two subspecies—*O. a. scorodumovi* and *O. a. cinreo-fusca* Schrenck, 1858. Investigations of E.L. Yakhontov and N.A. Formozov established that specimens of pikas caught by R. Maak and described by Schrenck as *Lagomys hyperboreus* var. *cinereofusca* ought to be referred to the northern pika, since the left bank of the Shilka should be considered the place of their collections, where only northern pikas live.

7. *Ochotona alpina nanula* Yakhontov et Formozov—Tuva pika. A small form of Altai pika: condylobasal length of skull less than 44 mm; length of upper tooth row 8.2 mm; width of rostrum usually less than 6.3 mm (cf. Table 6.1). The color of summer fur is reddish-brownish-brown, sides with rusty tinges; the belly is rusty-yellow. Distribution: south of Tuva—Tannu-Ola, Tsagan-Shibetu, Sangilen ranges as well as the extreme southeast of Altai—the upper reaches of the Chulyshman River and vicinity of Lake DzhuluKul'.

One more subspecies of Altai pika—*Ochotona alpina changaica* Ognev, 1940, is known from the territory of Mongolia (Khangai), which, in the morphological characters of the skull, occupies an intermediate place between the type subspecies and eastern Altai pika but, on average, has relatively larger length of the upper tooth row than in *O. a. alpina*. The color of summer fur in the Khangai pika is quite dull, of a yellowish tone.

According to the data of E.L. Yakhontov and N.A. Formozov, in Mongolia there is one more form of Altai pika of subspecies rank in Khangai, which has very large dimensions and sound signals different from the other subspecies.

Subspecies *Ochotona alpina argentata* Howell, 1928 has been described from China (Northern Gansu). However, the taxonomic status of this form needs vertification.

Biology

Habitat. Altai pikas may be found in the taiga zone, sparse forests, and in the subalpine and alpine belts. They may inhabit slopes of different exposures and inclination (to 50°). The main habitats of Altai pikas are stony mounds or "kurumniks" [rock-terrains] and the most favorable for them are the mounds with stones 0.5-1.0 m in diameter (Potapkina, 1971). On extensive mounds pikas live on their edges. The necessary condition for habitation of Altai pikas is the presence of vegetation, sufficient for foraging and collection of haypiles. In forest, Altai pikas can live under brushwood and between the roots of trees (Khmelevskaya, 1961). The most unfavorable for Altai pikas apparently should be considered areas with wet falls and low-snow winters, since in wet places the haypiles get damaged, and with insufficient snow cover the shelters of pikas are frozen (Potapkina, 1975).

In Kazakhstan on the northern slopes of the Ivanovsk Range, Altai pikas live at altitudes of 1,270-2,100 m above sea level in forest, subalpine and alpine belts, where large-stony mounds alternate with glades. In the forest belt, pikas prefer to live in fir-larch sparse forests (with an admixture of cedar, birch, mountain ash and European aspen) with an underbrush of currant and honeysuckle. In deciduous and dense fir forests, where the grass stand is poor and the stone mounds are rare, Altai pikas are not found. At the upper limit of forest and in the subalpine belt, juniper is common in Altai pika habitats; cedar or fir stands as well as rare larches are also found. In the alpine belt willows dominate among shrubs (Sludskii et al., 1980).

In the Altai territory Altai pikas live at altitudes from 400 (foothill forest steppe and mountain taiga) to 3,000 m above sea level (alpine belt). On the spurs of the Terektinsk Range, pikas live on small rock dumps overgrown with shrubs of spirea, pea shrub, and roses. On the eastern slopes of the Korgonsk Range (Mt. Tyudektu), Altai pikas have been found on rock dumps among sparse forests at altitudes of 1,700 m above sea level (Potapkina, 1978). Pikas avoid highly shaded places. On open level areas, Altai pikas are often found along the banks of water bodies. In the vicinity of Lake Dzhulkul' and in the upper reaches of the Yustyd River (southeastern Altai) pikas live in piles of moraine boulders and in small clusters of stones among dwarf arctic birches and alpine herbs. In high-mountain tundra, pikas often live under individual large boulders on slopes of southern and southwestern exposures. Altai pikas rarely live on the northern slopes (Potapkina, 1975). On the eastern slopes of the Torot Range (northeastern Altai) pikas are found in mixed forests, where they live among roots of cedar and under brushwood. On the western coast of Lake Teletskoe, Altai pikas live both in stone mounds (open or in forests) and outside of mounds—in cedar forests with dense underbrush and rich grass cover, as well as among individual stones on banks of small streams. Here Altai pikas are also found along forest fellings and along logging roads, where they live in birch stacks, and in masses of shoots (Khmelevskaya, 1961).

In Tuva, Altai pikas are the common inhabitants of the taiga; they also live in alpine meadows among stone barriers. In island forests of larch, pikas can enter the mountain steppe landscapes (Potapkina, 1978).

In the western Sayan, habitats of Altai pikas are most diverse. Thus, on the northern slopes of the Dzhoisk Range, Altai pikas form dense colonies in large stony mounds among green dense ceder forest at altitudes of 1,200-1,400 m above sea level. Pikas do not live on the extensive mounds, that are not overgrown with vegetation, on mountain peaks or in swampy mountain tundra. At altitudes less than 1,200 m, where small mounds overgrown with vegetation predominate, colonies of Altai pikas are sporadic. In this part of the range along streams is also found belt type habitations of Altai pikas. In the sparse forest belt pikas inhabit forest areas as well as open spaces with stone mounds of varying degree of turfiness. In this zone, at an altitude of 1,400 m above sea level on the gentle water divide, an unusual type of habitat of Altai pikas has been found—swampy cedar forest with moss cover rising above the tree roots. Here pikas use hollows between tree roots and lairs as shelters; moreover the density of their colonization is quite high (Naumov, 1974). On the southern slopes of the Aradansk Range, Altai pikas live in the upper slopes, in valleys of streams and in old burned-out forests with developed grass cover. In green dense cedar forests, Altai pikas do not form large colonies (Khlebnikova, 1978).

In the eastern Sayan on the Kryzhina Range in the Kyzyr-Kizirsk interfluve, Altai pikas are found in dark coniferous forests with well-developed grass cover as well as in the subalpine and alpine belts, where most often they occupy stone mounds overgrown with rhododendron, ledum, blueberry, sedges, ferns, reindeer moss, and different mosses (Kim, 1956, 1957).

In Khangai, in the sympatric zone with northern pikas, Altai pikas colonize the golets [bald peak] zone and open mounds in meadows below the limit of forest, whereas northern pikas prefer to colonize stony habitats of the forest belt. In the valley of the Urt-Tamir River, a sympatric zone of both species has been observed, but even here territorial separation between their family areas has been observed—Altai pikas occupy the central part of the extensive mounds, while territories of northern pikas are located along its periphery at the forest edge. In the Lake Khubsugul' area, Altai pikas are also found together with northern pikas, but here isolated colonies of Altai pikas surrounded by colonies of northern pikas are found in the golets [bald peak] zone (Formozov, 1986).

Population. The population of Altai pikas may differ significantly even in nearby areas. For instance, on the northern slopes of the Ivanovskii Range (Kazakhstan) during the warm season, in some areas Altai pikas were numerous, and in identical neighboring territories with good foraging conditions there were no pikas at all, although earlier they did live there (Sludskii et al., 1980). Such a phenomenon was observed also for regions of the Altai (Potapkina, 1975).

The population of Altai pikas is not uniform in different altitudinal belts. Thus, in Mongolia, their maximum number was noticed in the alpine belt (Bannikov, 1954). In the Kuznetskii Alatau the highest population density of Altai pikas was noticed in the golets [bald peak] zone—20-25 individuals/ha (Shubin, 1972). In the northeastern Altai the population of Altai pika is the highest in the dark coniferous middle mountains, where foraging conditions are better than in the upper and lower mountain belts and there are a large number of shelters (Smirnov, 1969). But even here, there were differences: the maximum population of pikas was in spruce-cedar taiga; and the minimum—in the birch-spruce forests and birch-European aspen forests (Table 6.2). In the course of the year, the population of Altai pikas changes, but in different habitats with different amplitude (Fig. 27).

In the western Sayan the maximum density of Altai pikas (to 60 individuals/ha) was reported in the subalpine belt with green dense cedar forests, abundant stone mounds, and subalpine meadows. In old, partially rejuvinated burned-out areas with developed grass cover, the population of pikas is somewhat lower—about 50 individuals/ha. Altai pikas inhabit

100

Table 6.2. Population density of the Altai pika in the northeastern Altai
(Smirnov, 1967)

Altitudinal belt and habitats	Year of census	Individuals/km^2	
		May 16-31	June 1- August 30
1. Low mountains (400-1,200 m above sea level)			
Pine-birch forests on stony slopes (outliers, valleys of large rivers)	1961	249	750
Pine-birch forests on banks of Lake Teletskoe	1963	192	384
Larch-birch forests on banks of Lake Teletskoe	1963	7	3
Kurumniks [Rock terrain] in European aspen-fir black taiga	1961-63	170	381
Cedar-fir black taiga		1,010	2,090
Birch-European aspen forests in extensive burned-out areas		78	165
2. Middle mountains (900-2,100 m above sea level)			
Birch-European aspen island forests	1962	195	453
Birch-spruce forests		144	450
Fir-cedar taiga		444	1,017
Spruce-cedar taiga		1,944	4,632
Spruce-fir-cedar taiga		870	2,496
Cedar taiga		930	2,466
Fir-Cedar subalpine forest		879	1,425
3. High-mountain (2,000-2,500 m above sea level)			
Dwarf arctic birch tundra		21	36
Stony tundra		123	219

the low-mountain belt of the northern mega-slope with the lowest density (1 individual/ha) (Khlebnikova, 1978).

Shelters. Most often voids between stones in mounds or crevices and recesses under individual stones serve as shelters for Altai pikas in all parts of their range; less often they occupy hollows in slashed trees and brushwoods. In winter, Altai pikas may hide in tunnels under snow.

In the western Sayan on stones semi-overgrown with moss, a nest of an Altai pika was found. It was situated under a slab covered with smaller stones at a depth of 1 m. The nest was a cup-like structure with a trough in the middle. It was constructed predominantly with thin leaves of grasses, thin long roots and a small quantity of moss. The area of the nest was approximately 12 × 4 cm; height, about 8 cm (Naumov and Lur'e, 1971). In another habitat in the western Sayan (swampy cedar forest), nests of Altai pikas were often located near tree roots, 70-100 cm from the trunk,

134

99

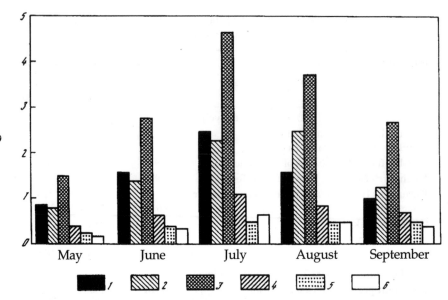

Fig. 27. Population of Altai pikas (in thousand individuals/km²) in dark coniferous middle mountains of the northeastern Altai in 1962. 1—cedar taiga; 2—sparse-fir-cedar taiga; 3—sparse-cedar taiga; 4—fir-cedar taiga; 5—birch-spruce forests; 6—birch-European aspen forests in old burned-out areas (from Smirnov, 1967).

sometimes under the trunk or 2-3 m from it in hollows of moderate dimensions; the tunnel in these hollows passed beneath. Here nests had a conusoidal form with the base area of 25 × 30 cm and with a depth about 6 cm at the tip of the cone. The bedding of the nest consisted of moss and a small quantity of leaves of grasses. Sometimes nests were located on a quite thick bedding of vegetative remnants, partly or completely rotten, which confirms the prolonged use of the very same place for the construction of the nest (Naumov, 1974).

In the forest zone of the Kuznetskii Alatau, in winter, Altai pikas build spherical nests 25-30 cm in diam. under the snow, using hay. These nests, from which issue many tunnels under the snow, are located 7-10 m from haypiles (Shubin, 1971).

At places of rest Altai pikas sometimes make special bedding. In the western Sayan, for such bedding pikas use ledum, cowberry, crowberry or twigs of cedar with the addition of twigs and leaves of rhododendron, cowberry, crowberry and pieces of cedar bark. The thickness of such bedding is from 1-2 to 4 cm, and unlike the soft nest, it is quite coarse (Naumov, 1974).

Altai pikas do not dig into more-or-less solid ground. Their burrows are found very rarely in areas with a thick moss cover and turfaceous or turfaceous-humus soils. In this case, burrows of Altai pikas are a system of long branched tunnels in loose peatlike moss lying below roots of crowberry and cowberry (Naumov, 1974).

Feeding. According to numerous observations (Kolosov, 1939; Kim, 1956, 1957, 1959, 1962*; Reimers, 1960, 1966; Shtil'mark, 1963; N.G. Shubin, 1963, 101 1971; Marin, 1984, and others), Altai pikas feed on almost all species of plants surrounding their colonies. For example, on the northern slope of the Ivanovskii Range (Kazakhstan) in mixed and larch sparse forest, nearly 100 plant species were recorded that were used by Altai pikas in their food, and in alpine meadows, about 70 (Sludskii et al., 1980).

The composition of diet of Altai pikas depends on the character of their habitats and time of the year. Thus in the forest belt of the northern slopes of the Ivanovskii Range, in April 1978, when the growth of plants had not yet begun, the main mass of winter reserve food was already consumed by pikas, and arboreal food predominated in their diet. At the same time, on the slopes of northern exposures, despite nonutilization of winter reserves, the pikas fed on buds, first leaves of shrubs and various grasses (Sludskii et al., 1980). During the warm period of the year, Altai pikas fed mainly on the green parts of higher plants. Flowers are absent in the diet of pikas in May-June, while seeds, berries, mushrooms and lichens are consumed in August-September (Table 6.3). Pikas more avidly feed on lichens after rains. Throughout the year vegetative parts of green plants 102 are constantly present in the food of Altai pikas. In the forest belt these plant parts are found in the stomachs of pikas most frequently in September, and in the alpine belt, in August. Feeding on and storing yellow and even dried leaves of willows, European aspen, aconite and dolgonog** has also been reported (Sludskii et al., 1980).

In the course of a day, Altai pikas feed several times. According to the data of I.P. Khlebnikova (1976), in the western Sayan in the second half of the day (from 16.00 hr until darkness) pikas venture out for feeding 1-2 times in an hour. Most often pikas nibble at the upper parts of plants by stretching their neck. When feeding on shoots or leaves of shrubs, pikas stand on their hind feet, resting their fore feet on branches. Like other species of pikas, they do not hold suspended food in their paws, but sometimes press the plant or its part on to the ground by their fore paws and nibble in bits. In a single act the animal eats 3-4 g of green grass or 0.7-1.2 g of dry vegetation (Khlebnikova, 1974). The weight of the well-

---

*Cited but omitted from Lit. Cit. in original.

**Local name; common name could not be confirmed—General Editor.

Table 6.3. Composition of the diet of Altai pikas (occurrence of food in %) on the northern slope of the Ivanovskii Range in Kazakhstan (from Sludskii et al., 1980)

Habitat, altitude above sea level	Month	n	Stomach contents			
			vegetative parts of plants		bark	flowers
			green	yellow		
Mixed and larch forest, 1,300-1,800 m	April	6	83.4	16.6	33.3	0
	May	3.2	100.0	6.3	3.1	6.3
	June-July	19	100.0	0	0	21.8
	August	47	100.0	0	0	0
	September	12	100.0	58.3	0	0
	October	12	100.0	16.7	0	0
Alpine meadow 2,000 m	July	15	93.4	0	0	33.3
	August	31	87.2	45.2	0	0

Habitat, altitude above sea level	Month	n	Stomach contents				
			seeds	berries	mushroom	lichens	fecal matter
Mixed and larch sparse forest, 1,300-1,800 m	April	6	0	0	0	–	50.0
	May	32	0	0	0	–	62.5
	June-July		0	0	0	–	52.6
	August	19	14.9	21.3	0	17.1	29.8
	September	47	0	8.3	8.3	0	50.0
	October	12	0	0	0	–	16.7
Alpine meadow, 2,000 m	July	15	0	0	13.5	46.7	
	August	31	3.2	9.7	0	12.0	16.1

Note: "–" denotes unreliability in identification of the given component. n—number of stomachs investigated.

filled stomach in adult Altai pikas constitutes 23-27 g, that is, from 6 to 7.1% of the body weight (Sludskii et al., 1980).

The haypiles made by them have a great significance in the feeding strategy of Altai pikas. In the beginning, the process of storage is rather ritualistic in character, since these animals consume the bulk of the food carried and the weight of such reserves does not increase (Naumov, 1974). In the forest belt of the Ivanovskii Range, the beginning of storage was noticed on May 23, and in the alpine belt, on May 19 (Sludskii et al., 1980). At this time, the animals begin to fetch plants and pile them under stones. Intensive storage of food, during which increases in the weight of the haypiles, begins in the forest belt in the first days of July, and in the alpine belt, in the middle of the same month. In other parts of the range a much later start of food storage is noticed in the alpine zone (Yurgenson, 1939). Altai pikas store food most intensively in August. Storage continues until the formation of a stable snow cover (usually until October). If the snow melts, Altai pikas continue to make haypiles. In the Kuznetskii Alatau, the periods of food storage are roughly the same as on the Ivanovskii Range (Shubin, 1971), and in the eastern Altai a much later (first days of August) start of storage is observed (Yurgenson, 1939).

In the western Sayan Altai pikas begin to store food soon after appearance of the young. In June from 1600 hr to dark, the number of loads per hour per individual constitutes, on average, 0.9, in July—3.2, in August—5.7, and in September—5.4 (Khlebnikova, 1978). August and the first half of September are characterized by the maximum collection activity of these animals, in these months the number of days with the maximum number of load increases (to 35-40/hr).

During collection of haypiles, Altai pikas move in hops, halting from time to time. If the plant is small, the animal fills its mouth with it and runs to the shelter. Large plants (to 80 cm long) are moved by pikas by dragging or breaking them into parts. For a day a single animal may collect up to 500 g of fresh food. The collection of plants may occur near as well as at quite considerable distance (50-60 m) from the shelter. Altai pikas can store food also after rain, during which they rarely cure the grass on the surface of stones and often carry it immediately to the place of storage (Shubin, 1971; Sludskii et al., 1980).

Altai pikas, as a rule, store their haypiles under large stones, in niches and recesses between stones; moreover, the largest haypiles are located usually in niches with a crown height of 20-40 cm. The plants are packed so compactly, that they can only be removed from there with great effort. The weight of the stored plants (without additional curing) may reach 27 kg. In the highly overgrown outliers of the eastern part of the Ivanovskii Range, haypiles were observed in the form of piles arranged under old weighty cedars or on stumps and branches (Sludskii et al., 1980). In the forest zone of the Kuznetskii Alatau, Altai pikas often make true piles near the trunks of firs, and less often they conceal their reserves under stones. The height of haypiles may reach 2 m, and the diameter at the base, 1.5 m. In large haypiles the animals make internal tunnels using stacks not only for food, but also as additional temporary shelters (Shubin, 1971).

In the alpine belt of the Ivanovskii Range, 70 of the 90 investigated haypiles were well-concealed under stone ledges and only 4 reserves lay in the open. A roughly similar picture was observed in the forest belt, although in the absence of convenient places under stones pikas can make their haypiles in the open on stones. A case was noticed when, despite a

103

Fig. 28. Haypile of an Altai pika in the forest zone of the northeastern Altai (photo by B.S. Yudin).

large number of large niches, the site of haypile was chosen under a thick trunk of cedar, which lay across the stone mound (Sludskii et al., 1980).

In the western Sayan in the upper reaches of Kantegir River (left tributary of the Yenisei [River]), haypiles of Altai pikas were observed in hollows of fallen trees and among their heaps. These reserves, small in volume, were covered on top with needles, branches of acacia and roses, which pikas did not use as food (Mel'nikov, 1974). According to the observations of I.P. Khlebnikova (1978), on the southern slopes of the Aradansk Range in the rejuvenated burned-out cedar forests, pikas made their haypiles mainly under kolodnik* (85%) or under crowns of trees and shrubs (15%). When thin brushwood firmly attached to the ground predominates in the area of habitation of pikas, a large part of that haypile is situated under the crowns of shrubs and trees (50%); 44% of haypiles are situated under brushwood, and the rest in open places (Fig. 28).

The number of haypiles in the territory of a colony depends on the number of animals and the availability of places convenient for storage. In some cases, the number of haypiles may reach 32. Where there are fewer convenient places and the number of pikas is not high, the number of haypiles varies from 1 to 14 (average—6.2) (Sludskii et al., 1980).

*Meaning not clear—General Editor.

Usually species of plants dominating around the colonies of Altai pikas predominate also in their haypiles. Most detailed information on the species composition of plants stored by Altai pikas is available in the works of P.B. Yurgenson (1939), A.M. Kolosov (1939), T.A. Kim , (1956, 1959), N.G. Shubin (1971), A.A. Sludskii et al., (1980), Yu. F. Marin (1984), I.P. Khlebnikova and A.I. Khlebnikov (1991).

104 The number of species stored by Altai pikas on the northern slope of the Ivanovskii Range in one area of habitation in mixed or larch-sparse forests varies from 21 to 39 (average 300), and in alpine meadow, from 13 to 26 (average—18.1) (Sludskii et al., 1980). In the Kuznetskii Alatau 62 plant species have been recorded that are consumed by Altai pikas. Here one haypile contained usually from 4 to 19 plant species, but 3-6 were dominant; moreso, the proportion of grasses was from 0.7 to 33.5% (Shubin, 1971).

In the central Altai (area of Lake Teletskoe), food reserves of Altai pikas recorded 106 species of vascular plants. Of these, 5 species were of ferns, 3 of horse-tails, 2 of club mosses, 5 of trees, 22 of shrubs, shrublets and lianas, and 119 of herbaceous plants (Yurgenson, 1938; Khmelevskaya, 1961; Marin, 1984). In the forest belt the basic haypile consisted of reed grass, grasses, sedges, raspberry, long-rooted onion, and water avens. In the subalpine belt predominant among haypiles are horsetails, reed grasses, false hellobore, shrubby willows, ground birch, Altai rhubarb, water avens, blue-berry, honeysuckle and saussurea. In the alpine belt the main mass of haypiles consisted of ground birch, sphagnum, fescue, long-haired anemone (*Anemonecrinita*), water avens, sweet vetch, and Frolov's sawwort (*sausurea*) (Marin, 1984). In some haypiles there was a predominance of poisonous plants: false hellebore, globeflower, water avens, peony and others (Khmelevskaya, 1961).

In the eastern Altai the food reserves of Altai pikas showed 59 plant species: 7 species of shrubs, 3 of semi-shrubs, 3 of mosses and 46 of herbaceous plants with predominance of 9 species, which constituted 68-75% by weight of all haypiles. The number of species in a haypile varied from 3 to 24. Haypiles differed in the occurrence of species. Thus reed grass was found only in haypiles made in sparse forest. Leopard's-bane (*Doronicum*) was more frequent, but in small quantities. In the Chui Altai pikas mainly cured grasses and sedges (Kolosov, 1939).

In the eastern Sayan (Mansk and Kuturchinsk Belogor'e) food reserves of Altai pikas showed 64 plant species belonging to 24 families. Among them were 40 species of herbaceous plants, 12 species of shrubs, 2 of small shrubs, 6 of deciduous trees, 2 of green mosses and 2 of lichens. The following families predominated quantitatively (by weight): Salicaceae (little-tree willow, goat willow, Russian willow), ferns (oak fern, Robert's wood-fern, lady fern) and Ericaceae (blueberry, mountain cranberry). In some food reserves twigs and leaves of ledum and Pontic azalea predominated (Kim, 1956, 1959).

Altai pikas collect some dominant plants reluctantly. Thus in the western Altai the food reserves of Altai pikas did not contain water avens, gentian, celandine, parnasia, thermopsis, juniper, Altai sibiria and hairy honeysuckle, although all these species grow in large quantity near habitations of pikas. The plants most frequently found in haypiles (long-rooted onion, willow weed, sweet vetch and blueberry) are often brought by pikas sometimes from far away (F. Samusev and I. Samusev, 1972).

At the very beginning of the period of storage Altai pikas collect mainly grasses. In the alpine belt of the Ivanovskii Range, starting from August they begin to collect twigs of willows, and in the forest twigs of small shrubs begin to appear in their haypiles. Altai pikas collect grasses in places rich in herbage, and predominantly in the second half of the storage period (from August). In the forest and alpine belts, Altai pikas collect some plant species not at the peak of flowering, but during the period of their defloration and fruiting. Such plants are, for example, sweet vetch, dryad, thermopsis and pea vine (Sludskii et al., 1980).

In the western Sayan (Aradansk Range) monthwise differentiated plant collection was also observed (Khlebnikova, 1976). In June, pikas here stored herbs of Altai saussurea, bird vetch, Gmelin's pea vine, few-flowered chickweed, Bunge's starwort, and white-flowered geranium. In the second-third decades* of June in the haypiles, in addition to herbs, there are plants of group of tall grasses: tall aconite, tall honeysuckle, vosmiyazychnyi** [eight-tongued] groundsel, low meadow rue, and great willow herb. Usually at this time plants are in the flower-bud stage or in an early phase of flowering. At the end of summer, pikas collect grasses, blunt-spiked reed grass [*Calamogrostis obtusata*], large-tailed sedge [*Carex macrouva*] and increase the frequency of curing of woody and shrubby plants: birch, European aspen, goat willow, honeysuckle, raspberry, black currant. In September, pikas cure sedges, often pulled out with the roots, fallen leaves of European aspen, birch, as well as crumbs of mosses, lichens and fungi/ mushrooms. According to the data of T.A. Kim (1956), on the Ardansk Range in the area of the Olen'ei Rechka [River] in 35 plant species collected in August, in haypiles of Altai pikas, false hellebore, aconite, reed grass, angelica, water avens and raznolistnyi [heteropetalous] cotton thistle predominated.

The keeping quality of hay in the food reserves of Altai pikas is different

---

*Ten-day period—Translator.
**Local name—General Editor.

at different places. In the belt of sparse forest and on mounds in the forest belt, where these haypiles are stored under stone slabs and better aerated, their keeping quality is good. Hay does not spoil also in the hollows of trees. In forests a significant part of the hay is stored under fallen tree trunks and in voids between stones overgrown with moss rots (Naumov, 1974).

Winter diet of Altai pikas consists mainly of the stored plants. Moreover, they feed on shoots and bark of trees in the neighborbood (larches, birches, mountain ashes and others) growing in the neighborbood and shrubs (currants, honeysuckles, pea shrubs), preferring young shoots. Sometimes, a considerable part of the branches is damaged on a single plant—in pea shrub for example, to 40% (Shtil'mark, 1963). According to the data of N.G. Shubin (1971), in the Kuznetskii Alatau, pikas in winter use bark and branches of fir and bird cherry in their food, and they particularly readily gnaw at mountain ash.

In the beginning of spring, Altai pikas are usually well-fed, and fat constitutes up to 5% of the body weight. The fat is deposited mainly in the region of the shoulder, on the neck, belly, ribs, girdles of the fore and hind limbs, and in the adominal cavity. The weight of fat in females decreases by the summer—by the time of their second pregnancy and in males decrease of fat reserves occurs in the spring-summer period (from May to the end of August); in December with the changeover of animals to a less active mode of life, the weight of fat reserves increases again (Sludskii et al., 1980).

Altai pikas apparently do not visit water holes; they receive a sufficient quantity of water in their body with their food. In winter, the animals can feed on snow.

Activity and behavior. The activity of Altai pikas outside their shelters depends on the weather conditions—mainly on the air temperature and insolation. Altai pikas prefer low (close to zero) temperatures, high humidity and seasonal freezing, but have a weak resistance to high temperature (Naumov, 1974). In experimental conditions, Altai pikas endured high temperature (from 25°C) with difficulty, and some of the animals died due to overheating. Hence on hot summer days, in Altai pikas there is a drop in activity—in the midday, they hide in shelters and are active only in the morning and evening. On cold foggy days the activity of Altai pikas rises— they are active throughout the sunny part of the day. When air temperature rises to 18°C, they become less surface-active, and they produce fewer sound signals. With high humidity, the activity of Altai pikas does not change significantly—they are quite active in strong fog, in weak drizzling rain and during weak snowfall (Naumov, 1974). On favorable summer days, the relatively high activity of Altai pikas outside their shelters begins with the appearance of first signs of dawn (2-4 hr before illumination of the areas of their habitation by the sun) and terminates with the onset of darkness. There is information about the nocturnal activity of Altai pikas (N.G. Shubin, 1963; Khlebnikov and Khlebnikova, 1972; Sludskii et al., 1980).

Even during hours of high activity, Altai pikas rest often, alternating running, foraging, curing of food and resting. Usually they rest sitting or lying on rocks, their hay stack or a fallen tree near the shelters; moreover, each individual has its favorite resting place. The average duration of rest of Altai pikas on the surface is only half of that in the nest. After young animals (2-4-week olds) are active for 10-15 min, they begin a period of total immobility, which lasts for 3 to 7 min. With age the duration of periods of rest and activity increases, and adult animals can rest on the surface from several minutes to 1-1.5 hr (Naumov, 1970). At low temperatures (below +8°C), Altai pikas sit shrunken, and at temperature +15-20°C they are relaxed with their head stretched forward. While resting they often scratch, groom, sometimes pull themselves up and yawn, and also eat their excreta (coprophagy). Altai pikas love to bask in the sun, during which they lie on stone, periodically changing body position, sometimes turning on their sides or even on their back (Khlebnikova, 1978).

In winter the activity of Altai pikas decreases and they spend a large part of time in shelters. Animals make tunnels under the snow between shelters and their haypile, and only occasionally appear at the surface. In winter contacts of animals are observed from adjoining family territories. In the event of death of the host of a territory, animals from neighboring territories use the haypiles (Khlebnikova, 1978).

In Altai pikas scars and scratches that are left behind after fights can be often seen. In the period of reproduction predominantly the males and, to a lesser extent, other individuals fight. In May-June injuries were noticed in 62.2% of adult males, but were absent in females. In August-September 33.3% of adult males, 35.7% of adult females, and 26% under yearlings had injuries from a fight (Sludskii et al., 1980).

Altai pikas have a characteristic and well-developed acoustic signaling with many types of call. The most common is the "chivi" ("chvi") signal, which constitutes 87% of the total number of signals recorded during observations on mounds and in forest (V.V. Labzin, in press). Exchange of these calls is most common between animals of one colony, which sometimes are composed in a series of repeat signals. The signal "chvi" differs well from the signal "tsik," which pikas make when they are greatly frightened. This type of call constitutes 4.5% of the total number of signals and is produced not only in real danger, but also in the event of a sudden appearance of birds not harmful to pikas (for example, thrushes and woodpeckers) or animals (for example, chipmunks and squirrels). Altai pikas usually produce an alarm call before entering their shelter after already hiding under a stone, but it happens that on giving the call they remain on surface. On the appearance of a predator (for example, an ermine), Altai pikas can hide in a shelter without previously giving out the call. Often Altai pikas give out an alarm call in response to such a signal of the neighbor without sighting the source of danger. Remaining in their shelter, pikas produce another type of call "te-se" or "tsi" (6.4% of the total number of signals). The functional significance of this call is not clear. Other types of calls are rare. The call "tsi-tsi-tsi" (1.4% of the total number of calls) is produced, apparently, under extreme agitation—during this call the animal takes a tense pose. The sharp and loud call "chirrr-tsi-tsi" is associated probably with reproduction, since outside the season of active reproduction this call is not heard. It constitutes 0.3% of the total number of calls recorded in this period (V.V. Labzin, in press).

According to the data of N.A. Formozov (personal communication), in the period of reproduction—from February to the middle of July—adult males sing a typical song, which consists of two parts. The first part is a sonorous trill, and the second—somewhat rapidly repeating pulses, which are identical to the signal made during danger. After half a minute's pause many individual impulses follow, but with a longer period. The frequency of giving calls depends on the population density and the character of the habitat. In favorable places of habitations and with high population density, the frequency of calls is higher (Khlebnikova, 1972).

Altai pikas show age-related and individual variability of voice tembre—in adult individuals the voice is choked and dry; and in young, thin and piercing (Khlebnikova, 1978).

Area of habitation. Altai pikas lead a family group mode of life. In each area there is a male and female, which use one territory, store food together and aggressively relate to individuals from neighboring areas intruding in their territory. In summer besides the male and female, the family area is occupied by young, sometimes of many broods. Individual territories are for loners. A single individual may possess a territory also in the event of death of another member of the family or during dispersal. Sometimes subordination of young animals to old ones is observed (Naumov, 1974; Khlebnikova, 1978). Family areas in old colonies are situated close to each other. The size and configuration of the area depend on a hierarchic status of resident families, as well as the ecological value of the area. The boundaries of areas, if at least one animal stays there after wintering, change insignificantly from year to year. The areas occupied by under-yearlings do not have permanent boundaries, however, the centers of familial areas, as a rule, remain as before (Khlebnikova, 1978).

According to some data (V.V. Labzin, in press), the area of individual (family) territories of Altai pikas constitutes, on average, 1,560 m^2 (n = 10) and varies from 570 to 3,040 m^2. According to other data (Khlebnikova, 1974, 1978), the sizes of familial areas of Altai pikas are considerably smaller—360 and even 150 m^2. The area of individual territories evidently depends on the age of the colony, habitat conditions and the quantity of available food.

The entire family territory area is not used uniformly by Altai pikas. There are places that the animals visit constantly—these are the entrance to the shelter, haypiles, toilets, foraging areas, resting places, observation points and signalling points. Such areas, as a rule, are connected to each other by beaten tracks [runways]. In foraging areas the system of tracks is more disorderly and less distinct than in the other places. There may be 2 to 7 foraging areas; these are either on the mounds and in small patches of grasses and shrubs, or beyond mounds in places with continuous plant cover. Some foraging areas, especially in the haypile construction period, are "mowed" by animals practically completely (V. Labzin, in press; Derviz, 1982).

Lavatories of Altai pikas are located either under stones or in open areas. Here only droppings are found, that have the form of thick globules about 3 mm in diam. In permanent toilets of old habitations, up to 20-40 and even to 100-300 l of droppings may collect (V.V. Labzin, in press). The other type of excreta is the wormiform droppings ("sausages") measuring several cm in length and about 3 mm thick: these are found in small heaps under stones and also on piles.

Between the neighboring families forming a single colony there are permanent contacts, and the more frequent of these are the visual-voice contacts. Animals call each other even in the absence of danger. Some areas of contiguous territories are apparently used by animals from family neighboring family groups simultaneously, which is testified by the capture of individuals of different family groups on the same runways (Khmelevskaya, 1961; Naumov and Lur'e, 1971; Derviz, 1982). However, during the period of curing food, particularly in the old resident colonies, the boundaries of family areas are defended by the inhabitants of these territories (Khlebnikov and Khlebnikova, 1972). On the defended part of the territory are located shelters, food reserves, and permanent places of observation and resting (Derviz, 1982). Between animals of neighboring areas with established boundaries often there are fights, and in the freshly occupied areas territoriality is less manifest. As a rule, the host of the area often chases away intruders from his territory, actively pursuing them (V.V. Labzin, in press).

Altai pikas, like the majority of species of the family, characteristically mark their territory with the secretion of their buccal glands. In males active secretion of these glands is observed from May to December; and in females, from August to December. Most often in Altai pikas the elements of marking behavior are observed in the period of dispersal of young and during straying of individuals into alien or unknown territory. Altai pikas also mark prominences of stones near their permanent shelters and food reserves with the secretion of their buccal glands. During marking a pika turns its head, and with the side of the neck rubs over the substrate, repeating this movement several times (Orlov, 1983).

Reproduction. In a year Altai pikas apparently have not less than two broods. Data on the presence of two broods are available for the eastern Altai (Yurgenson, 1938), the Kuznetskii Alatau (Shubin, 1971), the western Sayan (Khlebnikova, 1978), and the eastern Trans-Baikal area (Kuznetsov, 1929). There is a suggestion that Altai pikas can bear three broods in the reproductive season (Khmelnikova, 1961; Shubin, 1971; Naumov, 1974).

On the Ivanovskii Range and the eastern Altai, reproduction starts in the first decade of May (Yurgenson, 1938; Sludskiiet al., 1980); and in western Sayan, the beginning-middle of April (Naumov, 1974).

The second pregnancy happens when lactation is still on and occurs in the majority, if not in all, of females. On the Ivanovski Range, of the 15 females caught from June 1 to July 4, 14 were pregnant for the second time; moreover, only 2 had completed feeding the first brood, and the rest the were in the lactation stage. In August, of the 17 adult females caught, 5 were in the stage of anestrus, 1 in estrus and 11 in postestrus; 5 of the 15 females continued to feed young ones. In a later period pregnant and lactating females were not found. Spermatogenesis in males terminates in August, but in individual animals already at the beginning of July. Thus, already in August successful fertilization of females is less probable (Sludskii et al., 1980.

The second brood in Altai pikas is apparently larger than the first. Thus in the forest belt of the Ivanovsk. Range, the number of embryos in the first reproductive wave varied from 2 to 4 (average 3), and in the repeat pregnancy, from 3 to 6 (average 4.4) (Sludskii et al., 1980). In the Altai the average number of embryos in April was 2.5; in May, 3.2; in June, 3.6 (Potapkina, 1975). According to the data of N.G. Shubin (1971), the number of embryos in Altai pikas of the Kuznetskii Alatau in the first and second pregnancies is, on average, similar (4.1 and 3.9, respectively). In forests of the western Sayan, most often there are 3-6 embryos per female, with a maximum 7 (Shtil'mark, 1963). The small number of young ones in the brood of Altai pikas is probably related to the large sizes of embryos, each of which may reach 7-7.8% of the weight of the female.

In Altai pikas embryo mortality is low. According to observations in the

Kuznetskii Alatau (1961-1962), of the 167 examined embryos, just 3 (1.8%) were resorbed (Shubin, 1971).

Growth and development. The duration of pregnancy in Altai pikas has not been precisely established; apparently it lasts 30 days.

Newborn Altai pikas attain a length of 58-60 mm, in them vibrissae are well-developed and are up to 8 mm long. The weight of prenatal embryos reaches 21-22.4 g (Sludskii et al., 1980). According to the data of N.V. Khmelevskaya (1961), the weight of prenatal embryos is just 12.7-13.5 g, which apparently is erroneous. Newborn Altai pikas are completely covered with hair; the underside of the paws also has sparse lighter hairs; eyes and auditory meatuses are closed. Young of Altai pikas grow very fast, and the first young appear on the surface in May, and by August they have already reached the size of adult animals.

109 According to the data of R.L. Naumov (1974), in the western Sayan three fertilized underyearling females were caught between July 28 and August 1, 1971, which testifies to the possibility of participation in reproduction of underyearling females from the first brood. In Altai pikas sexual maturity mainly sets in the next year after birth.

Molt. In Altai pikas there are two molts in a year—spring and fall. In some parts of the range, Altai pikas retain the winter coat for up to nine months. The spring molt, particularly in females, is quite prolonged, which is linked with the period of reproduction.

In the western Sayan, the first signs of molting are observed in the end of May-beginning of June. Males begin to molt earlier than females. The spring molt is diffused in nature in the region of the withers, spine, sacrum and in the upper part of sides. In males molting begins in the area of the nose, around the eyes, and in front of the ears, and later passes over to the throat, neck and chest; later molting occurs in the region of the belly and back; the last to molt are hairs in the region of sacrum. Some areas on the back, belly and sides molt almost simultaneously. In the first half of July, males are found to have almost completely molted their summer fur. In females, as also males, molting begins in the region of the eyes, nose and ears, but simultaneously hairs molt around the teats, then they molt the remaining part of their chest and belly, followed by their withers and the anterior of the back; the last area to molt is the sacrum. Sometimes incomplete molting is observed in the region of the sacrum, when new winter hairs grow among the remaining old winter hairs; moreover in females this happens more often than in males. Molting slows down in the period of active reproduction and resumes after the termination of reproduction, and occurs more intensively. In Altai pikas from the western Sayan, an exceptionally high individual variability is observed in the character of the spring molt (Khlebnikova, 1978). In the Kuznetskii Alatau, males and females begin to molt simultaneously in spring—at the end of April-beginning of May, and the first completely molted animals begin to appear here from June 22 (Shubin, 1971).

The fall molt progresses faster than the spring molt and is not diffused; moreover, sex-related differences in the sequence and periods of molting are not noticed (Sludskii et al., 1980). In the Kuznetskii Alatau and the western Sayan, the fall molt in Altai pikas begins in September (Shubin, 1971; Khlebnikova, 1978) and in the Ivanovskii Range in Kazakhstan, in August (Sludskii et al., 1980). In the middle of October some individuals are completely covered in their winter coat. Fall molting starts from the buttocks and hind part of the back (Khlebnikova, 1978), or from the region of the ears and nose, then moves over to the shoulders, spine and the upper part of the sides, and only later to the sacrum and the lower part of the sides (Sludskii et al., 1980). The belly is the last part to molt.

In underyearlings the change of the juvenile hair coat begins from the head and occurs at a time when they still feed on their mother's milk with a body weight of 105-136 g. By mid-August young Altai pikas of first broods acquire the bright color of adult animals (Khlebnikova, 1978; Sludskii et al., 1980). Upon termination of the juvenile molt, in underyearlings begins the change of the summer coat to winter, which occurs at the end of August-September (Ognev, 1940). Thus, for young Altai pikas a quite fast change of fur is characteristic, first to adult summer coat and then to winter coat, that is, the process of molting extends almost uninterrupted.

Age and sex structure of population. The sex ratio in adult Altai pikas and underyearlings is roughly similar (Potapkina, 1975; Sludskii et al., 1980), but sometimes among them some predominance of males is observed—57.1% in underyearlings and 51.3% in adults. Among embryos also, a predominance of males is noticed—54.1% (Shubin, 1971) .

In the first half of May, the population of Altai pikas consists only of overwintered adult individuals, the average age of which is 1.68 yr. In early spring (March-April) the number of individuals older than one year may be more (to 47.2%) than in the fall, which can be explained by lower mortality of pikas that have already survived one winter (Orlov and Makushin, 1984). Underyearlings appear from the middle of May, and by the end of May they constitute 22.5% of the population. In June, underyearlings constitute 38.3% of the population; in July, 58.4%; in August, 82.5%; in September, 82%; in October, 80%. Some decrease in the proportion of young ones in the population after the termination of reproduction (September-October) is possibly related to higher mortality of young ones during their changeover to independent life and during the period of dispersal (Potapkina, 1975). According to the data of G.I. Orlov and G.I. Makushin (1984), among adult Altai pikas in populations of the

148

western Altai (Ivanovskii Range), underyearlings constitute 51.5%; two-year-olds; 33.3%, three-year-olds—12.9%, four-year-olds, 1.2%, and five-year-olds, 0.6%.

Adult Altai pikas usually undergo wintering with residents of the second brood . By spring their families usually consist of males and females of various ages, which, apparently decreases the possibility of inbreeding (Orlov and Makushin, 1984). Under unfavorable wintering conditions, there is a two-three fold reduction in population numbers; under more favorable conditions, up to 70% of animals survive (Khlebnikov et al., 1974).

Altai pikas apparently can live up to 6 years (Orlov and Makushin, 1984); however, in northern parts of the range their age most frequently does not exceed three years.

Population dynamics. The character of population fluctuations of Altai pikas has been well studied in the western Sayan (Khlebnikov and Khlebnikova, 1991). Multi-year observations have shown that the magnitude and level decreases in the population of pikas can be divided in three major types:

1. Depression in population. In this case there occurs considerable decrease in (by 50%) and disappearance of colonies over large areas. Such a fall in population of pikas was observed in the high-mountain zones of the Ergak-Targak-Taiga and Shepshir-Taiga ranges in 1970-1980.

2. Quite significant drop in population (up to 50-60%), but without disappearance of colonies over large areas. Such fluctuations were noticed for the majority of habitats of the southern megaslope. Here, the relatively high population of pikas (30/route km) was observed in 1960-1970, and from 1980 began a decrease in number, that by 1985 reached the 50-60% level. However, numerous colonies remained along the periphery of the investigated section (area 30 km²). If the reason for depression in population is probably an epizootic, then in the second case the contraction in population of pikas observed as a result of active grazing of horses regularly that destroyed fall reserves of pikas.

3. Local decrease of population, during which colonies disappear over small areas. This type of fluctuation of population is reported for all forest zones of the western Sayan, moreover, in some places restoration of numbers occurs for a short period, in others—its low level is maintained for a long time. Thus in 1971, in old burned-out forests of the southern megaslope, there were 78 individuals over 1.7 ha. In winter of 1972, up to 50% of the colonies were destroyed by sables, and by spring of 1972 the population became completely extinct. Restoration of these colonies occurred only after 17 years (Khlebnikova and Khlebnikova, 1991).

Enemies. There are many enemies of Altai pikas. In Kazakhstan these are the foxes, sables, ermines, solongoi* kolonk** and possibly weasels (Sludskii et al., 1980). In the Altai, these pikas are hunted by ermines, badgers and sable (Yurgenson, 1938; Afanas'ev, 1962). In the forest belt of the Kuznetskii Alatau and western Sayan, Altai pikas are one of the main objects of diet of sable. According to other data (Dul'keit, 1964), in winter in the western Sayan, sables practically do not catch Altai pikas, but in summer catch them casually.

Among birds, common buzzards, goshawks, black kites, kestrels, saker falcons, ravens and long-eared owls prey on Altai pikas (Shubin, 1971; Sludskii et al., 1980).

Competitors. On the whole, interactions of Altai pikas with other 111 mammals (not carnivores/predators) is neutral in nature. Food reserves of Altai pikas, particularly cured as haypiles, can be used in winter by ungulates. In summer the haypiles as well as runways are used by voles and shrews; moreover their number near the habitations of pikas is higher than in the neighboring forests (Naumov and Labzin, 1980).

In such places as the Gornyi Altai, the range of Altai pika overlaps with the ranges of Pallas's and Daurian pikas. However, the character of their habitat affinities are so different, that they are not found together. Consequently, competitive interactions between Altai pikas and Pallas's pika on one hand, and Daurian pikas on the other, are practically nonexistent. The situation is different in the sympatric zone of Altai and northern pikas (cf. sections "Distribution" and "Habitats"). Both these species have identical ecological niches and in places of combined habitation their competition is unavoidable. This explains the limited number of places where these species live together. The character of competitive interactions of Altai and northern pikas has so far been studied inadequately.

Diseases and parasites. Numerous fleas parasitize Altai pikas. In the Altai and the Kuznetskii Alatau, 10 species of fleas have been recorded: *Ctenophyllus subarmatus, Amphalius runatus, Rhadinopsylla altaica, R. dahurica, Paradoxopsylla* sp., *Frontopsylla hetera, F. elata, Ceratophyllus penicilliger, C. rectangulatus* and *Neopsylla mana* (Potapkina, 1967: Shubin, 1971). In the southeast Trans-Baikal area, 10 species of fleas have been found on Altai pikas: *Ctenophyllus hirticrus, Rhadinopsylla dahurica, Frontopsylla luculeanta, Ceratophyllus garei, C. armatus, Neopsylla bidentatiformis***, N. pleskei, Oropsylla silantiewi, Amphypsylla vinogradovi* and *Ophthalmopsylla kukuschkini* (Dubinin and Dubinina, 1951); only one species—*Rhadinopsylla*

Mustela (Mustela) altaica—General Editor.
**Or Siberian weasel, *Mustela (Mustela) sibirica*—General Editor.
***In Russian original, bidentatiformi—General Editor.

dahurica—is common to the Altai and Trans-Baikal area. Fleas are most numerous in spring and fall (up to 144 on one animal), but in summer their number is the least.

Among gamasid ticks the following parasitize Altai pikas: *Haemogamasus kitanois, Hirstionyssus isabeliinus, Poecilochirus necrophori, Laelaps hilaris, Parasitus* sp. (Potopkina, 1969; Sludskii et al., 1980). Of the ixodid ticks *Ixodes persulcatus* is predominant in the Altai. The maximum number of ixodid ticks was reported on pikas living in dark coniferous, middle mountains, whereas in the high mountains ixodid ticks are absent (Smirnov, 1967). On the Ivanovskii Range, in May-September 1977-1978, ixodid ticks were present in habitats of Altai pikas, but were not detected on animals (Sludskii et al., 1980). In the Borzinsk district of the Trans-Baikal area, ixodid ticks (*Ixodes persulcatus*, and *Dermacentor nuttalli*) were found as isolated individuals mainly on pikas living in isolated forest patches on mud volcanoes (Dubinin and Dubinina, 1951). In the forest belt of the Ivanovskii Range, in August and the first half of September, larvae of red mites were found with a frequency of 83.3-100% (n = 60). In the alpine belt at this time, they were found on only one individual (n = 27) (Sludskii et al., 1980).

On the Ivanovskii Range the louse *Hoplopleura ochotonae* was found on Altai pikas (Sludskii et al., 1980).

From the end of July to September, larvae of western warble flies parasitize Altai pikas, localizing mainly in the region of the sacrum and belly. In the Altai, Kuznetskii Alatau, Sayans and Trans-Baikal areas, larvae of the following species have been found: *Oestromyia laporina, O. potanini, O. rubtzovii, O. schubini, Portschinskia lowii, P. magnifica* (Kim, 1959; N.G. Shubin, 1963; Semenov and Potapkina, 1975). Up to 30-40 larvae of warble flies may be present on one individual (Shubin, 1971).

In the Altai (except the western Altai), the following helminths were found on Altai pikas: cestodes—*Schizorchis altaica, Rodentoleps straminea, Tarnia hydatigenae, Hydatigera* sp., *Paputerina candelabria*; nematodes—*Capillaria muris-sylvatici, Hepaticola hepatica, Tominx sadovskaja, Capillaridae* sp., *Heligmosomum dubinini, Cephaluris andrejevii*, Labiostomum vesicularis, Siphacia* sp., *Oxyspirura* sp. (Fedorov and Potapkina, 1975). In the southern Altai, Altai pikas showed *Schizorchis altaica, Capillaris muris-sylvatici, Heligomosomum dubinini, Cephaluris andrejevi, Dermatoxys schumakovitschi, Nematodixus aspinosus* (Gvozdev, 1967). These very helminth species were detected in pikas from the Ivanovskii Range (Sludskii et al., 1980). In the Trans-Baikal area (Berzinsk district), Altai pikas showed *Dermatoxys schumakovitschi, Cephaluris andrejevi* and *Helismosomum polygyrum* (Dubinin and Dubinina, 1951).

---

*In Russian original, *andreevi*—General Editor.

Practical Significance

Information on the economic significance of Altai pikas is somewhat contradictory. According to some data, with a high population Altai pikas can inflict substantial losses on the seedlings and outgrowths of woody species, particularly cedar, destroying from 63% (Loskutov, 1966) to 95% (Khlebnikova, 1974) of seedlings. Moreover, the most substantial harmful activity of Altai pikas is in old habitations. According to other data (Reimers, 1960), the negative role of Altai pikas is not high in cedar forests. In anthropogenic habitats—in logged forest and burned-out areas—Altai pikas, by destroying the fresh growth, inflict some damage to woody vegetation, but thereby they destroy also the grass cover by reducing the turf cover, which creates favorable conditions for forest rejuvination (Khlebnikova, 1972, 1974).

In years with low populations of rodents and non-yield of cedar nuts, Altai pikas become one of the diet objects of sables: at such time the percentage of their occurrence in sable stomachs increases to 78% (Khlebnikov, 1974) and even to 87% (Sokolov, 1965). However, in some places Altai pikas do not have significance in the diet of sables and do not exceed 8% (usually 1-2%) occurrence in sable stomachs (Dul'keit, 1964). Altai pikas can be an important part in the diet of ermines, minks and kolonok [Siberian weasels] (Dul'keit, 1964; Mel'inkov, 1974).

In the Altai in winters, with abundant snow, the food reserves of Altai pikas are eaten by stags, which at times consume their haypiles completely (Velizhanin, 1972). Sometimes piles of hay cured by Altai pikas are collected by locals to feed cattle (Sludskii et al., 1980).

Northern Pika
Ochotona (*Pika*) *hyperborea* Pallas, 1811

1811.   *Lepus hyperboreus* Pallas. Zoogeographia Rosso-Asiatica, p. 152, Chukchi Peninsula.

1858.   *Lagomys hyperboreus* var. *normalis* Schrenk. Reisen und Fors-chungen* Amur-Lande, I, p. 148.

1858.   *Lagomys hyberboreus* var. *ferruginea* Schrenk. Reisen. und Fors-chungen im Amur-Lande, I, p. 148-149, Kamchatka Peninsula, Khalzansk mountains.

1858.   *Lagomys hyperboreus* var. *cinero-flava* Schrenk. Reisen und Fors-chungen in Amur-Lande. I, p. 150. Khabarovsk territory, Udskoe ("Udskii Ostrog**").

---

*In Russian original, misspelled Forschunger—General Editor.
**Settlements in Kamchatka—Translator.

1852.   *Lagomys litoralis* Peters. Sitzungsb. der Gessellschaft Naturforsch. Freunde zu Berlin, p. 95, Chukchi Peninsula, Cape Chukotskii.

1903.   *Ochotona kolymensis* Allen. Report* on the Mammal. Collect. in North-East. Siberia, Bull. Amer. Mus. Nat. Hist., V. 19, p. 154-155. Middle stream of Kolyma River, Verkhnekolymsk.

1903.   *Ochotona hyberborea mantchurica* Thomas. A Collection of Mammals from Northern and Central Mantchuria. Ann. and Magaz. Natur. Hist., Ser. 8, p. 504-505. China, Inner Mongolia, Greater Khingan Range.

1913.   *Ochotona hyperborea coreana* Allen. et Andrews. Amer. Mus. Nat. Hist. Bull., Vol. 32, p. 429, North Korea.

1922.   *Lagomys kamtschatica* Dybowsky. Arch. Towar. Naukow. Lwo-wie, V. 3, p. 10 (nomen nudum). Kamchatka Peninsula.

1927.   *Ochotona hyperborea uralensis* Flerov. Pishchukhi Severnogo Urala [Pikas of the Northern Urals]. Ezhegodnik. Zool. Muzeya Akad. Nauk SSSR, V. 28, pp. 138-144.

1930.   *Ochotona hyperborea yesoensis* Kishida. Diagnosis of a new piping hare from Yeso. Lanzania, J. Arch. Zool., V. 2, pp. 45-47, Japan, Hokkaido.

1932.** *Ochotona hyperborea yoshikurai* Kishida. Lanzania, J. Arch. Zool., V. 4, p. 150, Sakhalin Island, central part ("Shirotoru").

1934.   *Ochotona hyperborea turchanensis* Naumov. Mlekopi-tayushchie Tungusskogo Okruga [Mammals of Tungusk District]. Trudy Polyarnoi Komissii Akad. Nauk., Vol. 17, p. 38, Krasnoyarsk territory, Nizhnyaya [Lower] Tunguska River, Uchami.

Diagnosis

Pikas of small and medium dimensions. Condylobasal length of skull less than 41 mm, height of skull in occipital part less than 11 mm, length of tympanic bullae less than 13 mm. Incisor fossa partitioned by protuberences of the intermaxillary bones. Length of the hind foot less than 28 mm. Color of summer fur on the back varies from light grayish-brown with pale rusty or gray-yellowish admixture to saturated—rusty-reddish-ochreous or dull brown-reddish-brown. Winter fur more gray or dirty gray with brown tone. Majority of vibrissae blackish-brown, their length 42-55 mm. Paws covered with whitish-gray, brownish or black hairs.

---

*In Russian original, misspelled Reort—General Editor.
**In Russian original, 1032—Translator.

Fig. 29. Northern pika *Ochotona hyperborea* from Tuva (photo by B.S. Yudin).

Description

113

The body length of adult individuals over the range varies quite considerably—from 133 to 190 mm and constitutes, on average (n = 219), 161.9 mm. Length of foot varies from 21 to 27 mm (average 23.9 mm). Length of ear, from 13 to 20 mm (average 16.5 mm).

The color of summer fur on the back varies considerably in different subspecies. The color of the sides is usually lighter and yellower, sometimes with a grayish tinge. The belly is whitish or grayish with a pale yellow admixture. The margin of the ears is whitish (Fig. 29). On sides of the neck there are small reddish-brown spots, formed by hairs shorter than on the remaining surface of the body. The paws are lighter grayish-brown or dark brown.

The color of winter fur is also subject to geographic variation. Winter fur in northern pikas is longer and denser than summer fur; often lighter [in color] than summer fur.

114 Young animals, the length of which does not exceed half the length of adults, have a dull gray coloration.

The skull is of small dimensions and is flattened in the parietal part.

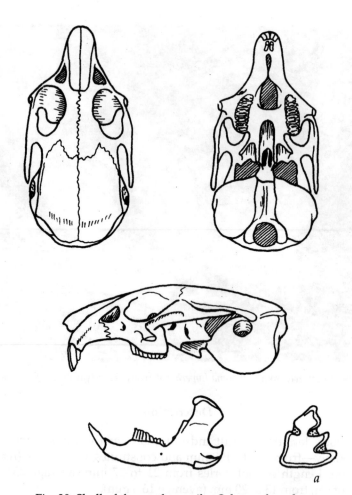

Fig. 30. Skull of the northern pika *Ochotona hyperborea*.
a—third lower premolar tooth of the northern pika.

The condylobasal length of the skull varies from 32 to 41 mm; zygomatic width, from 19.5 to 22.9 mm; interorbital width, from 3.2 to 4.9 mm; alveolar length of the upper tooth row, from 6.2 to 8 mm.

Frontal bones lack foramina. The incisor fossa is partitioned by the protuberances of the intermaxillary bones. The preorbital foramen is small and triangular in form. The nasal bones are broad in the anterior part and narrower—in the posterior. The orbits are small and round in form. The posterior zygomatic processes are quite long. The tympanic bullae are small. The mandible is low. The ascending ramus is short and quite broad. The angular notch [sinus] is situated high and most often rounded in form and deep. The angular process is broad (Fig. 30).

The third lower premolar tooth is quite variable; its anterior segment is rhombic in form and relatively large; entrant folds, separating the anterior segment from the posterior, are roughly uniformly deep; the posterior segment is variable in form, its inner margin with an entrant fold or without it (Fig. 30a).

The diploid complement has 40 chromosomes: 11 pairs of meta-submetacentrics and 8 pairs of submetacentrics; X-chromosome—submetacentric, Y-chromosome—small, acrocentric (Vorontsov and Ivanitskaya, 1973).

Systematic Position

In morphological characters the northern pika is closest to the Altai pika. In morphological and cytogenetic characters it is included with *O. pallasi* and *O. alpina* in one subgenus.

Geographic Distribution

The range of the northern pika extends from north to south from the tundra zone to Inner Mongolia and from west to east, from the cis-polar Urals to Chukchi, Sakhalin, Hokkaido, and the northern part of the Korean Peninsula. The general entirety of the range of the northern pika is disturbed by the existence of an exclave of this species in the cis-polar Urals, the formation of which is associated with the Late Pleistocene history of formation of the western Siberian lowland (Fig. 31).

The main part of the range of northern pika in the west is limited by the left bank of Yenisei or its upper reaches in the south to the north Siberian lowland in the north. Along the northern edge of the Putoran Plateau and the Anabar Plateau, the northern boundary of distribution of *O. hyperborea* reaches up to the lower reaches of the Lena and proceeds farther east along the arctic coast of the continent along 70-73° N. Lat. In the east the range of the northern pika extends along the seacoast from the Dezhnev Cape in the northeast of the Chukchi Peninsula to south of Primorsk [Pacific coastal] territory, Kamchatka and Sakhalin. The southern boundary of the northern pika passes over the territory of China and Mongolia. The extreme southwestern habitations of this species are found southwest of Tuva (cf. Fig. 31).

The Ural exclave of the northern pika is bounded at the present time by 62°40' N. Lat. in the south and 68° N. Lat. in the north. The southernmost localities of this species in the northern Urals are recorded on the slopes of the Sotchem-El'-Iz and Shuka-El'-Iz mountains on the territory of the Pechora-Ilychskii Preserve and on the Lyapin River (a tributary of the Sosva),

156

115

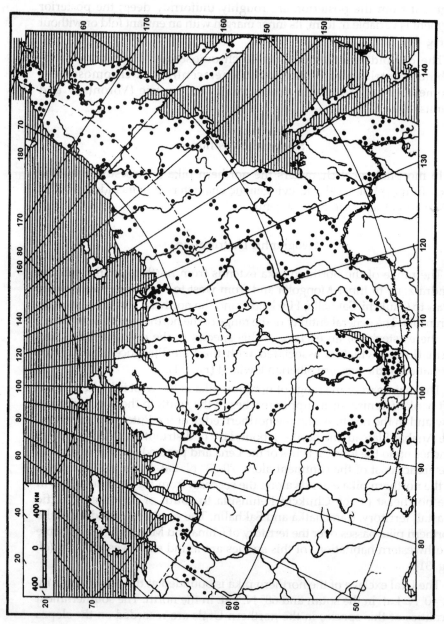

Fig. 31. Localities of northern pikas in Russia.

and the northernmost on the left bank of the Khadata-Yugan River, 20-25 km down Lake Khadatinskoe (Flerov, 1927; Gashev, 1971).

Northern pikas do not occur in the territory of western Siberia. The character of distribution of this species in the territory of central and eastern Siberia is so far unclear because of the absence of sufficiently complete faunistic information. Here the main localities of northern pikas are confined to river valleys and populated areas (cf. Fig. 31). However, currently available information about the localities of northern pikas makes it possible to fairly clearly define the general boundaries of the range of this species.

The northwestern boundary of the main range of nothern pikas passes along the western part of the Taimyr. Habitations of northern pikas are recorded here in the valley of the Khantaika River, and north of 69°30' N. Lat. they are apparently not found (Ognev, 1940).

Farther to the east the distributional boundary of the northern pika ascends to 69°30' N. Lat. (region of the Noril'sk mountains and upper reaches of the Volochanka River), and along the Khatanga River (Kotui) it reaches 71° N. Lat. (Naumov, 1934; Ognev, 1940; Romanov, 1941). In the basin of the Popigai River, the habitations of northern pikas are found along its left tributaries, the Rassokha and Fomich (up to 72°05' N. Lat.), and still northward—in the mouth of the Anabar River (Romanov, 1941; Krivosheev, 1971).

In northwestern Yakutia west of the Lena River, the northern boundary of the range of O. hyperborea passes roughly along 72° N. Lat. (Krivosheev, 1971). In the basins of the Olenek interfluve it occurs along the northern slope of the Moi Range (Syuryakh-Dzhangly), the southwestern slopes of the Chekanovskii Range and the valley of the Nizhnyaya [Lower] Maikangda River (Romanov, 1941). The northernmost habitation of northern pikas in this region is situated apparently in the Bulkur tract—72°48' N. Lat. (Romanov, 1941). In the lower reaches of the Lena, northern pikas were caught also above the island of Tit-Ara, in the middle and upper forks of the Oroktu-Ukhta River and in the vicinity of the Kysyr (Kapitonov 1961, Krivosheev 1971).

In eastern Siberia the northern distributional boundary of O. hyperborea passes along the western slope of the Verkhoyansk Range—in the upper reaches of the Sakhandzha River and farther along the Yana River it ascends northeast. Here northern pikas were caught on the northern spurs of the Kular Range—70°30' N. Lat. (Ognev, 1940; Krivosheev, 1971). Later, crossing through the Verkhoyansk Range, the boundary follows its eastern slope to the southeast, bending from the south of the Yana-Indigirsk lowland. Here northern pikas were caught in a small locality, Byyttakh, on the Yana and in the lower reaches of the Tuostakh River (Ognev, 1940;

Krivosheev, 1971; collections of ZIN). Along the Indigirka valley the northern boundary again ascends north and northeast bending from south of the Indigirsk and Abyisk lowlands. Northern pikas were caught by the mouth of the Moma River, 310 km from the mouth of the Indigirka (village of Shamonovo) and 80 km south of the town of Allakhin (collections of ZM MGU and ZIN). Data are available about localities of northern pikas in the lower fork of the Indigirka (vicinity of the village of Chokurdakh) and in the middle fork of the Berelekh and Khroma rivers (Krivosheev, 1971). Here, habitations of northern pikas reach as far north as 71° N. Lat. (collections of ZM MGU). There is no information about localities of northern pikas west of the Kolymsk lowland. Northeast of the Kolymsk lowland, the distributional boundary of northern pikas passes along the valley of the Alazeya River and ascends northeast to the valley of the Chukoch'ya River, where habitations of northern pikas are reported 30 km from the seacoast (Krivosheev, 1964, 1971). Along the seacoast northern pikas are found in the lower reaches of the Kolyma—between the tributary Stadukhina and the Kon'kovaya River in the valley of the Kon'kovaya River in the vicinity of the town of Pokhodskoe and northeast of the town of Sukharnoe (Ognev, 1940; Krivosheev, 1964, 1971; collections of ZIN).

From the lower reaches of the Kolyma, the northern boundary descends southeast and proceeds along the northern spurs of the Anyuisk Range. In particular, northern pikas were caught near Nizhnekolymsk, in the basin of the Koperveem River (E'nmenveem River) and in the upper reaches of the Malyi [Lesser] Anyui [River]. From the upper reaches of the Yarakvam River, the boundary ascends the northern macroslope of the Illirneisk Range in the north toward southern coast of the Chaunsk Inlet, where northern pikas were caught in the basin of the Pacheveem River, in the vicinity of Neitlin Mountain (collections of ZIN). Proceeding farther north, the distributional boundary of the northern pika passes along the eastern coast of the Chaunsk Inlet (collections of ZM MGU, ZIN and IE'RZh). Later the boundary proceeds east along the seacoast of central Chukchi, passing along the right bank of the Kuvet River, in the region of Cape Shmidt and in the lower reaches of the Amguema River (Yudin et al., 1976; Kostenko, 1976; Chernyavskii, 1984).

Information is absent about localities of the northern pika north of the Chukchi Peninsula. The extreme northeastern localities of northern pikas are Cape Dezhenev and the village of Ue'len—66°10' N. Lat. and 169°40' W. Long. (collections of ZM MGU, ZIN, and Geography Faculty of MGU). From here the distributional boundary of northern pikas joins the eastern and follows south along the coast. Northern pikas were caught in the vicinity of the villages of Lavrentia and Akkan', in the southeast Chukchi Peninsula—in the vicinity of the village of Yanrakynnot, and also along the coast of Provedeniya Bay, and eastward—on the islands of Arakamchechen and Yttyrgan (Belyaev, 1968; Formozov and Nikol'skii, 1979; Nikol'skii, 1984). The extreme southeastern habitations of northern pikas in the Chukchi Peninsula were recorded on Cape Chaplin—64°20' N. Lat. and 172°20' W. Long. (Gavrilyuk, 1966; Belyaev, 1968).

118

Passing along the coast of the Anadyr Gulf, the eastern distributional boundary of the northern pika turns west, and later north and passes south of the Chukchi Peninsula through the village of Sirenike, the E'rugveem River and along the northern coast of the Gulf of Krest through the village of E'gvekinot (Chernyavskii, 1984; collections of ZM MGU and ZIN). Later, the boundary passes through the basin of the Kanchalan River, the mouth of Anadyr' River, Dionisia Mountain, and later along the northeastern slope of the Koryaksk upland (Portenko, 1941; Yudin et al., 1976; Chernyavskii, 1984; collections of ZM MGU, and ZIN). There is no information about localities of northern pikas on the eastern slopes of the Koryaksk upland. Apparently, the eastern distributional boundary of the northern pika passes along the valley of the Anadyr', bending from south of the Anadyr' lowlands. Habitations of northern pikas in this region are recorded in the vicinity of the village of Tanyurer, near mouth of the Belaya River and 32 km upstream from its mouth (middle fork of the Anadyr'), in the vicinity of the village of Markovo and in the valley of the Ubienkovaya River (Portenko, 1941; collections of ZM MGU and ZIN). Later the boundary turns south, passes north of the Koryaksk upland—through the town of Penzhin, along the valley of the Apuka River, upper reaches of the Achaivayam River and in the region of Severnaya Glubokaya Gulf enters the coast of the Bering Sea (Ognev, 1940; Portenko et al., 1963; Chernyavskii, 1984; collections of ZM MGU and ZIN). Southward northern pikas were caught in the vicinity of Apuka, in the valley of the Pakhacha River and on the coast of the Olyutorsk Gulf (collections of ZM MGU and ZIN).

Northern pikas were caught north of Kamchatka on the coast of Korf Bay—in the vicinity of the village of Kultushnoe and in the lower reaches of the Vyvenka River (Gavrilyuk, 1966; Chernyavskii, 1964). In the central part of Kamchatka, on its eastern coast, northern pikas are found in the territory of the Kronotsk Preserve (Averin, 1948). In southern Kamchatka northern pikas were caught on the Ganal'skie Vostryaki Range (Formozov and Nikol'skii, 1979), and in the western part—on Khalzan Mountain, on the Pereval'naya River, in the upper reaches of the Moroshechnaya River and on the Bystraya River (Ognev, 1940; collections of ZIN). On the coast of the Penzhinskaya Inlet, northern pikas were caught in Kamenistaya plant and in Lovatakh (collections of ZM MGU).

Later the distributional boundary of the northern pika passes along the eastern slopes of the Kolyma upland, along the coast of the Nayakhanskaya Inlet, from the Omsukchan River to the Sea of Okhotsk, along the valley of the Takor River, upper and middle forks of the Bulun River (Kishchinskii, 1969; Yudin et al., 1976). Along the southwestern ranges of the Kolymsk Upland, the distributional boundary of northern pikas descends to the Babushkina Inlet on the Okhtosk Coast. Northern pikas were found on the coast of the Odyan Gulf, in the vicinity of Magadan and on the right bank of the Taui River, and in the region of Cape Severnyi (Belyaev, 1986*; collections of ZIN). Along the seacoast the boundary passes through the Eirineisk Gulf, Kul'ka Gulf and in the vicinity of Okhotsk (Yudin et al., 1976; collections of ZIN). There are no data on the habitations of northern pikas on the coast of the Sea of Okhotsk from Okhotsk to Ayan. If this is so, then the distributional boundary of northern pikas passes west from Okhotsk along the southeastern spurs of the Ulakhan-Bam Range in the region of the middle fork of the Aldan River, then turns south and passes through the Ayano-Maisk district in the vicinity of Nel'kan Mountain and passing through the Dzhugdzhur Range, enters the village of Ayan (Krivosheev, 1971; Alina and Reimer, 1975; Kostenko, 1976). Along the eastern spurs of the Dzhugdzhur Range, the eastern boundary descends south up to the Maisk Range, where habitations of northern pikas were recorded in the basins of the Maya and Uda rivers (Ognev, 1940; Kazarinov, 1973). East of the Uda River, northern pikas were caught along the coast of the Konstantin Gulf, on the island of Bolshoi Shantar, along the coast of Lake Orel' and southward—in the vicinity of Shaman Mountain and in the Taba Inlet of the De-Kastri Gulf (Ognev, 1940; collections of ZM MGU). Farther south, northern pikas were caught on the eastern slopes of Sikhote-Alin—in the vicinity of the village of Tumninskii (collections of ZM MGU).

In Sakhalin northern pikas are found in montane regions of the Shmidt Peninsula (at the extreme north of Sakhalin), in the Dagi Mountains, in the region of Cape Aleksandrovskii, on the Tym' River, in the Makarovsk district and in the vicinity of Dolinsk—the southernmost locality of the northern pika (Ognev, 1940; Timofeeva, 1962; V. Rozhnov, oral communication).

On the coast north of Primorsk territory, the northern pika has not been found so far. In the Primorsk territory the eastern distributional boundary of the northern pika apparently passes along the eastern spurs of Sikhote-Alin through the village of Vostok-2, along the upper reaches of the Kulumbe' River (basin of Iman) , along Abrek tract, along the territory of the Sikhote-Alin preserve and through Krasnorechensk (collections of ZM MGU, Geography Faculty of MGU and the Primorsk Antiplague Station; oral communication of M. Rutovskaya). In the southern part of the Primorsk territory, northern pikas were caught on the western slopes of Slkhote-Alin,

*Not in Lit. Cit., possibly a misprint for 1968—General Editor.

particularly in the vicinity of the village of Nizkie Pushki and on Khualaza Mountain (Khalaza) (Kostenko, 1976; collections of the Primorsk Antiplague Station). The environs of Khualaza is the extreme southeast locality of the northern pika (44° N. Lat. and 133° E. Long.). From here the eastern boundary meets the southern.

From south of Primorsk territory, the distributional boundary of the northern pika ascends north, passes along the valley of the Bikin River and in the vicinity of the village of Obor, and later turns west (Ioff and Skalon, 1954). The habitations of northern pikas were found in the Bol'shekhekhtsirsk preserve and south of Khabarovsk, in the Khingansk preserve and north of it (Chernolikh, 1973). Later the southern boundary bends around the Zeya-Bureya plain and proceeds north. The habitations of northern pikas were reported along the valleys of the Bureya and Mal'mal'ta rivers, and also in the vicinity of the village of Polovinsk (Krivosheev, 1984; collections of KGU). From here the boundary turns northwest and passes through the village of Dagmara in the basin of the Selemdzha River to the southeastern spurs of the Tukuringra Range (collections of KGU). Northern pikas were caught in the vicinity of the city of Zeya, in the valleys of the Motovaya and Bol'shaya E'rakingra rivers, on the right bank of the lower stream of the Gilyui and westward—in the vicinity of Solov'evsk (collections of ZM MGU, ZIN, BIN; oral communication of N.A. Formozov).

In the eastern Trans-Baikal area, habitations of northern pikas are reported in the basin of the Amazar rivers, along the left bank of the Shilka (15 km northwest of Ust'-Karsk), and in the vicinity of the village of Forkovo and city of Sretensk (collections of ZIN, ZM MGU, IGU; oral communication of N.A. Formozov). Along the right bank of the Shilka, the Altai pika already occurs. Following farther in a south-western direction, the southern boundary passes along the spurs of the Chersk and Yablonovyi ranges. The habitations of northern pikas were recorded on Sarankan golets [bald peak] in the valley of the Kumakhta River (right tributary of the Chita River), in the Aginsk steppe—south and north of Bol'zino and in the vicinity of Alkhanai Mountain (Ognev, 1940; collections of ZM MGU and ZIN). Later, the southern distributional boundary of northern pikas descends along the southwestern spurs of the Chersk Range and proceeds along the territory of the Sokhodinsk preserve. According to the unpublished data of N.A. Formozov, northern pikas are found in the southwestern part of the preserve—in the vicinity of Lake Nay'ya and Sokhodno golets [bald peak]. Eastern Trans-Baikal is the southernmost locality of northern pikas. Northwestward, pikas were caught along the Yamarovsk Range (collections of the China Antiplague Station).

Later the southern distributional boundary of the northern pika ascends north along the Zapadnyi [Western] Khamar-Daban. Habitations of northern pikas were encountered in the basins of Tumusun, Mergasan, Malaya and Bol'shaya Bystraya rivers and in the area of the Torsk basin (Tarasov, 1962). Along the Tunkinsk Range northern pikas were recorded in the Urungoi tract, in the vicinity of the village of Arashin, 1.5 km from the Inogda River, on Khubyty pass and along the Tubota River (Yudin et al., 1979; Formosov and Nikol'skii, 1979; collections of ICU, ZIN and BIN). West of Buryatia northern pikas apparently are absent. Later the southern boundary passes south and southwest from the upper reaches of the Azas and Kyzyl-Tashtyg rivers roughly to 92°30' E. Long. Pikas were caught in the upper reaches of the O-Khem River (tributary of the Bii-Khem River), in the upper reaches of the Balyktyg-Khem River and in the upper fork of the Naryn River (Yudin et al., 1979; collections of ZM MGU, ZIN, and BIN). Westward, in the basin of the Torgalyg River on the northern slope of the Zapdnyi [Western] Tannu-Ola Range, northern pikas were found together with Altai pikas (oral communication of N.A. Formozov). According to data available to us, the Torgalyg River is the extreme southwestern locality of

120   northern pikas in the territory of Tuva. From here the southern boundary turns northwest and meets the western boundary.

The western distributional boundary of the northern pika passes along the western slopes of the Dzhoisk Range—through Oninsk post, along the valley of the Byurkarak River (basin of Dzhebash) and sources of the Bol'shoi Arbaty (a tributary of the Abakan) (Kokhanovskii, 1962). In the valley of the Ana (Ona) river, according to the data of S.I. Ognev (1940), the range of the northern pika touches the range of the Altai pika. Farther northward the western boundary passes along the right bank of the Yenisei.

Along western slopes of the eastern Sayan, northern pikas were caught in the basin of the Kazyr River on the Kryzhina Range, on Mansh belogor'ye [white mountain] in the basin of the Sharovary River (right tributary of the Shinda River), in the area of the Mansk lakes and in the vicinity of Krasnoyarsk (Ognev, 1940; Nikol'skii, 1984; collections of ZIN). Following farther northward, the western boundary of the range of the northern pika passes first along the left bank of the Yenisei, through the villages of Kolmogorovo and Vorogovo, and then—along the right bank—through the town of Sulomai on the Poodkamennaya Tunguska, in the vicinity of the town of Bakhta, in the vicinity of Turukhansk, and ascends to the Khantaika River on the western Taimyr (Ognev, 1940; collections of ZM MGU, BIN and ZIN; B.I. Sheftel', oral communication).

Outside of Russia the northern pika is widely distributed in Mongolia—in the Khentei, Khangai, Cis-Khubsugul' [Lake] area, the Greater Khingan; in China it is known from Inner Mongolia—the Greater and Lesser Khingan, Manchuria; in Japan northern pikas occur on Hokkaido Island; on the Korean Peninsula—in the mountains of North Korea (Fig. 32).

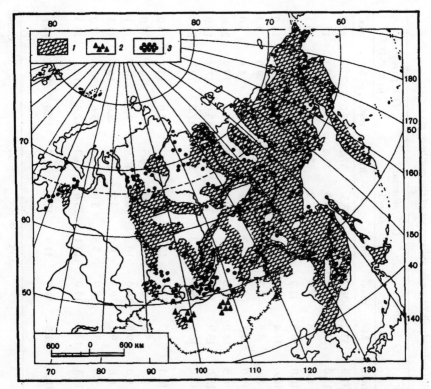

Fig. 32. Range of the northern pika. *1*—schematic distribution according to S.I. Ognev (1940); *2*—localities of the northern pika in Mongolia (Sokolov and Orlov, 1980); *3*—localities of northern pikas in Russia.

Geographic Variation

Apparently 7 subspecies of northern pika occur in the territory of the USSR.

1. *Ochotona hyperborea hyperborea* Pallas, 1911 (syn. *Lagomys litoralis* Peters, 1882)—Chukchi northern pika.

The type subspecies of northern pika is characterized by small dimensions. The condylobasal length of the skull is less than 36 mm; diastemic width of upper molar teeth—less than 12 mm; length of tympanic bullae—less than 12 mm; height of the skull in the region of the rostrum—more than 5.8 mm; length of the orbits—more than 10.8 mm; width of the rostrum—more than 5.5 mm (Table 7.1). The color of the summer fur on the back has a predominance of yellowish-gray tones with distinct black brown ripples; the upper part of the head is brownish-brown, sides pale yellowish; belly gray, sometimes with a small admixture of yellowish tinge.

Table 7.1. Skull dimensions (in mm) of adult individuals of seven subspecies of the northern pika

Indices	O. h. hyperborea (n = 24)			O. h. minima (n = 14)			O. h. shamani (n = 10)			O. h. ferruginea (n = 284)		
	min.	M	max.	min.	M	max.	min.	M	max.	min.	M	max.
Condylobasal length	32.8	34.6	37.0	32.0	33.2	36.2	34.7	35.5.	37.2	35.0	37.2	42.0
Length of rostrum	13.6	14.7	16.0	13.0	13.5	15.0	13.6	14.5	15.5	13.0	15.3	17.1
Length of upper tooth row	6.8	7.1	7.8	6.2	6.8	7.0	7.0	7.2	7.5	6.8	7.6	8.1
Length of orbit	10.7	11.3	12.0	10.0	10.8	11.3	11.0	11.2	12.2	11.0	11.7	13.4
Length of tympanic bulla	11.0	11.6	12.0	10.6	11.4	12.4	12.1	12.6	13.2	11.0	12.2	13.5
Width of tympanic bulla	8.0	8.7	9.5	8.0	8.5	9.3	8.3	9.3	10.0	8.6	9.0	9.4
Interorbital width	4.0	4.7	5.0	4.0	4.5	4.9	4.2	4.7	5.0	4.0	4.6	5.0
Zygomatic width	18.0	18.9	19.9	16.5	18.0	19.8	18.3	19.2	20.1	19.5	19.9	21.8
Width of cranium	15.6	16.5	17.1	15.0	15.9	17.8	16.5	16.9	17.6	15.0	17.4	18.9
Width in the region of tympanic bullae	17.3	18.2	19.2	17.0	18.0	19.8	18.3	19.5	21.0	17.0	19.9	21.5
Post-orbital width	12.2	12.8	13.3	11.3	12.3	13.0	12.5	13.2	14.2	12.0	13.2	15.0
Width of rostrum	5.4	5.8	6.8	5.1	5.6	6.0	5.5	6.0	6.5	5.4	6.0	7.0
Occipital height	9.8	10.2	10.9	9.0	9.5	10.0	10.0	10.3	10.9	9.0	10.7	11.1
Height in frontal region	9.0	9.6	10.5	8.6	9.1	9.9	9.0	9.7	10.5	9.6	10.2	11.5
Height of rostrum	5.3	5.9	6.5	5.0	5.3	5.9	5.5	5.9	6.5	5.6	6.2	7.0
Height in the region of tympanic bullae	12.2	12.8	13.5	11.3	12.4	14.0	13.1	13.6	14.3	12.8	13.6	15.3
Diastemic width of upper molar teeth	11.0	11.6	12.0	10.6	11.3	11.9	11.9	12.1	12.7	12.0	12.5	13.1

Contd.

123

Table 7.1 contd.

Indices	O. h. davanica (n = 7)			O. h. stenorostrae (n = 12)			O. h. uralensis (n = 12)		
	min.	M	max.	min.	M	max.	min.	M	max.
Condylobasal length	40.0	40.9	42.0	34.2	35.4	35.5	32.0	34.6	36.3
Length of rostrum	14.8	16.5	17.0	13.7	14.1	14.8	13.0	13.9	13.9
Length of upper tooth row	7.8	8.1	8.3	6.7	7.2	7.5	7.0	7.1	7.5
Length of orbit	12.0	12.1	12.3	10.6	11.2	12.0	10.2	10.6	11.0
Length of tympanic bulla	12.1	12.4	12.7	10.9	11.4	12.1	11.0	11.7	12.5
Width of tympanic bulla	9.2	9.9	10.5	8.5	8.8	9.3	8.0	8.6	9.0
Interorbital width	4.3	4.7	5.3	3.7	4.0	4.3	5.0	5.1	5.2
Zygomatic width	20.8	21.5	21.8	18.1	19.0	20.1	18.1	19.1	20.0
Width of cranium	17.8	18.9	19.5	16.0	16.6	17.1	15.5	16.4	17.2
Width in the region of tympanic bullae	21.0	21.7	22.5	18.3	19.1	19.7	17.5	18.7	19.6
Post-orbital width	13.3	13.9	14.8	12.2	12.8	13.6	12.0	12.8	13.5
Width of rostrum	5.4	6.1	6.6	5.0	5.2	5.4	6.0	6.1	6.5
Occipital height	10.2	11.0	11.2	9.6	10.0	10.7	10.0	10.2	10.7
Height in frontal region	10.7	10.8	10.5	8.8	9.4	9.8	9.0	9.6	10.0
Height of rostrum	6.3	6.7	7.0	5.4	5.7	5.9	5.8	6.0	6.2
Height in the region of tympanic bulla	13.7	14.4	14.7	12.5	12.9	13.3	11.9	12.8	13.5
Diastemic width of upper molar teeth	12.8	13.2	13.8	11.3	11.9	12.4	12.0	12.2	12.7

Winter fur has a predominance of light gray color. The Chukchi northern pika occurs in the Chukchi Peninsula—from the sea-coast in the east to the right bank of the Amguema River in the west.

2. *Ochotona hyperborea minima* ssp. nov.—Anadyr northern pika.

The smallest subspecies of *O. hyperborea*. The condylobasal length of the skull is usually less than 35 mm, diastemic width of upper molar teeth—less than 2 mm, zygomatic width—usually less than 18.5 mm, height of the skull in the occipital region—usually less than 9.8 mm, width of the rostrum—less than 6 mm, interorbital width—less than 5.5 mm (Table 7.1).

The Anadyr northern pika is distributed in the basin of the Anadyr River—from the Chukotsk (Anadyr) Range in the east to the Oloisk Range in the west.

3. *Ochotona hyperborea shamani* ssp. nov.—Lower Indigirka northern pika. A small subspecies of *O. hyperborea*. The condylobasal length of the skull is usually from 35 to 36 mm, height in the frontal region—usually less than 10 mm, interorbital width—less than 4.9 mm, diastemic width of the upper molar teeth—usually more than 12 mm, zygomatic width—usually more than 18.5 mm, width of the rostrum—more than 5.5 mm, length of tympanic bullae—more than 12 mm, length of the orbit—usually more than 11 mm.

The lower Anadyr northern pika is known from the vicinity of the villages of Shamanovo and Kondakovskoe—the lower reaches of the Indigirka River (right bank), and lives possibly also on the Alazeisk Plateau.

4. *Ochotona hyperborea ferruginea* Schrenk, 1858 (syn. *Lagomys hyperboreus* var. *cinereoflava* Schrenk, 1858; *O. kolymensis* Allen, 1903; *O. hyperborea turchanensis* Naumov, 1934)—Rusty northern pika. A northern pika of medium and large dimensions. Condylobasal length of the skull is more than 35 mm, height in the frontal region—more than 10 mm, width of the rostrum—more than 5.5 mm, length of the orbit—more than 11 mm, zygomatic width—usually less than 21 mm, width in the region of the tympanic bullae—usually less than 21 mm (cf. Table 7.1).

The color of the summer fur is quite variable. Populations inhabiting the Okhotsk seacoast are characterized by a light yellowish-gray color on the back with an admixture of ochreous color on the back and with lighter. sides. The summer fur on the back in Kamchatka populations is yellowish-reddish-brown with an admixture of ochreous tones, the sides are yellow-ochreous, and the belly somewhat lighter than the sides. The winter fur is a monochrome, yellow-brownish-brown on the back and yellow with ochreous tinge on the sides. The color of summer fur of rusty northern pikas inhabiting the basin of Yana, Indigirka, Kolyma and the Koryaksk upland is light reddish-brown or brown on the back, grayish-brown on the upper part of the head, light reddish-brown with a yellowish tinge on the sides and gray with light rusty tinge on the belly. Winter fur is a dark gray with a chestnut tinge. Rusty northern pikas inhabiting the Amur and Ussuri territories, in the Trans-Baikal area, and the Sayans, differ in having summer fur with dark coloration—brownish-brown with a black ripple on the back and a rusty-brown with grayish tinge on the sides; winter fur in these pikas is grayish-brown. Rusty northern pikas in the basins of the Nizhnaya [Lower] and Podkamennaya Tunguska rivers have summer fur which differs in having bright rusty-brown coloration—on the back the fur is an intense rusty with an admixture of hairs with black tips and on the spine passes a dark, almost black, stripe; the sides are a rusty-red without an admixture of black; the belly is yellowish-rusty. Winter coloration is lighter and grayer, particularly in the region of the head and neck; the belly is very light with a weak rusty-ash bloom.

22

The rusty northern pika has a wider distribution—it inhabits the Kamchatka, the Koryaksk uplands, the basins of the Kolyma, Yana, Indigirka, Lena and Aldan rivers, ranges bordering the Sea of Okhotsk, the Sikhote-Alin, Amur and Ussuri territories, the eastern and southern Trans-Baikal area, the eastern and western Sayans, the basins of Podkamennaya and Nizhnyaya Tunguska rivers, and the Putoran Plateau.

5. *Ochotona hyperborea davanica* ssp. nov.—Northern Baikal northern pika. A large subspecies of northern pika. The width of the skull in the region of the tympanic bullae is usually more than 21 mm, zygomatic width usually more than 21 mm (cf. Table 7.1).

In summer ochreous-brown tones predominate with a dark, almost black ripple on the back of northern Baikal pikas. The sides are lighter than on the back, are ochreous and without dark speckles; in the cheek region ochreousness is the most bright. The belly is grayish-ochreous. The winter fur is twice as long as the summer fur and is brownish-gray.

O. h. davanica is known from north of the Baikal Range (Davan Pass); besides the Baikal Range it inhabits possibly also the northern Baikal upland.

6. *Ochotona hyperborea stenorostrae* ssp. nov.—Tuva northern pika. A northern pika of moderate and small dimensions. The condylobasal length of the skull is usually more than 35 mm, interorbital width—less than 4.4 mm, height of the skull in the frontal region—usually less than 10 mm, the width of the rostrum—less than 5.5 mm (cf. Table 7.1).

In summer the back is dark ochreous with black and sandy ripples, and along the spine sometimes there is a dark stripe. The upper part of the head is less bright, the sides are ochreous, the belly dark gray with ochreous bloom or ochreous-gray, and the throat is dark gray. The winter fur is brown with dark gray speckles.

The Tuva northern pika lives on the western Tannu-Ola and eastern Tannu-Ola ranges, and also on the Akademik [Academician] Obruchev Range.

7. *Ochotona hyperborea uralensis* Flerov, 1927—Ural northern pika. A quite small subspecies of *O. hyperborea*. The length of the orbits is less than 11 mm, diastemic width of the upper molar teeth—more than 12 mm, interorbital width—more than 4.9 mm, and width of the rostrum—more than 5.9 mm (Table 7.1).

123

The summer fur on the back is yellowish-rusty with black-brown ripples, the sides are a lighter, ochreous-yellow, and the belly is grayish.

The Ural northern pika lives in the Cis-Polar and northern Urals.

In 1932 a subspecies of northern pika *O. hyperborea yashikurai* Kishida was described from Sakhalin. Inadequate collection material does not so far make it possible to confirm or reject the existence of this subspecies.

The following subspecies of northern pika are known from the territories of adjoining countries: *O. hyperborea yesoensis* Kishida, 1930 (Japan, Hokkaido), *O. hyperborea coreana* Allen and Andrews, 1913 (North Korea), and *O. hyperborea mantchurica* Thomas, 1909 (China: Inner Mongolia—Manchuria, Bolshoi [Greater] and Malyi [Lesser] Khingan).

Biology

Habitat. The most typical habitats of northern pikas are the stone mounds situated at various heights and most often arising as a result of weathering and translocation of bedrocks. The disjunct distribution of this species is apparently associated with the geological structure and geomorphological history of the given territory. There is a suggestion that in the mounds formed by different rocks dissimilar conditions are created for the formation of the soil layer, which leads to differences in the productivity of the plant cover and, consequently, to dissimilar conditions for habitation of northern pikas. Thus, northern pikas are less numerous or absent in mounds of limestone and some forms of sandstone, which disintegrate faster, turning into a heap of small slabs. Northern pikas are most numerous in stable mounds of moderate sizes (from 0.5 to 2.5 m across) of fragments of crystalline schists, gneisses, gabbro and granites (Revin, 1989). In plain taiga and tundra habitats, northern pikas prefer habitats with banks of fallen trees or accumulations of driftwood.

124

In northeast Siberia northern pikas live in taiga and tundra zones, both in the plains and in the mountains. In mountains, which do not usually rise over 2,000 m above sea level, northern pikas are found up to the upper limit of the goltsovoi [bald mountain] zone (Yudin et al., 1976). In taiga as well as in tundra, northern pikas remain in stone mounds comprising stones of medium and large size. The preferred places are mounds alternating with turfy areas covered with herb, shrub or woody vegetation. Usually the most thickly inhabited are the lower parts of mountain slopes with rich vegetation. Northern pikas are found also on the peaks of ranges in the goltsovoi [bald mountain] zone with scanty, predominantly lichen vegetation. In Anadyr valley northern pikas occupy steep banks with stony slopes and mounds, and also in the adjoining thickets of willows and alder-cedar elfin woodstand. In the plain part of the middle and lower course of the Kolyma, in the valleys of the Omolon and Kegali rivers, northern pikas live along banks and on the islands in heaps of driftwood. On the raised rightbank of the middle and lower course of the Kolyma, sporadic colonies of northern pikas were noticed in larch-dense green forest massifs. In the lower reaches pikas enter open tundra areas with sedge-cotton grass hummocky marshes (Yudin et al., 1976). In the western part of the Okhotsk-Kolymsk upland (Berendzhinskii Range) the maximum number of northern pikas was recorded in the flat areas of mounds—at the foot of slopes and in a trough with herbaceous vegetation and rare shrubs of clove currant and cedar elfin wood-stand. Here northern pikas avoid colonizing dense thickets of cedar elfin wood-stand (Krivosheev, 1989).

In the coastal part of the Koryaksk upland, northern pikas do not ascend above 500-600 m, and in the inner parts of the uplands—above 1,000-1,300 m above sea level. Here the colonies of the northern pikas are often found in stone mounds and less often in flood plain thickets of alder, poplar and *Chosenia*. Large-stone mounds with an area not less than 100 m^2 alternating with meadow patches or clusters of elfin woods are particularly favorable for habitation by northern pikas (Portenko et al., 1963).

125   In Yakutia, northern pikas inhabit montane regions, valleys of rivers in plains, and forests in plains. In mountains northern pikas occupy afforested and open-stony slopes with large-stone mounds—from the foot and coastal embankments to peaks of water divides with high-mountain tundra. If there are large-stone mounds and sufficiently dense vegetation, pikas occupy slopes of any exposure and steepness. With the presence of suitable habitats, they penetrate the subalpine belt and zone of mountain tundra. Thus, in the area of the Verkhoyansk Range, pikas are found at altitudes up to 2,000 m, and in the area of the Chersk Range—at altitudes up to 1,500 m. In the basin of the Adycha River along areas of large-stone mounds, pikas enter the goltsovi [bald mountain] zone, but with the presence of shrub and herb-vegetation there (Krisvosheev, 1971). Talus situated above the forest limit and devoid of shrub and herb vegetation, as a rule, is avoided by pikas. However, on the Kharaulakhak Range colonies of northern pikas have been recorded in mountain tundra, where the nearest shrub thickets are to be found at a distance of 5 km (Kapitonov,

1961). Colonies of northern pikas are also found in plain forests, but with the presence of buried stones and hollows, in the soil cover, formed by frozen cracks covered with moss and brushwood; such colonies are usually less numerous. Sometimes northern pikas colonize burned-out forests along hummocky marshes, on banks of lakes and in larch forest massifs. Along the valleys of rivers in plains, pikas enter the tundra zone almost to the seacoast—here they occupy steep banks with driftwood and clusters of driftwood (Krivosheev, 1971).

On the Lena Plateau colonies of northern pikas are confined to talus on slopes of deeply cut river valleys. Near the mouth of the Nizhnyi [Lower] Dzhege River (left tributary of the Tokko River, 300-400 m above sea level) isolated colonies of pikas have been observed in loamy, weakly fixed mounds surrounded by pine forests (Revin, 1989). On the Olekmo-Charsk Upland, northern pikas most often inhabit lower parts of mountains, where they prefer to colonize small (to 0.5 ha) large-stone mounds in river valleys, on mountain terraces, and in rock terrain surrounded by high tundra forest. On extensive mounds, where the central part has small-blocky structure, colonies of pikas are concentrated near the upper edges of talus, near the outcrops of bedrock or in their lower part. On the Charoudsk Plateau favored places of habitation of northern pikas are the stony valleys of brooks, the bottom and lower parts of talus, that are surrounded by forest. Pikas are found here not only in talus, but also in wind-fallen woods and thick shrub thickets under the forest canopy (Revin, 1968).

In the Khabarovsk territory (Khingan and Zeya preserves), northern pikas are found on mounds among mountain larch; larch-birch and spruce forests (Chernolikh, 1973). In the basins of the Uda and Maya rivers, and also along the eastern spurs of the Dzhugdzhur Range, northern pikas are most frequently found in the upper reaches of rivers and streams with rock ledges and mounds. On the coast of the Sea of Okhotsk, pikas usually occupy coastal rocks and extremely rarely colonize the plain areas (Kostenko, 1974).

In the Primorsk [Pacific Coastal] territory pikas inhabit mainly medium- and large-blocky mounds with different degrees of overgrowth and situated basically in the zone of spruce-fir mountain forests and in high mountains, where there are tundra areas among thickets of cedar elfin wood. In Sikhote'-Alin pikas are found in mounds along steep banks of montane rivers in the zone of cedar-broad-leaved forests, and in years of high population—in broken trees along gravelly and sandy bars and in rocks of the seacoast (Kostenko, 1976).

In Kamchatka habitats of northern pikas differ little from Primorsk habitats, but here pikas are found additionally at the base of old lava flows surrounded by sparse Japanese stone pine, elfin and thinned-out herbaceous vegetation (Averin, 1948).

126 In Sakhalin colonies of northern pikas are confined to large-stone mounds with thickets of blueberry, cowberry, shrubs and Japanese stone pine, as well as in mixed forests, where hollows between tree roots serve as shelters. Pikas are also found in alder thickets on banks of springs (Timofeeva, 1962, 1963*).

In southern Cis-Baikal and in the Trans-Baikal area, northern pikas are found in mountains from the forest belt to the goltsovoi [bald mountain] zone, and their colonies, as in a large part of the range, are often confined to talus or outcrops of bedrock (Tarasov, 1962). In the forest belt northern pikas live mainly in the middle and upper parts, where the main mass of talus is situated. Pikas are also numerous in the lower part of the goltosovoi [bald-mountain] belt. Here they colonize shrubby and moss-lichen tundra, where thickets of willow, ledum, alder and ground birch are encountered. In this part of the range, the most optimal habitat for northern pikas is talus in the sub-goltsovoi [bald mountain] belt, where the grass stand is rich, and Japanese stone pine and thickets of different shrubs are common (Shvetsov et al., 1984). In the western Cis-Baikal area (Primorsk Range) in years of high populations, northern pikas colonize not only mountain talus in the forest belt, but also banks of streams and dumps of forest wastes along rivers and roads, and also on mountain tops (Reimers, 1966; Shvetsov and Litvinov, 1967).

In the Cis-Khubsugul' area, Khangai and Khéntéi, besides the habitats typical for the species, under favorable conditions northern pikas can occupy open places—mounds among burned-out forests and forest fellings, and in the area of Lake Khubshgul'—stone dumps on the lake bank (Shetsov et al., 1984**).

In the region of the middle Siberian Plateau in basins of the Malaya Tunguska and Podkamennaya Tunguska rivers, northern pikas remain predominantly along talus atop ranges and also in river valleys, where they inhabit heaps of deadwood (Naumov, 1934).

In the Polar Urals in the valley of the Khadata-Yugan River (at the northern distributional boundary of the Ural northern pika) small colonies of pikas were noticed in thickets of Manchu alder with an admixture of mountain ash, individual trees of larch and a poor grass cover. To the south, with the increase in diversity of plant groups and rise of vertical zonation, the affinity of pikas to the zone of the upper limit of forest and sub-goltsovoi [bald mountain] zones becomes clear. Here, almost always they occupy

---

*Not in Lit. Cit.—General Editor.
**Not in Lit. Cit. Presumably reference is to Shvedov et al., 1984—General Editor.

large-stone dumps, that are small in area and overgrow with mountain ash (Gashev, 1968).

Population. The population of northern pikas depends on the character of the habitat, time of the year and climatic conditions. In the northwest of their range (vicinity of the village of E'gvekinct) in relatively favorable habitats (large-stone mounds on slopes of southern exposure with numerous meadow patches), in July 1967, no less than 15 animals/ha were counted (Chernyavskii, 1984). A roughly similar high number of northern pikas was observed in this region also in August 1970 (Yudin et al., 1976). A still higher population density (20 individuals/ha) was observed in analogous habitats in the middle fork of the Omolon River (Chernyavskii et al., 1978). In small areas of talus in the upper reaches of the Yarakvaam River (western Chukotka) in July 1965, the density of northern pikas did not surpass 3-5 individuals/ha, and in larch sparse forest on the Omolen River in July 1972—0.05-0.1 individuals/ha (Chernyavskii, 1984). The highest population density of northern pikas (up to 100 individuals/ha) was recorded in the Olekmo-Charsk upland (Revin, 1968).

In the southern Cis-Baikal area, the least number of northern pikas (0.25-0.5 individuals/ha) was recorded for the goltsovoi [bald-mountain] belt, although in the lower part of this belt, where vegetation is developed much better, the population of pikas is higher—4-7 individuals/ha; the maximum population density of pikas in this region (5-15 individuals/ha) was observed in talus on the sub-goltsovoi [bald mountain] belt. In the upper part of the forest belt of the Khamar-Daban Range, in 1972-1975 the population density of pikas was 4-6 individuals/ha, and on mounds in mixed and pine forests of the southern slope of the Malyi [Lesser] Khamar-Daban Range—3 individuals/ha (Shvetsov* et al., 1984).

127   In northern Khéntéi (territory of the Sokhondinsk preserve) the maximum number of northern pikas (25 individuals/route km) was recorded in talus in the river flood plains of the taiga zone. Pikas were fewer in the bald mountain belt (on average, 9 individuals/route km) and in the taiga zone outside of the river valleys (7-8 individuals/route km). Least of all (1-2 individuals/route km), pikas were found in the belt of mixed and deciduous forests (Baranov, 1984).

Shelters. The character of shelters of northern pikas depends on the type of habitats. In large-stone mounds, recesses between stones and niches serve as shelters for pikas. The digging activity of northern pikas is minimal, and the nests are situated deep in the mound under stones, in cracks and hollows. Sometimes one individual may have several nests (Flerov, 1927).

---

*Cf, footnote on p. 171—General Editor.

In places with a shortage of natural shelters, northern pikas can dig burrows. Thus, in the northern Far East, 50 km from the village of Seimchan, a colony of northern pikas was found with a complex system of tunnels and burrows (Yudin et al., 1976). The system of tunnels was located 7-12 m below the talus and was represented by 10 to 25 cm pits in the soil, which were formed as a result of many years use of the very same paths by these animals. The underground tunnels were made in those places, where foraging areas and shelters were separated by a sharp boundary. For instance, in the tundra zone in the village of Lavrentia, a system of burrows of northern pikas was detected at the junction of talus and the area of low tundra vegetation. Such a system of burrows ensures risk-free collection of food. In the valley of the Seimchan River, brood burrows dug in steep banks of rivers and under tree roots were detected (Yudin et al., 1976). Burrows of northern pikas with simple structure were found in the Kolymsk upland (Kishchinskii, 1969) and on small rubbly mounds in the Trans-Baikal area (Kuznetsov, 1929). In Manchuria, northern pikas in summer make spherical nests of grasses under the bushes or on hummocks (Ognev, 1940). In the northern Urals, in high-trunk taiga, pikas often dig underground tunnels 6-9 cm in diam., the length of which reaches 1-1.5 m. These tunnels are weakly branched, but have several outlets, which are usually connected to each other by well beaten paths [runways] (Flerov, 1927).

Feeding. Northern pikas use usually the same species of plants that grow in the vicinity of their habitats as food and for their cache; most often these are massive species of herbaceous plants, shrubs and semishrubs. In many regions, the diet of northern pika includes pileate fungi, and in montane regions—lichens. In summer and fall, practically everywhere their food includes berries and seeds.

The composition of food changes with season. Thus during an analysis of stomach contents of northern pikas from the mountain-taiga belt of the upper fork of the Kolyma [River], it was established that green vegetative plant parts (85% frequency) predominate in the summer diet of pikas. Quite frequently (in 30% of stomachs) stones of crowberries and black currants were found, and in 15% of stomachs, fungi were present (Yudin et al., 1976). On the Charuodsk Plateau in spring, shrubs and shrublets predominate in the food of northern pikas, and in summer—herbaceous plants (Table 7.2). Here the stomachs of northern pikas showed remnants of animal food (Revin, 1989).

In summer, diet of pikas in the basins of the Olekma and Yana rivers includes the green parts of herbaceous plants, shrubs and shrublets, and also a considerable proportion consists of the needles of larch, juniper and stone pine, stems and berries of cowberry, land lichens and pileate fungi

174

Table 7.2. Composition of the diet of northern pikas from the Charuodsk
Plateau in different months (April–September), in volume
per cent of stomach contents (from Revin, 1989)

Type of food	April (n = 32)	May (n = 31)	June (n = 24)	July (n = 16)	Aug. (n = 7)	Sept. (n = 35)
Plant food						
Shrublets, shrubs	72.8	67.8	31.3	6.3	–	30.8
Herbaceous plants	15.0	20.8	61.7	83.8	100.0	58.7
Mosses	–	–	6.6	–	–	2.0
Lichens	12.2	10.8	0.4	3.1	–	1.7
Fungi	–	–	–	3.1	–	6.8
Animal food	–	0.8	–	3.7	–	–

*n—number of stomachs investigated.

Table 7.3. Frequency of different food items in stomachs of northern pikas
(n = 52) in the upper reaches of the Lena (Voronov, 1964)

Type of food	Number of stomachs with the given food	Stomachs with the given food as a % of all filled stomachs
Green vegetation (grass)	46	86.2
Cedar seeds	17	32.7
Berries (honeysuckle, blue-berry, bog bilberry)	6	11.5
Leaves of woody trees (alder, birch)	4	7.7
Seeds of herbaceous plants	2	3.8
Flowers of arctic bramble	1	1.9
Fungi	1	1.9

(Krivosheev, 1971). Near villages in the northern Far East, the main food
of northern pikas is fireweed, which dominates among herbaceous plants
in the river flood plains and in anthropogenic landscapes (Yudin et al.,
1976). In the Cis-Polar Urals, northern Khéntéi and the upper reaches of
the Lena, the summer diet of pikas, besides green parts of plants, has a
large quantity of cedar seeds (Gashev, 1971; Baranoov, 1984; Table 7.4).

In winter, northern pikas feed mainly on plants from their haypiles, and
also on bark and terminal shoots of shrubs. The composition of haypiles
is determined by the ratio of species of plants found in direct proximity to
shelters of pikas. Most often the number of plant species in piles does not
exceed 5-10, though there are piles made of only one species or that contain
more than 10 plant species. Instances have been recorded where common
species of plants are practically absent in haypiles. Thus, in the northeast

of their range, the rather common plant species are the Daurian larch, stone pine, *Chosenia*, Mongolian poplar, and reindeer moss, but these are rarely stored by pikas.

The beginning of food storage of northern pikas varies in different parts of the range and at different altitudes; it depends on the weather conditions, that is, this behavior is not stable among years. According to the observations of N.S. Gashev (1971), in the Urals, northern pikas start storing food coincident with the onset of budding (July 1-18), and massive storage occurs during the period of flowering and fruiting of these plant species, which are frequently found in the haypiles of pikas.

In the forest zone in the northern Far East, northern pikas begin to store food in the second decade of July, while in the tundra zone (the village of Lavrentia) haypiles are not made at this time. In the northeast of the range, August is everywhere the period of active food storage. In the beginning of the second decade of September, in the middle fork of the Seimchan River, food storage has almost terminated (Yudin et al., 1976). In the lower reaches of the Lena the beginning of food storage occurs in the middle of July (Kapitonov, 1961), in Leno-Khatanga territory—in the first half of June (Romanov, 1941), in Olekma basin—in the last decade of July (Revin, 1968), and in the upper reaches of the Lena—in the first days of August (Voronov, 1964). Thus, in Yakutia, in more northern regions, storage of food starts earlier than in the southern regions.

In the high mountains of the southern Cis-Baikal area in 1972-1975, northern pikas began to store food on June 15-20. In the high mountains of the Turkins in the goltsovoi [bald peak] belt, first haypiles were noticed on July 18 (1975) and in the sub-goltosovi [bald peak] zone—on July 22-24, in the Khamar-Daban Range in the goltsovoi [bald peak] zone—on July 13, and in sub-goltsovoi [bald peak] zone—on July 18; that is, at higher altitudes pikas begin to store somewhat earlier (Shvetsov* et al., 1984). The termination of storage usually coincides with the onset of snow cover.

It has been reported that different species of plants, and also their parts, are stored by northern pikas at different times. Herbaceous plants are stored, as a rule, during their flowering period, while shrublets and shrubs—in the period of fruit maturation. In the alpine belt of the Koryask upland in the first half of August, storage of mainly willow and sweet vetch [Hedysaruin] are reported; active collections of manzanita [*Arctostaphylos*] and among procumbent shrubs—false hellebore and dwarf cornel—begins by the end of August. In September, herbaceous plants are almost not collected—at this time pikas store mainly the leaves of dwarf shrubs (bog bilberry, cowberry, and dwarf cornel), and in later fall—fallen leaves of

---

*Cf. footnote on p. 171—General Editor.

trees and shrubs (Portenko et al., 1963). In the northern Urals in July, haypiles contain mountain ash, fireweed, blueberry, bog bilberry, cloudberry, arctic bramble and dwarf arctic birch; and in September appear cowberry and fallen leaves of mountain ash and dwarf arctic birch; in October, fallen leaves of mountain ash predominate in collections (Gashev, 1971).

Northern pikas collect plants in haypiles without preliminary curing by dessication. In stone habitats they place their haypiles under large slabs, sometimes in recesses between small stones. Haypiles are found in funnel-like broadened entrances to the burrow and in specially dug pits (Kapitonov, 1961). In plain larch forests pikas make haypiles under overhanging trunks of trees, in hollows between roots, under stumps and brush-wood. Pikas living on riverbanks store hay in stumps of driftwood, under tree roots and in niches under steep banks (Krivosheev, 1971).

The size of haypiles and their number depend on the presence and volume of hollows convenient for storing reserves. In the northern Far East (Chukchi, Anadyr basin) haypiles of northern pikas are usually not large (the weight of each pile is 200 to 500 g), and they are located at a small distance (0.5-1 m) from each other (Yudin et al., 1976). According to the data of V.G. Krivosheev (1964) in the taiga of the Kolyma lowlands, haypiles of pikas may reach 6 kg. In Olekma valley haypiles are just of 300-500 g (Krivosheev, 1971). In the lower reaches of the Lena, depending on the places haypiles are stored, the weight of piles varies from 100-500 g to 2-3 or even 5 kg; the lesser the weight of haypiles, the more their number in the area of habitation (Kapitonov, 1971).

South of Yakutia, reserves of northern pikas are small and play only an auxillary role in their winter diet. Here, over an area of 50 × 50 m, 3-4 pikas make not more than 20 haypiles with overall weight of 6-7 kg (Revin, 1989). Pikas make relatively small haypiles in the northern Urals, where there are 2 to 6 piles for every family, the weight of which varies from 100 g to 8 kg (Gashev, 1968).

In the southern Cis-Baikal area in the Ulan-Burgassy Range, haypiles of northern pikas have been found, the height of which was 30 cm and the base area 1.4 × 0.7 m, the weight of which exceeded 10 kg (this pile consisted mainly of stems of blueberry). However, such large haypiles of northern pikas in the Baikal basin are found extremely rarely. In this part of the range, one pika, on average, makes 4 kg of hay. Roughly similar quantity of haypiles are made by one pika in montane forests of Khéntéi and Khangai (Shvetsov[*] et al., 1984).

Data on the composition of winter haypiles of northern pikas are

[*]Cf. footnote on p. 171—General Editor.

130 numerous and cover practically the entire spectrum of habitats, particularly in the north and northeast of their range (Kapitonov, 1961; Tarasov, 1962; Krivosheev, 1964, 1971; Kishchinskii, 1969; Gashev, 1971; Yudin et al., 1976; Baranov, 1984; Shvetsov* et al., 1984; Chernyavskii, 1984; Larin and Shelkovnikova, 1987, 1991).

In captivity, as apparently also in nature, northern pikas willingly drink water. Even in the presence of succulent foods in their diet, pikas must have water. According to the observations of N.S. Gashev (1971), in captivity northern pikas cannot survive for more than five days without water.

In northern pikas, as compared to other species of the family, the phenomenon of coprophagy has been studied more completely (Pshennikov et al., 1990). The fecal mass of northern pikas is represented by black and reddish-brown "sausages" up to 95 mm long and 0.477 g in weight. The discharge of soft excreta occurs in the period of rest and is characterized by a definite rhythm, but unlike in hares, the fecal mass is not always immediately ingested, but collected in specific places, usually protected against rain. In summer and fall up to 40% of pikas eat this fecal mass in a small quantity and irregularly, while in winter the entire mass of soft feces, including also that stored earlier, is consumed. In chemical composition the soft fecal mass differs from solid excreta and also from plants eaten by pikas. It is rich in phosphorus, ash elements and proteins, the amino acid composition of which is comparable with products of animal origin, and the content of cellulose in it is low, half that found in the solid excreta. Thus, the high biological value of the fecal mass allows pikas to utilize plant food more effectively.

Activity and behavior. Northern pikas are active mainly in the brighter part of the day. The character of their activity depends on the time of the year, weather conditions, habitat peculiarities and the type of colonies.

In the northeast part of their range, in June and September, the surface activity of northern pikas has two sharply expressed peaks. The first rise of activity is observed from 04.30 to 12-13 hr, the second from 16-17 hr to 20-21 hr. In July-August, when the mortality of young pikas increases, the two-peak nature of activity becomes less manifest, and the overall daytime activity increases in connection with storage of food. In the vicinity of Magadan, during the period of July white nights, northern pikas are active from 04.30 hr to 23.00 hr, and near the village of Lavrentia, to 24.00 hr (Yudin et al., 1976). Summer observations on pikas in the Kulu River basin during the period of their daytime activity (from 8 to 19 hr) showed that in sunny weather adult individuals spend a large part of their time foraging and storing hay (about 45%); and young, for rest and cleaning (about 50%).

*Cf. footnote on p. 171—General Editor.

Young hide under rocks more frequently than adults (Krivosheev and Krivosheeva, 1991).

In river valleys the period of activity of northern pikas is shorter than on talus on mountain slopes (because of the late dawn and early onset of darkness. In places, where foraging areas are far from shelters, the activity of pikas is mainly crepuscular—they come out to feed at 4-7 hr in the morning and 20-23 hr in the evening. In sparse colonies the surface activity of pikas is usually low in spring and in the beginning of summer (Portenko et al., 1963; Yudin et al., 1976).

Northwest of the Verkhoyansk Range, northern pikas were found on the talus surface throughout the day; however, in spring and fall their maximum activity was recorded in midday; in summer, in the cooler time of the day, in morning and evening. In overcast calm weather pikas are usually active, but during rain and in windy weather their activity falls sharply (Kapitonov, 1961).

In winter, northern pikas appear on the snow surface only on calm sunny days, when they come out to warm up in the sun or to run from place to place. On such days pikas make many tunnels and paths from one mound to another under the snow, sometimes to a distance of 100-150 m. However, pikas apparently do not forage on the snow surface (Kapitonov, 1961).

131   According to the observations of A.G. Bannikov (1954), northern pikas living in Mongolia are active only in the brighter part of the day. According to other data (Shvetsov and Litvinov, 1967; Shvetsov* et al., 1984), the activity of northern pikas of Mongolia and the Cis-Baikal area is round the clock. Here the maximum surface activity is observed from April to June (the mating period), and from the middle of July to September (period of storing food). During the period of reproduction the most active are sexually mature females and males; usually pregnant females become more cautious and less mobile. Usually peak activity is during the morning and dusk hours, although periodically these animals come out to feed also at night. In August the daytime activity increases, particularly in clear and dry weather (Shvetsov* et al., 1984).

In northern pikas vocal signaling is well developed. As a rule, the vocal signal is produced by northern pikas in the direct proximity of a reliable place of hiding. In habitats with a small number of shelters, northern pikas squeek considerably less often. Caught unawares away from shelters, animals stop dead soundlessly, firmly pressed to the ground. In rock slides along river valleys and in nuclear colonies in shale mounds, lowering of vocal signaling has been observed in northern pikas (Belyaev, 1968; Yudin et al., 1976).

---

*Cf. footnote on p. 171—General Editor.

During the period of food storage, vocal signaling in northern pikas becomes considerably activated. Then, particularly with high numbers, calls of northern pikas are heared almost continuously. Not only animals sitting on rocks squeek, but also those that are feeding or are in shelters. In the period of storing hay, northern pikas show two types of vocal signaling—calls and signals forewarning danger (Nikol'skii and Srebrodol'skaya, 1989). Calls, as a rule, are provoked either by the squeeks of animals from the neighboring areas or by the proximity to haypiles of the host of the area. During calls the animals most often (in 79% of cases) collect on those rocks, under which the largest haypiles are stored. Having run to the highest point on the rock, a pika turns to the side of the partner, raises the front part of its body high and remains in the same pose to the end of the call, testifying to the high level of excitement. In the nature of modulation of frequency of constituent sounds, calls differ well from signals forewarning of danger. Vocal signaling of northern pikas addressed to neighbors in the period of food storage, apparently, fulfills the function of territorial demonstration, thus supporting a definite spatial structure of a population (Nikol'skii and Srebrodol'skaya, 1989).

In the mating period the acoustic activity of northern pikas is not only high, but also more diverse in the quantity of signals produced. Thus, as compared to other seasons, at this time three characteristic forms of signals are noticed, two of which are produced only by males, with one of them (short and dull trill "krrrr") males meet females before pursuit and courtship demonstration. The other signal (loud acoustic demonstration of males) consists most frequently of 11-14 short monotone cries produced at almost uniform intervals. Sometimes this begins with a short trill "Krrrr." In individual territories of males often there is a definite place, where it spends a considerable part of its time, and which may be considered the center of its activity. Judged from the combination of calls, males and females differentiate each other by sound. In case, when individual territories of animals are highly overlapping, males (as well as females) often reply to calls of specific individuals of the opposite sex. However, with increased sexual activity preference for a specific partner is almost not manifest in returning calls (Nikol'skii and Srebrodol'skaya, 1989).

Individual variability has been noticed in the sound of northern pikas (Portenko et al., 1963). Moreover, in different situations animals produce sounds of different strength [intensity] and modulation. According to the data of N.A. Formozov and A.N. Nikol'skii (1979), the signals forewarning danger in northern pikas from the Urals, Chukchi and Kamchatka are identical between themselves and differ from signals of pikas from the Trans-Baikal area, that is, there exists geographic variation in the nature of some acoustic signals of northern pikas.

132

When northern pikas are quiet, they sit with drawn-in paws and a sunken head. In such a pose the animal may remain for a long time. With the advent of danger, the pika rises on its fore limbs and the head is held high. If the danger increases, the pika rises still higher on its fore limbs without lifting them from the ground, raises its head and starts to call. During the call its mouth is wide open and the ears are laid back. Then the pika, continuing to call, sits on its fore limbs and stretches the head and shoulders. Finally the pika becomes silent and hides. The smallest movement of the enemy is enough for the animal to take to scampering. Such behavioral features at the moment of danger are characteristic also of other species of pikas (Formozov, 1981).

Areas of habitation. Northern pikas live in more or less large colonies, which consist of individual families. A family group consists of a pair of adult animals and their progeny. In the northern Urals family areas, as a rule, are situated at a distance of not less than 120 m from each other, that is, they do not overlap. The dimensions of family areas of northern pikas in this part of the range vary from 400 to 600 m^2 (Gashev, 1969). The boundaries of family areas are marked with urine on rocks and stumps and also are defended by warning calls. Intruders are usually actively chased away.

Animals are very much attached to their territories, because they contain the permanent location of haypiles and shelters. Within family areas individual territories are marked, the dimensions of which are much smaller than that of family areas, since individual territories overlap to a large degree. Thus, while observing tagged northern pikas in the sub-goltsovoi [bald mountain] part of the Okhotsk-Kolymsk upland, it was established that during summer the area of individual territory of four adult individuals changed and in males constituted 3,978-4,370 m^2; in females, 2,155-5,557 m^2. In the same year's brood, individual territories are located, as a rule, at the periphery of the family territory, and their area is small—425-1,345 m^2 (n = 22). In social interactions pikas of one family group most often revealed segments of nonaggressive behavior (68% of the 32 recorded contacts). Among the elements of aggressive behavior most frequently observed were mutual chasing, attacking and boxing by adult males and females (18.7% of cases), and less often (9.4%)—chasing young animals by adult females (Krivosheev, 1987; Krivosheev and Krivosheeva, 1981).

In the northern Urals massive dispersal by northern pikas has not been observed. At the end of summer, when young leave parental nests, they break into pairs and settle not far from the territories of parents. In exceptional cases, with a shortage of available territories having convenient shelters and places for making piles, young animals inhabit neighboring mounds. Animals of later broods spend winter together with their parents and feed from the common haypile, in the making of which they participate actively (Gashev, 1969).

Reproduction. The periods of reproduction and fecundity of northern pikas in different parts of the range have their own peculiarities. In the tundra zone of Chukchi and in the basin of the Omolon River the reproductive period of northern pikas is extended. In this region reproduction begins in mid-May. Pregnant and nursing females are found here up to the end of July. The size of testes in males also testifies about the termination of the reproductive period in this part of the range by the beginning of August (Chernyavskii, 1984).

In the basin of the middle fork of the the Kolyma [River], reproduction of northern pikas begins at the end of April and continues to the middle of August (Krivosheev, 1971). In central Yakutia mating of northern pikas was recorded in the later half of April (Safronov and Akhrimenko*, 1982). In southern Yakutia (the Olekmo-Charsk upland), mating starts at the end of April; the process of induction of overwintered females in reproduction extends roughly for a month—from the middle of April to the middle of May. Here individual females are ready to reproduce on April 15-20, but in the majority the estrous condition sets in by the end of April-beginning of May. In the second half of May, barren females constitute not more than 15%. This part of females is ready to reproduce only in June, when the majority of females have already borne the first brood (Revin, 1989). Massive births of young occur in the second decade of July, and in the third decade of July a vast majority of females terminate reproduction. In the first half of July degeneration of testes occurs in males. Thus, the reproductive period in populations of northern pikas of the Olekme basin extends for 3-3.5 months (Revin, 1968).

In the high mountains of the southern Cis-Baikal area, the reproductive period of northern pikas extends from the middle of May to the middle of August; and in the forest of Khangai and southern Khéntéi, mainly from April to July (Shvetsov** et al., 1984). In the northern Urals mating in northern pikas begins in May, and the mean duration of reproductive cycle is about two months (Gashev, 1969).

In different parts of the range and, possibly in different years, female northern pikas bear 1 or 2 broods during the summer. According to the data of V.I. Kapitonov (1961), N.S. Gashev (1971), and F.B. Chernyavskii (1984), in the lower reaches of the Lena, in the northern Urals and in Chukotka, there is only one litter per year in northern pikas. However, over a large part of the range, overwintered females reproduce twice (Table 7.4).

133

---

*Not in Lit. Cit.—General Editor.
**Cf. footnote on p. 171—General Editor.

Table 7.4. Dynamics of reproduction of northern pikas in some parts
of the range (from Krivosheev and Krivosheeva, 1991)

Region, year	Number of females investigated	Participated in reproduction			Author
		Over wintered		Current year's brood	
		May-June	July-August	July-August	
Olekmo-Charsk upland (1961-1964*)	175	100.0	40.0	–	Revin, 1968
Kolymsk upland (1963-1964)	22	100.0	17.0	–	Kishchinskii, 1969
Lower Kolyma basin (1960-1964)	66	71.0	22.0	9.5	Krivosheev, 1971
Kulu River basin (1981-1985)	65	83.3	28.6	–	Krivosheev and Krivosheeva, 1981)

*In Russian original, 1964—General Editor.

South of Yakutia repeat mating occurs apparently soon after the first births, and lactation starts simultaneously with embryonic development of the young of the second brood, which appears in July (Revin, 1968). In a southward direction the proportion of post-partum reproducing pikas increases, and in the taiga zone of the southern Cis-Baikal area some females may bear three broods during the summer (Reimers, 1966). In the central and southern parts of the range, apparently only the overwintered females take part, while in the north the increase in population of northern pikas occurs partly also on account of reproduction by the current year's individuals (Table 7.4).

According to data of embryo analysis and postnatal scars, broods of northern pikas may have 1 to 11 young, and the average litter size over the range changes from 2.0 to 5.9. It is seen from Table 7.5 that fecundity of northern pikas is lowest in the intercontinental regions of eastern Siberia, and it increases to the north, northeast and south.

Viability of the number of young in the spring and summer broods has been studied only for the Olekmo-Charsk upland population of northern pikas (Revin, 1968). In this part of the range, the average size of spring and summer broods is roughly the same. The mean value of litters by the current year's females, according to observations in the basins of the Yana and Kolyma, is 4.6 (7 females investigated), with fluctuation of the number of young in one litter from 2 to 7—this is somewhat less than the number of young in overwintered females from this same region (Krivosheev, 1971).

Table 7.5. Fecundity of northern pikas in different parts of the range
(from data of analysis of embryos and placental scars)

Region	Number of females investigated	Average size of litter	Range	Authors
Chukchi Peninsula	6	4.8	–	Yudin et al., 1976
Lower reaches of the Anadyr	5	5.5	–	Chernyavskii, 1984
Omolon basin	12	4.8	3-8	Chernyavskii, 1984
Lower reaches of the Kolyma	20	5.9	3-11	Krivosheev, 1971
Middle Kolyma	12	4.5	3-7	Labutin et al., 1974
East of the Kola upland	6	3.5	–	Kishchinskii, 1967
West of the Kola upland	22	3.2	–	Krivosheev and Krivosheeva, 1991
Lower reaches of the Lena	7	4.8	2-6	Kapitonov, 1961
Olekomo-Charsk upland	77	2.2	1-4	Revin, 1969
Upper reaches of the Lena	16	2.0	–	Voronov, 1964
Southern Cis-Baikal area	–	3.5	1-7	Shvetsov* et al. 1984
Middle Urals	–	–	2-5	Gashev, 1971

134

In northern pikas embryo mortality is usually low—from 1.9 to 5.5%, but in unfavorable years may reach 25% (Gashev, 1971; Revin, 1989). Resorption of embryos in the spring, and also in females bearing large litters is more likely.

Growth and development. In northern pikas gestation extends 28 days (Gashev, 1971).

Prenatal embryos in northern pikas from Hokkaido may attain a weight of 12 g and a length of 50.5 mm, and newborn animals weigh up to 9.6 g and have a length of 60 mm (Raga, 1960). Newborn animals from the northern Urals have a weight of 5-8.5 g and a length of 51 mm. Eyes and auditory meatuses are closed at birth; on the digits, the claws are black; incisors and premolars are in the cutting stage. According to observations in vivarium conditions (Gashev, 1971), five days after birth auditory meatuses open in northern pikas, while eyes open on the seventh to eighth day. Fourteen days after birth young leave the nest, and after another two days they begin to independently dig earth, drink water and even begin to whistle; at this time their weight reaches 32-33 g. In nature at the beginning of July, individual young ones of earlier broods, appear that have attained

*Cf. footnote on p. 171—General Editor.

the size of adult animals. In late-born pikas, by fall the growth halts and resumes in the spring of the next year. Young attain sexual maturity in the year of their birth and, in some cases, take part in reproduction.

Molt. In northern pikas there are two molting seasons—spring and fall. The start of the spring molt and its periods depend on the geographic latitude, altitude above sea level, weather conditions, and also on the sex and age of animals. In the northern Khéntéi in the belt of larch and mixed forests, the spring molt begins at the end of March; but in the taiga—at the end of April; it continues correspondly, to the beginning of June or second decade of July. In the goltsovi [bald mountain] zone the periods of spring molt are extended; fall molt starts early, such that in 40% of individuals in these habitats, the spring molt fades into the fall molt (Baranov, 1984).

135 In the northeast of the distributional range (the basin of the Omolon River and the Koryaksk upland), spring molt starts in May and its period is highly extended (roughly up to August) and is not synchronous in different individuals. Here, the end of the spring molt may coincide with the start of the fall molt (Portenko et al., 1963; Chernyavskil, 1984). In the lower reaches of the Lena until the middle of August, pikas are caught with remnants of winter fur in the sacral region (Kapitonov, 1961). And in the taiga zone in the upper reaches of the Lena, molting starts only at the end of June and extends to first decade of September (Voronov, 1964). The spring molt of the Ural subspecies of the nothern pika extends from June to August (Gashev, 1969), and from May to July in pikas from the middle Siberian Plateau (Naumov, 1934).

As compared to males the spring molt in females of northern pikas begins late and terminates apparently also at a later time. Moreover the data available on the character of molting of northern pikas from different parts of the range confirm its individual variability.

According to the data of N.P. Naumov (1934), the change of winter fur in northern pikas starts on the neck—in a semicircle up to forepaws and simultaneously around eyes. Then molting covers the anterior part of the snout up to the ears, descending from the neck to half the spine and later— along the sides up to the rump. Last to molt are the nape and lower part of the spine.

The fall molt occurs in a condensed period and is more harmonious than the spring molt. Almost over the entire range, northern pikas change summer fur to winter during the course of September (Kapitonov, 1961; Baranov, 1984).

The fall molt, like the spring molt, has individual characteristics but occurs apparently in the reverse order. Summer hairs are retained longest of all on the chest, belly, limbs and at the end of the snout (Kapitonov, 1961).

185

In young northern pikas juvenile fur initially changes to adult summer fur and then to winter; moreover the fall molt in young, as compared to adults, is somewhat delayed.

Sex and age composition of population. The sex ratio in different age groups of northern pikas is roughly similar (Table 7.6) although, according to observations in southern Yakutia (Revin, 1989), among young at the age of 1-4 months, females predominate (1:0.83); and among over-wintered— males are roughly in the same proportion. However, the actual ratio of adult males and females in a population is difficult to establish by the capture method, since in different seasons the activity of females and males may differ significantly; this is particularly apparent in the period of nursing young ones and in the mating period when males always predominate in catches.

Since over a large part of the range northern pikas have two peaks of reproductive activity, at the end of summer their populations have young of two age groups—the first and a second brood. In southern Yakutia the main increment in population occurs on account of the second brood; moreover here only the overwintered individuals participate in reproduction. By the end of summer, young constitute about half of the population. In August one-year olds constitute 37.5% of the population, and in September—44.4% (Table 7.7).

By fall the increase in the proportion of one-year-olds occurs as a result of the deaths of some juveniles. The proportion of two-year and older individuals in the populations of southern Yakutia is insignificant, and by fall decreases to 8.4% (Revin, 1989). Northern pikas do not live more than three years. In populations from the Kolymsk lowland and Yana basin, the proportion of resident individuals increases mainly on account of the first brood, and young from the second brood constitute not more than 13% of the population (Krivosheev, 1971). By fall, despite a second reproduction by some overwintered females and possible participation of juvenile individuals in reproduction, the proportion of surviving individuals in populations, as also in south of Yakutia, decreases somewhat (Table 7.6).

In conditions of southern Yakutia, where a short reproductive period is characteristic of northern pikas, the later onset of sexual maturity (nonparticipation of juvenile individuals in reproduction), and also low fecundity of females, the annual increment in population numbers is insignificant—in a majority of habitats it constitutes 200%. At such a level of reproducibility, any unfavorable conditions can cause a considerable decrease in population numbers.

Population dynamics. Observations on northern pikas in the goltsovoi [bald mountain] zone in the vicinity of Magadan showed that no significant changes in population occurred in course of three years,

Table 7.6. Sex and age composition of populations of northern pikas in different regions of Yakutia (Krivosheev, 1971)

Region	Month	Entire population		Overwintered			Surviving		
		n	Surviving %	n	Males %	Females %	n	Males %	Females %
Olekmo-Charsk upland	April-May	171	–	171	56.7	43.3	–	–	–
	June-July	131	18.3	107	55.1	44.9	–	–	–
	August-September	101	48.5	52	48.0	52.0	73	54.8	45.2
Kolymsk lowland	April-May	11	–	65	50.0	50.0	–	–	–
	June-July	21	52.4	–	–	–	–	–	–
	August	–	–	–	–	–	46	52.2	47.8
	September	79	44.2	–	–	–	–	–	–
Yana basin	June-July	26	53.9	–	–	–	–	–	–
	August	–	–	36	62.5	37.5	32	44.4	55.6
	September	42	42.9	–	–	–	–	–	–

Table 7.7. Age structure of a population of northern pikas in south of Yakutia
(Revin, 1989)

Month	n	Including, %		
		Juveniles	One-year olds	Two-year olds
April	45	–	80.0	20.0
May	131	–	84.7	15.3
June	86	3.5	80.2	16.3
July	43	51.3	40.3	8.4
August	72	52.8	37.5	9.7
September	36	47.2	44.4	8.4

136

although in July-August of the exceptionally warm year 1962 the number
of pikas was somewhat higher than in other years, and after the cold
winter of 1962/63 a decrease in their numbers was observed (Yudin et al.,
1976). No significant fluctuations were recorded in numbers of northern
pikas also in the western part of the Okhotsk-Kolymsk upland (Krivosheev,
1989). Over the observation area, animals were absent in 1983, when
numbers of northern pikas shrank over the entire territory of the Okhotsk-
Kolymsk upland, and the maximum population density was noticed in
1982 and 1984 (Fig. 33).

Considerable fluctuations in numbers of northern pikas were observed
in 1959-1960 on stony mounds in hills near the mouth of the Apuka
137 (Koryaksk upland). In August and September 1959, the number of pikas

137

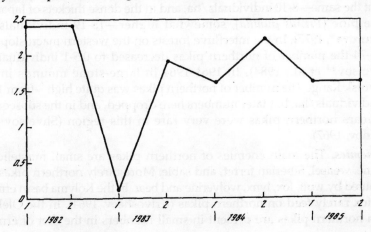

Fig. 33. Change in population density (individuals/ha) of northern pikas
in the first (1) and second (2) half of the summer of 1982-1985 in the basin
of the Kulu River (from Krivosheev and Krivosheeva, 1991).

there was 100 individuals/ha, and after the severe fall of 1959 and the subsequent snowless winter the number of northern pikas shrank sharply (Portenko et al., 1963).

Quite a high number of northern pikas was observed in September 1956 in the middle fork of the Ortoku-Ukhta River (northwest of the Verkhoyansk Range), where about 30 individuals/ha (Kapitonov, 1961) lived. A similar number of northern pikas was found in some areas in the basin of the Olekma and on the western edge of the Yukagirsk Plateau (Krivosheev, 1971). In southern Yakutia, on the Olekma-Charsk upland (near the mouth of the Charoda River) in old burned-out forests with rock slides and stony mounds in April-May 1963, the number of northern pikas was very high—50-75 individuals/ha. A still higher number of northern pikas was recorded at the end of July 1964 on rock dumps and on slopes of supraflood plain terraces in the mouth of the Sykra River—about 100 individuals/ ha. In places with weakly dissected relief or on edges of stony mounds in the basin of the Tokko River, in the middle of June 1961, the number of northern pikas was low—10 individuals/ha (Revin, 1968). On the whole, for Yakutia an insignificant fluctuation of numbers from year to year is characteristic for northern pikas (Krivosheev, 1971).

For forest and montane zones of the Baikal basin, quite a considerable fluctuation in number of northern pikas is characteristic, which occurs synchronously on different ranges. In subgoltsovoi [bald mountain] mounds of the Baikal Range above Davan Pass in August 1975, the number of northern pikas was 8-10 individuals/ha. On the Ulan-Burgassy Range on individual mounds in taiga (September 1972), the number of pikas was about the same—8-10 individuals/ha, and in the dense thickets of Japanese stone pine [*Prinus pumila*], somewhat higher—15-19 individuals/ha (Maturova*, 1977). In the interfluve forests on the western macroslope, in 1973-74 the number of northern pikas decreased to 0.5-1 individual/ha (Shvetsov** et al., 1984). In 1961-1964 in large-stone mounds in the Primorsk Range, the number of northern pikas was quite high—from 10 to 30 individuals/ha, but later numbers here dropped, and in the subsequent 3-4 years northern pikas were very rare in this region (Shvetsov and Litvinov, 1967).

Enemies. The main enemies of northern pikas are small mustelids—ermine, weasel, Siberian ferret, and sable. More rarely northern pikas are devoured by wolf, fox, lynx, wolverine and bear. In the Kolyma basin ermine and fox rarely feed on northern pikas (Krivosheev, 1964). In the Olekma
basin northern pikas are present in small numbers in the diet of ermine

---

*Not in Lit. Cit.—General Editor.
**See footnote on p. 171—General Editor.

and Siberian ferret. According to the data of V.V. Larin and T.A. Shelkovnikova (1991), on the Putoran Plateau pikas are devoured by wolves (34.7% encounters in excrements in summer, and 25.3% in winter), ermine (33.3% in summer and 25.0% in winter), sable (50.0% in summer and 44.4% in winter), and in winter, in addition, polar foxes (27.6% encounters in excrements), brown bear (5.0%), wolverines (2.3%) and lynx (50.0%). In places where northern pikas are numerous, they may be one of the main objects in the diet of sable. Thus, in the area of the Aldano-Uchursk Range, northern pikas were found in 1.0-3.7% of investigated sable stomachs, and in the basin of the Olenek River—in 10% (Krivosheev, 1971). In the upper Lena taiga in 1967-1969, encounters of northern pikas in the stomachs of caught sables, as also in their excrements, constituted from 19.3 to 50.0% (Mel'nikov and Tarasov, 1971). In eastern Kamchatka northern pikas are devoured by weasel, ermine and fox (Averin, 1948).

Among the feathered predators northern pikas are hunted by horned owl, golden eagle, hawk, rough-legged buzzard, Ural owl, merlin and skua (Romanov, 1941; Kapitonov, 1961; Portenko et al., 1963; Revin, 1968; Krivosheev, 1971).

139 *Competitors.* Some species of voles, which use winter resources of northern pikas, can be occasional competitors of northern pikas for food in different regions. In spring the hay cured by northern pikas may be consumed by marmots (Kapitonov, 1961). In the Sokhondinsk preserve a decrease in the population of long-tailed Siberian susliks is explained as the result of their trophic competition with northern pikas (Baranov, 1984). A stiffer competition exists between northern and Altai pikas in the regions of their combined habitation (cf. section "Biology of Altai Pika").

In the central Putorans, of 52 species of flowering plants, mosses, lichens, horsetails and fungi eaten by northern pikas, 39 species form the food of wild reindeer, and 16 species are utilized by bighorn sheep. However, the dynamics of utilization of habitats and plant groups in pikas, sheep and reindeer has its own peculiarities, which levels the competition (Larin and Shelkovnikova, 1987).

Diseases and parasites. Among ectoparasites of northern pikas, 10 species of fleas, 6 species of gamasids, 2 species of trombiculids and 1 species of ixodid mites, 3 species of lice (Table 7.8) have been detected. Larvae of warble flies *Oestromya fallax* have been found in pikas from Kamchatka (Averin, 1948), and in the lower reaches of the Lena (Kapitonov, 1961); *O. leporina*—in the Kolyma upland (Kishchinskii, 1969); and *Oestroderma schubini*—in the Kegali valley (Yudin et al., 1976). Sometimes northern pikas are heavily infected with larvae of warble flies: thus in August 1979, in the Sokhondinsk Preserve in the upper reaches of the Bukukun River, 60% of pikas were infested with larvae of warble flies

190

Table 7.8. Ectoparasites of northern pikas in different parts of their range

Species	Region of distribution	Author
Sucotria		
Ctenophyllus subarmatus	Tuva	Letov and Letova, 1971
C. armatus	Trans-Baikal	Ioff and Skalon, 1954
	Tuva	Letov and Letova, 1971
	Magadan region	Yudin et al., 1976
Ceratophyllus tolii	Lower reaches of the Lena	Kapitonov, 1961
	Magadan region	Yudin et al., 1976
C. advenarius	Magadan region	Yudin et al., 1976
C. pseudodahurica	Magadan region	Yudin et al., 1976
Amphalius runatus	Lower reaches of the Lena	Kapitonov, 1961
	Tuva	Letov and Letova, 1971
	Northern Urals	Gashev, 1969
Oropsylla silantiewi	Lower reaches of the Lena	Kapitonov, 1961
Rhadinopsylla altaica	Tuva	Letov and Letova, 1971
Paradoxopsyllus scorodumovi	Tuva	Letov and Letova, 1971
Wagneria tecta	Tuva	Letov and Letova, 1971
Ixodoidea		
Ixodes angustus	Magadan region	Yudin et al., 1976
Gamasoidea		
Hirstionyssus isabellinus	Magadan region	Yudin et al., 1976
	Lower reaches of the Lena	Kapitonov, 1961
Euryparasitus emarginatus	Magadan region	Yudin et al., 1976
Haemogamasus ambulans	Magadan region	Yudin et al., 1976
	Lower reaches of the Lena	Kapitonov, 1961
H. ivanovi	Kolyma valley	Krivosheev, 1971
Eulaelaps stabularis	Magadan region	Yudin et al., 1976
Hypoaspis sp.	Magadan region	Yudin et al., 1976
Trombiculidae		
Trombiculus sp.	Magadan region	Yudin et al., 1976
Leptotrombidium schlugerae	Magadan region	Yudin et al., 1976
Siphunculata	Magadan region	Yudin et al., 1976
Hoplopleurra acanthopus	Magadan region	Yudin et al., 1976
H. emarginata	Magadan region	Yudin et al., 1976
Enderleinellus sp.	Lower reaches of the Lena	Kapitonov, 1961

(Baranov, 1984) and in one specimen, caught in September 1969 in the Kegali valley, 54 larvae were detected under the skin (Yudin et al., 1976).

Round worms of the family Oxyuridae have been found in the blind gut and adjoining parts of the small intestine in northern pikas. In the Verkhoyarsk region two species of cestodes, *Schizorchis altaica* and *Taenia tenuicollis*, and two species of nematodes *Cephalurus andrejevi* and *Dermatoxys schumakovitchi* have been found (Gubanov, 1964). In pikas from the northern Urals, in addition, *Labiostomum vesicularis* and *Alveococcus multilacularis* have been detected (Gashev, 1969).

In the basins of the Kolyma and Olekma rivers, and also in the lower reaches of the Lena, the infectivity of northern pikas with helminths fluctuates from 57 to 74%; pikas inhabiting talus on the right bank of the Kolyma are quite strongly infected with endoparasites, but pikas inhabiting larch forests of flood plains are almost free of them (Kapitonov, 1961; Krivosheev, 1971).

Practical Significance

The forestry significance of northern pikas has not been studied. Northern pikas most intensively damage mountain ash, European birdcherry, alder and cedar, and to a lesser extent—willow, fir, birch and spruce. However, in some places damage to fir by northern pikas is quite extensive. In the upper reaches of the Lena, by destroying cedar seeds, northern pikas can inflict some loss to the crop (Voronov, 1964). In the forest zone of the Cis-Baikal area, according to the data of N.F. Reimers (1960), the adverse impact of northern pikas on restoration of cedar is insignificant. At the same time, northern pikas, being consumers of phytomass and an object in the diet of some carnivorous animals, play a definite positive role in the mountain ecosystems.

In Yakutia, from 1941 to 1955, there existed commerce in northern pikas. At that time from 700 to 6000 hides were processed annually (Krivosheev, 1971; Revin, 1989). At present, commerce in northern pikas has been discontinued.

Pallas's Pika
Ochotona (Pika) pallasi Gray, 1867

1867.   *Ochotona pallasi* Gray. Ann. Mag. Nat. Hist*, V. 20, p. 220. Kazakh-stan, Dzhezkazgansk region, Balkhash district.

---

*In Russian original, Yist.—General Editor.

1911. *Ochotona (Ogotona) pricei* Thomas. Ann. Mag. Nat. Hist. Vol. 8, p. 760. Mongolia, basin of Kabdo River, west of Lake Achit-Nur.

1939. *Ochotona pricei opaca* Argyropulo. Novaya forma gobiiskoi pishchukhi iz Kazakhstana [New form of Gobi pika from Kazakhstan]. Izv. Kaz. FAN SSSR, No. 1, p. 31-33.

Diagnosis

Pikas of large and medium dimensions. Condylobasal length of skull more than 40 mm. Interorbital space narrow—usually less than 4 mm. Incisor fossa not partitioned by protuberances of intermaxillary bones. Frontal bones lacking foramina.

Color of summer fur on back varies from ochreous rusty-brownish to more dull gray-brownish with a yellowish tinge. In winter the back is pale gray with a pinkish-rusty tinge or dark gray with a brown tinge. On the neck behind the ears, a dark reddish-brown and dark ochreous spot is formed by short hairs. Vibrissae are mainly dark brown, their length 60 mm. Paws whitish.

Description

Body length of adult individuals from 170 to 224 mm (average—197* mm), similar in males and females. Length of foot from 26 to 34 mm (average—29.2 mm). Length of ear from 22 to 30 mm (average—25.3 mm). Body weight in males 127 to 240 g (average—186.3 g); in females—149.5 to 307 g (average—206.5 g).

The color of summer fur on the back varies from ochreous-rusty-gray to dull grayish-brown. The head is somewhat brighter than the back. The belly is dirty whitish or ochreous-yellow. Ears are the same color as the back. A light border on the ear margin is expressed weakly (Fig. 34). Along the sides of the neck there are spots of dark reddish-brown or dark ochreous color. Paws are whitish above, sometimes with a pale grayish or ochreous tinge.

Winter fur is denser and longer than summer fur. The color of the back is pale gray with a Pinkish-rusty tinge, sometimes dark gray with a brownish tinge. Sides of the body and head in lighter individuals are lighter than the back. The belly is ochreous-whitish or ochreous-sandy. Vibrissae attain a length of 60 mm. They are dark brown in color and the longest of these have white tips. The claws are lighter, longer and sharper in winter; dark brown, relatively short and blunt—in summer.

---

*In Russian original 397—an obvious misprint—General Editor.

Fig. 34. Pallas's pika *Ochotona pallasi*. Altai, Chui steppe
(photo by V.V. Kucheruk).

Skull of moderate dimensions. The condylobasal length of the skull varies from 40 to 48.9 mm; the zygomatic width—21.9-26.0 mm; interorbital space—2.6-5.0 mm; alveolar length of upper tooth row—8.1-10.0 mm.

Skull weakly bulged. Frontal bones lack foramina. Incisor fossa is partitioned by protuberances of intermaxillary bones. The preorbital foramen is small and triangular. Nasal bones are broadened in the anterior part and narrowed in the posterior. Orbits are large and oval. The upper margin of the orbit forms a small ridge in the region of the frontal bone. Posterior zygomatic processes are short and broad. The tympanic bullae are of moderate dimensions, quite considerably projecting from the sides (Fig. 35).

The mandible is not massive, but quite high in the dental section. The articular process is high, broad, and almost not shifted back. Angular section is broad. Angular notch is oval in form and not very deep (cf. Fig. 35).

The third lower premolar tooth has a large anterior segment, which is rhombic in form, but often with a reduced antero-outer margin. The posterior segment is quite broad and short; on its inner margin sometimes there is a fold (cf. Fig. 35a).

The os penis is relatively long and thin; its length is more than 8.5 times the width (Aksenova and Smirnov, 1986).

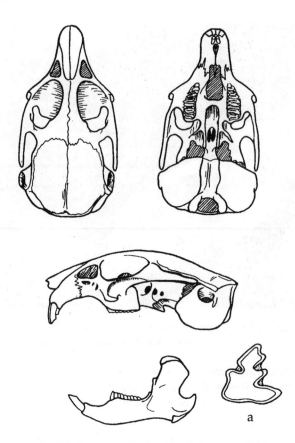

Fig. 35. Skull of Pallas's pika *Ochotona pallasi*.
a—third lower premolar tooth of Pallas's pika.

The diploid complement has 38 chromosomes: 10 pairs meta- and sub-metacentrics; and 8 pairs of subteleocentrics; X-chromosome—submetacentric, Y-chromosome—the smallest in the complement (Vorontsov and Ivanitskaya, 1973).

Systematic Position

In morphological and cytogenetic characters, Pallas's pika is close to northern and Altai pikas, and forms one subgenus with them.

Geographic Distribution

The range of Pallas's pika consists of two disjunct parts—Kazakhstan and Mongolia. The Kazakhstan part covers southern part of the Kazakh upland, northern part of Betpak-Dala and a narrow strip along the eastern Lake Balkhash area. The Mongolian part of the range of Pallas's pika includes the Chui steppe and adjoining regions of the southeastern Altai, south and west of Tuva, Mongolia and the Gobi Altai, southern Khangai and trans-Ili Gobi.

43   The Kazakhstan part of the range of Pallas's pika lies in the central and eastern parts of the Dzhekazgansk* region, east of Semipalatinsk and northeast of the Taldy-Kurgansk regions (Fig. 36).

The western distributional boundary of this species in the eastern Dzhezkazgansk region has not been adequately defined. According to the data of A.M. Belyaev (1933), habitations of Pallas's pikas are found on the northwestern edge of the Betpak-Dala and in the southern Kazakh upland—from the valley of the Sary-Su River and the Ulutau mountains to the valley of the Tokrau River. However, faunistic studies conducted in the Ulutau mountains by B.A. Kuznetsov (1948) and A.A. Sludskii (Ismagilov, 1961), did not confirm these reports. Thus, the northwestern distributional boundary of Pallas's pika apparently should be drawn along the northwestern edge of the Betpak-Dala—through the Koktas River to Mt. Munglu, later along the Shazagoi River (90 km north of the Bulatau mountains) and the Ata-Su River (Bondar', 1956; Ushakova, 1956; Ismagilov, 1961). Farther north the distributional boundary of this species

142

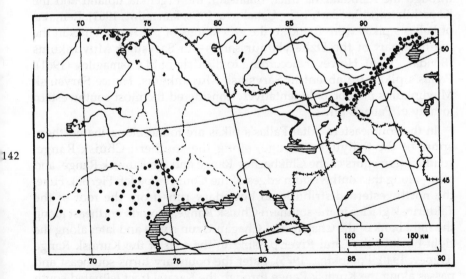

Fig. 36. Localities of Pallas's pika in Kazakhstan and Russia.

*In Russian original, Dzhezgazgansk—General Editor.

passes through the northern foothills of Kyzyltau (Ismagilov, 1961). The Zhana-Arka station (right bank of the Sary-Su) may be considered as the extreme northwestern locality of Pallas's pika in this region, and between the Sary-Su and Ata-Su rivers Pallas's pikas apparently are absent (Gvozdev, 1962).

Later the boundary turns southeast and proceeds along the Kazakh mlekosopochnik [low-hilly area] to the southern edge of the semidesert zone of the Sary-Arka. Pallas's pikas were recorded in the mountains of Kopal, Arab, Shunduk and Airtau to Mt. Kzyl-Tas (Andryushko, 1952; Ushakova, 1956; Shubin, 1959). From Mt. Kzyl-Tas (the northernmost locality of Pallas's pika in this part of the range) the boundary descends south and southeast and passes along the right bank of the Tokrau River, and from the mouth of the Dzhenishke River—along the left bank of the Tokrau [River] through the Arkarly and Baktauta mountains and farther east along the northern bank of Lake Balkhash to the Shubartau mountains (Afanas'ev and Varagushin, 1939; Andrushko, 1952). The Kyly and Arkaly tracts may be considered as the extreme eastern locality of Pallas' s pikas in the eastern Lake Balkhash area (Ismagilov, 1961).

On the whole, the southern boundary of the Kazakhstan part of the range passes along the northern bank of Lake Balkhash. From the Ishan-Koran tract in northern Betpak-Dala, the boundary turns northwest and passes through the Kalikhadzha tract, Bulat-tau, the Ergenkta upland and the Kogashik tract (Bondar', 1956; Ushakova, 1956; Ismagilov, 1961). Remains of Pallas's pikas were found in the eructations of predatory birds in the western part of the Dzhetykonur sands—in Sarysuisk Muyunkums (Bondar', 1956). However, according to the data of M.I. Ismagilov (1961), Pallas's pikas are absent in Sarysuisk Muyunkums. Hence Sarysuisk Muyunkums may be only tentatively considered the most southwestern locality of Pallas's pikas.

In the southeastern Altai Pallas's pikas are found along the northern spurs of the Sailyugem Range, along the southern Chuisk Range, southeastern spurs of the Chikhachev Range and the Kuraisk Range, and also—along the southwestern edges of the Chui steppe (cf. Fig. 36). Here, the northwestern distributional boundary runs from the foot of the Taldurinsk glacier in the southern Chuisk Range, ascends northeast along the right bank of the Chagan and Chagan-Uzun rivers and later along the right bank of the Chui River passing to the foot of the Kuraisk Range (Ognev, 1940; Potapkina, 1975). Later the boundary turns southeast and passes along the Kuraisk Range through the Karatal tract (situated south of the discharge of the Chagan-Uzun River into the Chui) and Ortalyk (Firstov, 1957; Potapkina, 1975). The boundary then proceeds to the northern edges of the Chui steppe in the region of the village of Kosh-Agach

(Ognev, 1940). Isolated small colonies of Pallas's pikas are recorded on the left bank of the Bar-Burgazy River—roughly 10 km from its emergence in the Chui steppe (Yudin et al., 1979).

From the southeastern spurs of the Chikhachev Range, the eastern distributional boundary of Pallas's pika turns west and passes from Durbe't Daba Pass along the northeastern spurs of the Sailyugem Range along the Chui tract to the village of Tashanta and the emergence of the Ulandrik River in the steppe (Lazarev, 1971; Yudin et al., 1979). The southern and southwestern distributional boundary Pallas's pika passes along the northern part of the Sailyugem Range, the spurs of which form the southern edge of the Chui steppe, and passes through the upper reaches of the Ulandrik, Chagan-Burygazy, Tarakhta and other rivers (Lazarev, 1971; Yudin et al., 1979). In the western part of the Sailyugem Range, the southern boundary passes through northern spurs of the Tagan-Bogdo Range and emerges in southern regions of the Ukok Plateau (Demin, 1960). Habitations of Pallas's pikas in this region are at the extreme southeastern limits of the Altai part of the range.

In Tuva, Pallas's pikas live in desertified stony steppes of the eastern spurs of the Altai (Mt. Mongun-Taiga and the Tsygan-Shibetu Range), in the foothills of the western and partly eastern Tannu-Ola (cf. Fig. 36). Tuva, like the southeastern Altai, is a peripheral area of the Mongolian part of the range of Pallas's pika. Here, this species enters mainly along intermontane depressions and valleys of rivers flowing from Mongolia. West of the Tuva part of the range, Pallas's pikas were recorded in the valleys of the Shara-Kharagai, Mogen-Buren, Bukhé-Murén and Tolaity rivers (Letov and Letova, 1971; Ochirov and Bashanov, 1975; Yudin et al., 1979). In the eastern Mongun-Taiginsk region, habitations of Pallas's pikas are found along slopes of the mountain massif of Mongun-Taiga, in the valley of the Mugur River, on the northern slopes of spurs of the Tsagan-Shibetu, in the upper reaches of the Barlyk River, lower reaches of the Tolailyg River and along the valley of the Kargy River (Mokeeva and Meier, 1969; Letov and Letova, 1971; Zonov and Evteev, 1972). Eastward, Pallas's pikas inhabit the southern slope of the western Tannu-Ola Range, particularly along valleys of the Sagly, Kaady-Khalsy and Khanda-gaity rivers (Vlasenko, 1954; collections of ZM MGU and ZIN). Colonies of Pallas's pika were recorded east of the Ulatai River and in the interfluve of the Kadvoi and Irbitei rivers (Vlasenko, 1954; Proskurina et al., 1985; collections of BIN). In the northeastern part of Ubsunursk basin, Pallas's pikas inhabit stony steppes in the valley of the Kholu River (Yudin et al., 1979). On southeastern slopes of the eastern Tannu-Ola, Pallas's pikas were caught in the environs of the village of Khol'-Ozhu (collections of BIN). Easternmost colonies of Pallas's pikas in the territory of Tuva are present apparently in the southwestern part of the E'rzinsk region (collections of MGU).

145 Pallas's pikas are widely distributed in Mongolia, the Mongolian and Gobi Altai, Khangai and trans-Ili Gobi (Fig. 17b) .

Geographic Variation

Two subspecies of Pallas's pika live in Russia and Kazakhstan.

1. *Ochotona pallasi pallasi* Gray 1867 (syn. *O. pricei opaca* Argyropulo, 1939)—Kazakhstan Pallas's pika.

The type subspecies is characterized by medium and small dimensions. The length of rostrum is usually less than 15 mm and the length of the orbit less than 18 mm (Table 8.1). Summer color is gray-brown with a ochreous or yellowish tinge. Winter fur is of the same tone, but lighter.

Kazakhstan Pallas's pika occurs in the southern part of the Kazakh upland, in the northern part of Betpak-Dala and in the eastern Lake Balkhash area.

Table 8.1. Skull dimensions (in mm) of adult Pallas's pikas
O. pallasi pallasi and *O. pallasi pricei*

Indices	O. p. pallasi (n = 66)			O. p. pricei (n = 11)		
	min.	M	max.	min.	M	max.
Condylobasal length	40.0	42.1	44.9	40.6	44.2	48.9
Length of rostrum	16.2	18.0	19.1	17.8	19.4	21.8
Length of upper tooth row	8.1	8.9	9.6	8.2	9.3	10.0
Length of orbit	13.1	14.2	15.0	14.0	15.3	17.0
Length of tympanic bulla	12.9	14.0	15.1	13.1	14.4	16.2
Width of tympanic bulla	10.0	11.1	12.5	9.8	11.2	12.5
Interorbital width	3.1	4.0	4.9	2.6	4.0	4.5
Zygomatic width	21.9	23.0	24.8	22.0	23.7	26.0
Width of cranium	17.5	18.9	20.0	17.0	19.7	22.0
Width in region of tympanic bullae	21.0	22.7	24.4	21.1	23.6	26.6
Postorbital width	14.0	15.0	15.9	14.0	15.0	16.5
Width of rostrum	5.2	6.3	7.4	6.0	6.6	8.0
Occipital height	11.8	12.7	13.5	12.0	13.2	14.7
Height in frontal region	12.0	12.7	13.9	12.0	13.1	14.0
Height of rostrum	6.5	7.3	8.0	6.5	7.7	8.2
Height in region of tympanic bullae	14.0	15.4	17.0	14.0	15.5	17.0

144

2. *Ochotona pallasi pricei* Thomas, 1941—Mongolian Pallas's pika.

Mongolian Pallas's pika is characterized by large dimensions. The length of the rostrum is usually more than 18.5 mm and the length of the orbits usually more than 15 mm (Table 8.1). The color of summer fur is paler than in the type subspecies; an ochreous tinge is weakly developed in the back coloration.

Mongolian Pallas's pika occurs in Tuva, in the Altai (region of the Chui steppe) and in Mongolia.

In 1912, Thomas described *Ochotona hamica* from the southeastern part of the Tien Shan (Hami, China), which Chinese authors (Feng and Zheng, 1985) consider a subspecies of Pallas's pika. These very authors have identified one more subspecies of Pallas's pika—*O. pallasi sunidica* from the territory of Inner Mongolia. The taxonomic status of this form needs verification.

Biology

Habitat. Pallas's pika prefers to colonize places where natural shelters are abundant—among extensive or local piles of stones with rich vegetation around them. They avoid damp places.

In Kazakhstan, Pallas's pikas are most frequently met with in stony habitats formed by broken outcrops of quartzites, hard limestones and some varieties of granite, alternating with areas, of softer soil cover. In the Kazakh upland, such places serve Pallas's pikas as survival habitats (Shubin, 1959). Colonies of Pallas's pikas are often met with along slopes of uplands, ridges and in ravines. Soil composition does not exert considerable influence on the number and distribution of colonies of Pallas's pikas. Colonies are found on both hard clay and strongly rubbly soils, and also on chernozems. Usually Pallas's pikas colonize places with an abundance of shrubby vegetation, often making burrows at the base of shrubs. In the northeastern part of central Kazakhstan (northern Lake Balkhash area), Pallas's pikas are common in rocks and piles of large blocks in granite massifs. Occasionally they are found in valleys with streams and rivulets; here burrows were observed under clumps of junipers, at the base of broken blocky structures and in low bluffs of clay soils (Andrushko, 1952). Pallas's pikas are rarely found along slopes of mountains with soft soil. Unique places of habitation of Pallas's pikas are the winter quarters of cattle, which are enclosures with walls made of stones, the gaps between which are shelters for pikas, and the rich vegetation growing in the manured soil serves as their food.

In high-population years Pallas's pikas can settle in places devoid of rock outcrops. However, in low-population years, and particularly in the

146 zone of high humidity, they strictly remain in stony habitats. Such places are the survival habitats for Pallas's pikas (Tarasov, 1950). In the arid zone, for example, in the area of the Bosaga station, their survival habitats are the dry ridges and slopes of hills, covered with shrubs (Smirnov, 1974).

In the southeastern Altai, habitats of Pallas's pikas are confined to rubbly finely turfaceous, mountain steppes with occasional stones or small piles of coarsely fragmented rocks on spurs of the Sailyugem and southern Chui ranges, and also to feather grass-gravely desertified steppes on the southern edges of the Chui steppe. They occupy different habitats, particularly, stony mounds and outcrops of bedrocks, and also different parts of desertified steppes. They are also met with in more humid places— along the banks of bluffs almost touching water or in willow thickets (Yudin et al., 1979).

In Tuva, Pallas's pikas occur in desertified stony steppes of the Ubsunursk basin, where they colonize predominantly rock outcrops on the southern slopes of mountains (Yudin et al., 1979). They are also met with in semi-deserts and suppressed vegetation (Yanushevich, 1952; Vlasenko, 1954) and in diverse grass, cinquefoil-wormwood-herbaceous— desertified steppes (Zonov and Evteev, 1972). Along the southern slopes of the Tannu-Ola Range, Pallas's pikas occur on granite rocks—in places with many cracks and stones (Galkina et al., 1977). In winter in Tuva, Pallas's pikas are numerous in wormwood steppe with rock outcrops and stones. Usually they are frequent in grass-herb steppes and occasionally in desertified steppes. Pallas's pikas avoid inundated glades, flood-plain forests, willow groves along ravines and stone mounds in high-mountain meadows (Zonov et al., 1983).

In the Khangai mountains in Mongolia, colonies of Pallas's pikas are confined to stony, grassy and grass-saltwort semideserts (Bannikov, 1954). They are particularly frequent in places with outcrops of bedrocks, on detritus and outliers. Such places are particularly characteristic of the northern limit of distribution of Pallas's pika in the Khangai. In the Gobi 147 Altai, Pallas's pikas are found mainly along dry streams (sairas), ravines and bottom of narrow gorges in the mountains (Bannikov, 1954). In the southern part of their range, Pallas's pikas remain in sairas—where the plant cover is richer, and do not colonize rocks and gravels. In lake basins (Uryuk-Nur, Achit-Nur), Pallas's pikas are found mostly along the plain, usually near gravels and rocks. Here they select areas with a richer vegetation. In the Mongolian Altai (for instance, hear Lake Tolba-Nur) Pallas's pikas occur among stony detritus and near rocks (Bannikov, 1954). According to the data of A.N. Formozov (1929), the most common places of habitation of Pallas's pikas in Mongolia are rocks and outliers, that have been subjected to water and wind erosion. Pikas dig burrows under such rocks. Sometimes Pallas's pikas are found also in the desert, where there are no rocks, and in this case their burrows are constructed under thickets of pea shrub. In the valley of the Bzapkhyn River, colonies of Pallas's pikas were found along sandy mounds of Sdersu and in the open gravelly places with isolated thickets of pea shrub. Daurian pikas also live here. In the Ikh-Bogdo mountains, Pallas's pikas live in the upper zone—in places inhabited by Siberian ibex and marmots.

The altitudinal distribution of Pallas's pikas in the Kazakh upland is restricted by the height of hills—pikas colonize their slopes from the foot to an altitude of 1,000 m above sea level, and sometimes even higher (Shubin, 1959). In the southeastern Altai, colonies of Pallas's pikas are encountered at altitudes from 1,700 to 2,500 m above sea level, but the optimum for them lies between 2,000 and 2,300 m (Lazarev, 1971). In Mongolia, Pallas's pikas ascend mountains to an altitude to 2,900-3,200 m; at an altitude of 3,000 m, for example, their colonies were met with on the southern slope of Ikh-Bogdo (Bannikov, 1954), and at the altitude of 3,100 m—on the Adzhi-Bogdo Range (Chugunov, 1961).

Population. Numbers of Pallas 's pikas depend on the presence of places with the shelters favorable for them, the degree of ruggedness of the terrain of the locality and food resources. The most stable and high density of habitation of pikas (10-30 individuals/ha) is found in habitats with dissected and mosaic landscape; in such places the amplitude of seasonal fluctuation of numbers is low and the effect of poor weather conditions is less perceptible. In the Kazakh upland, the highest density of colonization of Pallas's pikas (41 individuals/ha) was found on the outlier mountains, and also in habitats with outcrops of quartzite and hard limestone, and the lowest—in habitats with granite rocks (3.1 individuals/ha) and on less exposed hills (2.7).

In years with favorable weather conditions, pikas can disperse beyond the limits of rock habitats to the plain areas, where they dig burrows. For instance, in 1951-1952 in the areas of the Bosaga and Kiik stations, the number of Pallas's pikas was very high, and their colonies, which occupied a vast territory, were situated not only in rock habitats, but several kilometers from rocks and stones. In the following two years the number of pikas in this area remained high only within limits of rock habitats, and in the plains animals perished (Sludskii et al., 1980).

In Tuva, Pallas's pikas are most numerous on rocky mountain slopes—here, their density reaches 18 individuals/ha. Less densely colonized are beds of spring streams (6.7 individuals/ha), melkosopochnik [low rounded hills) (8 individuals/ha), grass-wormwood and wormwood-grass steppes (7.5 and 5.5 individuals/ha, respectively). In river flood plains, stony-rubbly steppes and on moraines, Pallas's pikas are found very rarely, and here their density does not surpass 0.9 individuals/ha (Boyarkin, 1984).

   In the mountains of Mongolia (Mongolian Altai) the maximum density of Pallas's pikas (39 individuals/ha) was recorded in wormwood associations of the intermontane valley of Mt. Chandmankhairkhan* (Table 8.2). In the Mongolian Altai the most densely colonized areas are mountain peaks. On the middle slopes the number of burrows increases, and at the foot of the mountains the number of burrows per unit area decreases again, but remains higher than on the peaks (Chugunov, 1961).

Table 8.2. Number of Pallas's pikas in different habitats in Mongolia (from Chugunov, 1961)

Range	Place of census	Habitat	Date	Number of pikas/ ha	Size of census area, ha
Taishir	10 km west of Dutindaba Pass	Middle of slope; herb-grass steppe on forest edge	23.08.1956	8.0	0.25
Khasagt	Mt. Dolan	Peak; sedge-grass vegetation	23.05.1955	4.0	1.0
	Mt. Dolangiin-nuru	Foothill; grass-wormwood association	9.09.1956	2.0	0.5
Mongolian Altai	Eastern part of the Gichige' in-nuru Range	Peak; sedge-association	27.07.1956	1.3	0.75
		Foothill; grass wormwood association	27.07.1956	16.0	0.5
	Mt. Chandman-khvirkhan	Intermontane valley; wormwood association	31.07.1956	39.0	1.0
	Valley of the Sagsaingol stream	Intermontane valley; herb-wormwood association	5.09.1956	0.57	7.0

146

148

   Shelters. Burrows of Pallas's pikas are quite complexly built, and they are located usually under stones, in rock crevices or at the base of thickets. In the southeastern Altai and southwest of Tuva, the most widespread is the diffused type of habitations, in which individual burrows are separate from each other. However, prolonged existence of colonies and an increase

---

*In Table 8.2, Chandmankhvirkhan—General Editor.

in the number of entrances may lead to uniting burrows in a continuous massif, occupying an area up to several hectares. Usually the system of tunnels of one occupied burrow covers an area from 3 to 160 m², and the number of entrances varies from 5 to 42 (on average, 18). The diameter of entrances varies greatly, and their form may be round, but often it is oval—extended in a horizontal or vertical direction; the maximum distance between the edges reaches 12 cm, the minimum—5 cm or somewhat more. In one burrow there are 4 to 12 feeding chambers with a diameter up to 40 cm. There is usually one nesting chamber; its dimensions are roughly the same as those of feeding chambers; usually, the nest is situated at a depth of about 70 cm. In those cases when colonies of Pallas's pikas are situated in rocky habitats, burrows are built more simply (Lazarev, 1974; Okunev and Zonov, 1980).

In Kazakhstan Pallas's pikas usually find shelters and build burrows in deep niches, voids and crevices in rocks, against large stones and under stone blocks. They avoid moist places. Burrows, as a rule, are of two types—temporary and permanent; in temporary burrows pikas hide themselves during sudden appearance of an enemy. Permanent burrows have from 5 to 12 entrances with a diameter of 5-9 cm (Fig. 37). Each burrow may cover an area of up to 10-15 m². Entrances are connected to each other by beaten tracks [runways] 5-6 cm wide. Tracks also lead to places used for foraging, to temporary burrows, and to neighboring colonies; their expanse sometimes reaches several tens of meters.

Almost all entrances are connected to each other by underground tunnels up to 10 cm in diameter. Usually there are many blind alleys 40-60 cm long. The shorter of these are often used as lavatories. Each permanent burrow has up to 2-3 nesting chambers 15 to 24 cm in diameter with a

148

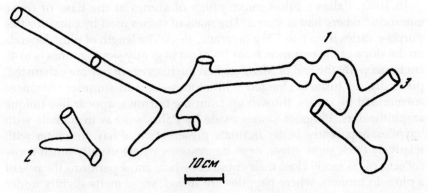

Fig. 37. Schematic drawing of a burrow (plan) of Pallas's pika in the vicinity of the Bosaga station in Kazakhstan. 1—nesting chamber, 2—branch, 3—entrance (from Shubin, 1959).

bedding of dry leaves of grasses, less often—of twigs of wormwood and other plants. The thickness of such a bedding is usually 2-4 cm (Shubin, 1959).

The depth of burrow may reach 10-20 cm in stony ground among roots of shrubs and 40-50 cm in softer ground. If the ground is high in rubble, Pallas's pikas build simple burrows—with 2-3 entrances, from which run tunnels to the nesting chamber, located usually 60-80 cm from the entrances. Sometimes burrows are very small with a single entrance and a nesting chamber. In temporary burrows, nesting chambers and lavatories are absent, and the number of entrances is low (Shubin, 1959; Smirnov, 1974).

In the Kazakh upland Pallas's pika may settle in burrows of sousliks (Smirnov, 1985), and in Mongolia (the Gobi Altai)—abandoned burrows of marmots. Here sometimes up to 81% of abandoned burrows of marmots are occupied by pikas (Chugunov, 1961). In the southeastern Altai, besides burrows of marmots, young during dispersal often occupy shelters of flat-skulled voles (Eshelkin, 1978).

Near burrows of Pallas's pikas, as a rule, there lie small heaps of dug up earth, through which deep tracks pass. If shelters are situated in stony habitats, pikas collect small stones, piling them around the burrow in heaps or ridges, the length of which may reach 1 m. Sometimes together with stones pikas deposit heaps of camel and horse dung, and also twigs of shrubs. Burrows of Pallas's pikas were found, protected from the stream flowing nearby, by a large ridge made of rubble—this was clearly a water diversion structure. Some investigators (Formozov, 1929; Bannikov, 1954) suggest, that circular ridges of stones serve Pallas's pikas to protect their burrow entrances against wind, which probably is not the primary reason.

In Tuva, Pallas's pikas make plugs of stones at the base of those entrances* where hay is stored. The mass of stones used by pikas for this purpose varies from 5 to 170 g (average, 45 g). The length of plug depends on the slope of the entrance: from 10-20 cm in gently sloping tunnels to 40 cm in steep. By the end of March, when the reserves of hay are exhausted, pikas dig up tunnels clogged with stones—and in summer entrances surrounded by stones, thrown up from the burrows, appear like unique amphitheaters. Plugs of stones made by Pallas's pikas in tunnels with haypiles, apparently, better facilitate preservation of hay. In winter with relatively thick snow cover, or in depressions where usually much snow collects, pikas rarely close their entrances—here snow performs the role of a plug. In tunnels, where haypiles are stored, snow melts slightly under

---

*In Russian original, vidov = species or types which is a misprint for vkhodov = entrances—General Editor.

the impact of the warm air passing through the burrow, and as a result a nice plug is formed (Zonov and Evteev, 1972).

In winter time Pallas's pikas often build burrows under snow, which may be located a few tens or hundreds of meters from rocks, where pikas remain for the summer. These burrows are made in depressions, where more snow collects and the feeding conditions are more favorable. Areas, devoid of snow, are avoided by Pallas's pikas in winter (Shubin, 1959).

By the end of fall usually two pikas—a male and a female—live in one burrow. Some pikas die in winter. Thus in the end of February-beginning of March, 1972, each burrow (of the 70 investigated) had 1.5-1.8 individuals; in February 1973, on average, there was 1 individual: in March 1969-1972, on average, there were 1.6 individuals per burrow (Okunev and Zonev, 1980).

Feeding. Usually Pallas's pikas feed on plants growing in the direct proximity of their colonies, hence the species composition of the plants eaten by them and stored for winter is different in different parts of the range. Moreover, individual selectivity of food has been noticed—in one and the same habitat in different haypiles there may be a predominance of different species. However, such selectivity is characteristic only of pikas living in habitats with rich plant cover. In places without abundant food, usually background species predominate in haypiles. Selectivity of foods decreases or disappears altogether in years when numbers of Pallas's pikas are increasing, when, excluding plant parts unsuitable as food, the entire vegetative stand is devoured (Sludskii et al., 1980).

In Kazakhstan Pallas's pikas feed in spring on dry grass left behind from winter, and also on the ephemers and ephemeroids growing at this time. Later they begin to eat young shoots of wormwood, grasses and herbs. In summer to these plants are added tender twigs of shrubs of spirea, winter fat, juniper, rose and others (Sludskii et al., 1980).

In the southern part of their range in Kazakhstan already in May-June Pallas's pikas begin to store food (Ismagilov, 1961), and in the central part— from June (Andryushko, 1952; Shubin, 1959). In places with rich vegetation, pikas begin storing food later. In the mountains of southwest Tuva and in the Gobi Altai, storage begins in the end of July-beginning of August, but already in the beginning of July it is possible here to find small haypiles of grasses not far from burrows (Chugunov, 1961; Okunev and Zonev, 1980). The haying period extends usually for 1-2 months.

Usually the current year's young are the first to start making haypiles. In stony habitats tufts of grasses are stored in crevices or niches. In areas of level steppe, the plants are dried near the entrance to the burrows; some plants are placed in the burrow, and some on the surface. Sometimes such haypiles are in the shape of elongated (to 80 cm) piles, the height of which may reach 35 cm. In windy weather Pallas's pikas place small stones, soil clods, twigs, cow dung and horse dung on the open haypiles. When the wind dies out, pikas remove some of the stones and a new layer of plants is arranged only after the previous one becomes dry (Andrushko*, 1951). In Tuva pikas usually push the cured hay into the tunnels of burrows and special chambers (Okunev and Zonov, 1980).

The mass of hay in the haypiles of Pallas's pikas inhabiting the Kazakh mlekosopochnik [low rounded hill] areas varies from 200 g to 20 kg (Shubin, 1959). In the southern Kazakhstan part of the range, and in habitats with poor vegetation, the mass of haypiles, on average, is more than in places with rich vegetation and in the north of the range. In the Kazakh upland in habitats with outcrops of quartzites and hard limestones, and also on relictual small mountains (with a population density of pikas at 41 individuals/ha), there were 28 to 30 haypiles per hectare, and the total mass of hay in them ranged from 89 to 111 kg. In places with granite rocks, where the population density of pikas was low (3.1 individuals/ha), there were 2 piles per hectare with a total mass of 2.6 [kg]. A similar number of haypiles with a mass of 5.2 kg was recorded on less exposed hills where the population density of pikas constituted 2.7 individuals/ha (Sludskii et al., 1980). In Tuva the haypile mass of Pallas's pikas varied from 1.8 to 14 kg per burrow (Okunev and Zonov, 1980). Foraging individuals or pikas living amidst rich herb cover do not make haypiles.

Seasonal change of species of plants eaten and stored is a characteristic of Pallas's pikas. Thus, in Kazakhstan giant fennels and other fast withering plants are found in cured haypiles and in their diet only in the first half of summer. In fall (from September) Pallas's pikas store capitate wormwood and Bukhtarminsk libanotis (Shubin, 1959). In the Altai, initially, herbs are stored, followed by dragonhead, and the upper layer of haypiles predominantly consists of wormwoods (Yudin et al., 1979).

The species composition of plants in haypiles of Pallas's pikas, and also their summer diet generally, depends on the vegetation which is around their shelters. In Kazakhstan from 7 to 66 species of plants have been recorded in haypiles; in each haypile not more than 9 species predominate. A detailed list of plants stored by Pallas's pikas is available in the works of A.N. Formozov (1929), A.V. Afanas'ev and P.S. Varagushin (1939), P.P. Tarasov (1950), D. Tsybegmit (1950), A.N. Andrushko (1952), A.G. Bannikov (1954), and I.G. Shubin (1959).

In winter Pallas's pikas feed mainly on the hay from their stores, but in Tuva and in the Altai often underground parts of plants are used for food, uprooting them from beyond limits of their burrows (Zonov, 1974; Okunev and Zonov, 1960). Thus in the Kosh-Agachsk district (southeastern Altai) in 1969-1971, pikas fed on hay and dry grass throughout winter, and from January started digging out underground parts of plants—tubers of prostrate summer cypress, bulbs of bear's garlic and small garlic, roots of wormwood, blue grass and sedge. In addition, they actively collect lichens on the soil surface (Derevshchikov, 1971).

In has been recorded, that, in spring after a winter with favorable weather conditions, plumpness of Pallas's pikas and their mass is not lower than in fall, whereas during severe winters pikas loose weight, even when adequate food was stored (Sludskii et al., 1980).

151 In captivity during cage rearing, adult individuals of Pallas's pikas consume up to 150 g of green grass or 31-40 g of hay in a day (Derevshchikov, 1971; Sludskii et al., 1980). Animal food was not found in stomachs of Pallas's pikas.

Since Pallas's pikas most often live in places lacking open water, they evidently are satisfied with the moisture contained in the food consumed by them.

Activity and behavior. The nature of daily activity of Pallas's pikas changes according to the season and depends on the weather conditions and number of individuals. It is also possible that their activity has its own peculiarities in different parts of the range.

In Kazakhstan the summer activity of Pallas's pikas is apparently round-the-clock in nature. Thus, in the Kazakh upland these pikas were repeatedly found at night or their calls were heard (Afanasev and Varagushin, 1939; Sludskii et al., 1980). The nocturnal activity of Pallas's pika is confirmed by their remnants in the pellets of eagle owls (Shubin, 1962). Possibly, the nocturnal activity of Pallas's pikas increases at the end of summer and during the period of dispersal of young, and also in years of high numbers. Thus, in August 1970, when the number of Pallas's pikas south-east of the Altai was quite high, in the central part of the Chui steppe, a young female was caught in the night; more so it was at least 10 km away from the nearest colony of this species (E.Yu. Ivanitskaya).

Pallas's pikas living in the Gobi Altai have a polyphasial activity in summer with a peak of activity at 17.00-18.00 hr and the lowest activity between 16.00 and 17.00 hr (Chugunov, 1961). According to the data of A.G. Bannikov (1954) and P.P. Tarasov (1950), Pallas's pikas are active only during daylight.

During the mating period the activity and mobility of Pallas's pikas increases sharply. They can be seen at any time of the day. On termination of mating the activity of Pallas's pikas decreases, but they are found outside the burrows throughout the daylight part of the day. They feed in the morning and evening, and in midday often lie on stones warming under the sun. The duration of such rest may constitute from 1 to 2 hr. During the period of curing food, the activity of Pallas's pikas has two peaks—at 08.00-10.00 hr (maximum activity) and 18.00-20.00 hr., and in the hot time of the day their activity decreases sharply. As soon as the vegetation dies out and storing food ceases, the activity of pikas decreases outside the shelters. In fall they come to the surface at any time of the day, but for short periods. With the onset of considerable cooling, Pallas's pikas come out of shelters only on warm sunny days (Okunev, 1975).

In winter in Tuva, where there are larger reserves of hay in the burrows, Pallas's pikas use every non-windy day to feed at the ground surface. On the southern slopes of mountains almost devoid of snow, pikas appear from the burrows 20-30 min after sunrise. Two peaks have been observed in their activity—from 10.00 to 13.00 hr and from 16.00 to 18.00 hr. About 10-20 min before sunset these pikas stop exiting from the burrows. Thus, in winter the main factor influencing the outing time of Pallas's pikas on the surface is the duration of day light. On shady slopes the activity of pikas outside the shelters is very low and the peak activity is at 14.00 hr— the warmest time of the day. In the presence of firm and thick snow cover, pikas feed only on their haypiles, and then even on the southern slopes of the mountains their maximum activity on the surface is at 14.00 hr. In winter, even in good weather, pikas do not venture out of the burrows farther than 100 m. Snowfall in the absence of wind almost does not affect their activity (Okunev, 1971; Zonov and Evteev, 1972; Zonov, 1974).

In *O. p. pallasi* females outside their defended territories do not fight with each other upon chance meetings. On the defended territory females are aggressive to each other, but indifferent in relation to males. A male always drives out females from his defended territory, but does not touch them at the periphery of his individual area. In Pallas's pikas aggressive behavior of the type subspecies is manifest in the form of chasing, pursuit, sometimes quite persistent; however rarely is it accompanied by pushes and bites.

Members of the other subspecies, *O. p. pricei*, are extremely aggressive in relation to each other, which is manifest in persistent chases with pushes and bites, and prolonged boxing. All pikas are aggressive when meeting a neighbor in their own territory or near it (Proskurina et al., 1985).

The aggressiveness of pikas increases in summer during the period of dispersal of young and decreases by autumn with the stabilization of boundaries of their individual areas. In southwestern Tuva, the elements of aggression appear in Pallas's pikas also during the period of storing food, when pikas, irrespective of their sex and age, chase away encroachers from their areas. The maximum aggressiveness is shown to individuals living 50 m or more from their areas. During fights pikas stand on their hind feet and hit each other on the head with their fore feet. There are no serious injuries due to this behavior. Fights continue for 3 to 15 sec and, as a rule, the host of the territory chases away the intruder (Tarasov, 1959; Okunev and Zonov, 1980). The aggessive behavior of Pallas's pikas may bear a ritualistic character and is manifest as a demonstration of a threating pose at the border of an ndividual's territory. This form of aggressive behavior is manifest usually in colonies with established spatial structure (Deviz et al., 1983).

Behavior of solitary Pallas's pikas differ from behavior of those from large colonies. They lead a more cryptic mode of life, do not leave the burrows farther than 10-15 m and rarely squeek (Sludskii et al., 1980).

According to the data of M.I. Ismagilov (1961), Pallas's pikas are monogamous. However, later studies, particularly in the territory of Kazakhstan, testify, to the polygamy in Pallas's pikas (Sludskii et al., 1980). Males live separate from females and do not take part in rearing the young. According to the observation of P.K. Smirnov (1972), females do not allow males to approach their offspring.

Before parturition females make a pile of grass or hay 20-25 cm high and 30-35 cm in diam., at the top of which they make a depression and line it with softer plants. The diameter of this nest reaches 12 cm, its depth—from 10 to 15 cm. When young are born, females cover them on top with a layer of grass. If the air temperature is low, the grass layer is 5-6 cm, and if it is fairly warm, then the thickness of this layer is not more than 1-2 cm. In cold weather babies remain huddled together, and when it is hot—they lie stretched separate from each other. Females usually remain away from their litters and approach the nest only at the time of feeding. Usually the female feeds babies at intervals of 3-5 hr, each feeding lasting 5-10 min. Females periodically change the bedding and remove excreta of babies. If babies crawl out of the nest, she pushes them back. The duration of lactation is about 18 days, and if second brood appears, then the period is 2-3 days less (Smirnov, 1972).

In Pallas's pikas vocal signals are well developed. In the mating period, in Kazakhstan, it is possible to hear courtship songs of females, consisting of two parts: first—short, interrupted and piercing "pi-ik", and second—issued after a short pause of 1-2 sec, sounding the same note as "pi-ik", but vibrating and relatively prolonged (5-6 sec) trill (Smirnov, 1972). In other species of pika and in Pallas's pikas from the Altai, the female song was not heard during the reproductive period. The courtship songs of males of Pallas's pikas can be heard already by the beginning of February, and of females—even earlier. Trills are produced in the day and the first half of night. At the end of summer and beginning of fall, it is possible to hear only rare songs of animals in which the reproductive period is extended. During singing animals take a characteristic pose, in which the front part of the body is stretched upward almost vertically, the head is held back, and the open mouth acquires a round form. After fertilization females stop singing (Smirnov, 1972, 1976, 1988).

In Pallas's pikas inhabiting Tuva, a unique vocal signal has been recorded—the trill. It plays the role of territorial marking. Trills are produced both by females and males only on the defended territory. In August 1974, 70 of 186 recorded trills, that is, 39%, were related to appearance of another individual and accompanied by antagonistic behavior. In young Pallas's pikas such a trill appears only after dispersal and occupation of a specific burrow (N.A. Formozov, in press).

Area of habitation. The dimensions of individual territories of Pallas's pikas depend on the type of habitat, time of the year and number of individuals. Individual territories in *O. pallasi pallasi*, living in habitats rich in vegetation, reach 600-1,300 m², and in females (average 800 m², n = 9) and 4,200-5,200 m² in males (n = 2). The inner defended area of the individual territory in females is 20-25%, and in males—5%, which in both cases is about 200 m². The territory of one male may overlap the areas of many males, and the territories of females only partially overlap each other; the defended territories do not overlap (Proskurina et al., 1985).

A different character of distribution of individual areas was observed in *O. pallasi pricei* inhabiting habitats less rich in vegetation. The dimensions of the individual territories of males and females are roughly similar. In diffused colonies in the southeastern Altai, their area varies from 370 to 720 m², constituting on average of 525 m² (Eshekin, 1978), and in Tuva somewhat less—from 100 to 600 m² (n = 8), average—230 m² (Proskurina et al., 1985). The entire area of the territory is defended by the host, and territories themselves do not overlap (Fig. 38). In those cases when there is overlap (7% of all investigated territories), the overlapping areas are the place of constant fights between neighbors (Proskurina et al., 1985).

In winter the areas of individual territories of Pallas's pikas are determined by the distances between shelters and places suitable for foraging, and constitute from 42 to 306 m² (average—128.7 m²). The maximum distance, to which Pallas's pikas leave the burrow in winter, extends 60 m. During periods of deep depression of a population, Pallas's pikas decrease the area of their individual territory (Okunev and Zonov, 1980).

Pallas's pikas mark their territories with secretions of specific skin glands located on the sides of the neck behind their cheeks (Tarasov, 1958). Active marking of territories begins in June-July, and by August the entire territory of a colony becomes divided into areas, the boundaries of which are marked not only with secretion from glands, but also with specially

153

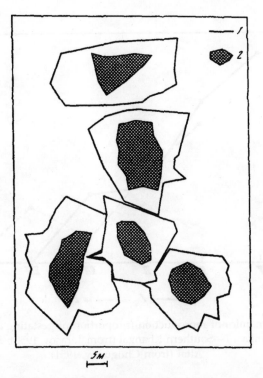

Fig. 38. Schematic drawing of the distribution of individual areas of Pallas's pikas in the southern Altai. *1*—boundary of the area, *2*—territory, occupied by burrows (from Eshelkin, 1978).

made pyramids of their excreta. These pyramids from 1 to 7 cm high are made by pikas on the highest places along the periphery of their territory. If there is not suitable elevation, pikas make special earthen heaps, that are also marked. When the dung serving as a marker dries, pikas throw it away and make a new heap at the same place (Eshelkin 1978; Zonov et al., 1989).

The grown young of Pallas's pikas relocate from the parental burrow sometimes to considerable distances. According to the observations in Kazakhstan, dispersal of young has been observed to 8-10 km. The current year's young of last litters are forced to disperse to especially greater distances (Sludskii et al., 1980). During dispersal Pallas's pikas may swim across rivers 15 m wide (Boyarkin, 1984).

Reproduction. In Kazakhstan reproduction starts in the first decade of April. On the southern slopes of hills, reproduction begins 10-15 days earlier than on the northern slopes (Sludskii et al., 1980). In Tuva according to some data (Ustyuzhin, 1971), mating starts in February, and according

212

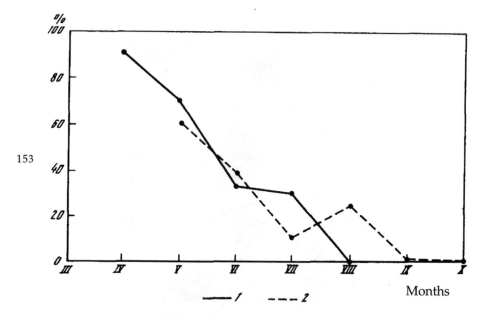

Fig. 39. Dynamics of reproduction (proportion of gestating females in %) of Pallas's pikas. 1—Southern Khangai (from Tarasov, 1950); 2—Mongolian Altai (from Chugunov, 1951).

to another (Boyarkin, 1984)—in April. In Khangai the start of reproduction is extended, but usually the reproductive cycle starts in the first decade of April (Tarasov, 1950).

In Kazakhstan the reproductive period lasts 2-3 months and terminates in May-June; in Tuva—in the first decade of August (Ustyuzhin, 1971) or in July (Boyarkin, 1984). In Mongolia the reproductive period of Pallas's pikas is extended and terminates in some parts of the range in July, and in others—in September (Fig. 39).

During the reproductive period females of Pallas's pikas bear up to 2-3 broods. In the Mugun-Taiginsk district of Tuva, first gestation in females occurs in April-May, the second terminates in June, the third in July, and in individual pikas—in the first decade of August (Krylova, 1974). West of Tuva a third brood is not reported. Many females (60%) bear two broods each, and about 11% of females do not participate in reproduction (Boyarkin, 1984). In the Kazakh upland the intensity of reproduction is very high in Pallas's pikas. All over-wintered females bear two broods each, and the presence of a third brood depends on the weather conditions and character of the habitat (Sludskii et al., 1980). In captivity females of Pallas's pikas usually bear two broods each, but sometimes, three (Smirnov, 1972).

The current year's males do not participate in reproduction in the first summer. Females of the first brood begin to reproduce at the age of 1-1.5 months (Lazarov, 1968; Krylova, 1974). According to the data of Yu.D. Chugunov (1961), in the Mongolian Altai females of Pallas's pikas participate in reproduction, not having attained the body mass of adult individuals: of 455 females investigated, in 0.2% of the confined and 0.4% of the gestating females body weight was less than 50 g, and 0.65% of the confined and 1.3% of the gestating weighed not more than 100 g. In Tuva the current year's females may already bear offspring in July (Table 8.3), although their participation in reproduction was not observed each year (Krylova, 1974).

In central Kazakhstan the number of young in a litter varies from 2 to 13 (average 8), fecundity of pikas decreases somewhat toward the end of the season: 7 females, gestating for the second time, had, on average 9 embryos each, and in females gestating for the third time—from 5.7 to 6.3 (Sludskii et al., 1980). According to observations in Tuva, in females of Pallas's pikas at the age of 1 year and 3 years the number of embryos varied from 4 to 7, and in two-years old—from 3 to 8 (Krylova, 1974). In the current year's individuals and females having attained the age of 3 years, the number of embryos, on average, was lower than in females surviving one winter (Tables 8.3 and 8.4).

In Khangai, old females had, on average, 6 embryos each, and young—4.8. Fewer embryos are observed in April (average 4.1), which apparently is linked to the insufficiency of food. In May the number of embryos reaches 7, then it decreases again—in June—to 6, and in July—to 5.8 (Tarasov, 1950). In the Mongolian Altai the number of embryos in Pallas's pikas varies from 2 to 10 (Table 8.5).

Growth and development. The duration of pregnancy in Pallas's pikas is roughly 25-27 days (Shubin, 1956; Smirnov, 1972).

Babies are born naked, blind, with short (about 2 mm) vibrissae. The weight of newborns is 8-9 g. Incisors are well-developed, digits are separated, and claws are quite long. On the second day of life a short hair coat appears, on the 5th-6th day auditory meatuses open, and on the 9th-10th day the eyes open; by this time the body weight doubles, molar teeth are cut, the growth of the tympanic bullae intensifies, and the hair coat becomes luxuriant.

No sooner than the babies begin to see, they become highly mobile. By the end of lactation young change over to feeding on plants. By this time their body weight reaches 50-60 g. The growth of juvenile pelage terminates. One and one-half to two months after birth, juveniles attain the size and weight of adult animals (Smirnov, 1972).

Molt. In Kazakhstan the spring molt in the majority of Pallas's pikas begins in May and continues until July. In males the molt begins 10-15

214

Table 8.3. Reproductive indices of female Pallas's pikas in Tuva
(Kara-Bel'dyr River, June, 1973) (from Krylova, 1974).

Age	Number of females			Number of embryos	
	Total	Gestating, %	Confined, %	min.-max.	Average
Current year's	53	24.5	1.9	3-7	5.0
Adults	55	73.7	25.5	3-10	6.5

Table 8.4. Fecundity of Pallas's pikas of different age in the Mugun Taiginsk
district of Tuva in July-August 1972 (from Krylova, 1975)

Age of females, years	Total females	Embryos, average	Placental spots, average
1	12	5.2	5.5
2	13	6.1	6.3
3	6	5.5	6.0

Table 8.5. Number of embryos in Pallas's pikas in different parts of the range

Place of observation	Number of females with number of embryos									
	1	2	3	4	5	6	7	8	9	10
Southern Khangai (Tarasov, 1950)	2	1	14	39	62	79	35	16	5	–
Mongolian Altai (Chugunov, 1961)	–	8	12	58	91	81	70	20	21	3

days earlier than in females, and occurs more intensively, terminating 20-30 days earlier than in females. The duration of the molt is influenced by weather conditions. For example, in 1953 and 1960, under conditions of early and mild spring, molting terminated in June, that is, a month earlier than usual (Shubin, 1972).

As a rule, the spring molt starts from the head and the anterior part of back, followed by the sides and the posterior part of body. The molt progresses in patches. Tufts of long gray hairs, retained particularly longer in the region of the sacrum, are retained together with the already emerged short brown summer fur (Shubin, 1972).

The fall molt starts from the middle of the back, proceeds to the anterior part of back, head, belly and the sacrum; limbs are the last to molt. In 1959, in Kazakhstan, in individual pikas, molting was observed in July, and by August. Pallas's pikas had a winter hair coat on their backs. By the end of August-beginning of September, 2-3 weeks after the termination of fall molt, growth of supplementary hairs was noticed, that occurred in the same sequence as the fall molt. Possibly, the appearance of supplementary hairs was caused by severe weather conditions (Kryltsov, 1962; Shubin, 1972).

Molting in the current year's individuals starts at the age of 25-30 days. Luxuriant and relatively long gray color of juvenile fur begins to fade gradually, being replaced by the adult summer hair coat. The change of juvenile fur occurs for 10-20 days. The back molts first and is followed by the sides and the belly. In some babies the juvenile pelage grows faster in the lower part of the sides and on the belly, and in the majority it appears simultaneously on all body parts. The undergrowth of new hairs in the juvenile molt in Pallas's pikas starts sometimes even before the complete formation of juvenile fur. Hence in the youngest individuals during the undergrowth of primary hairs on sides and back starts the growth of secondary hairs in the lower part of the sides (Kryltsov, 1962). Termination of the juvenile molt occurs in individuals having attained a weight of 90-120 g. With a body weight of 130-150 g, when young pikas externally differ little from adults, a new change of pelage begins in them. It occurs usually in the hottest period of summer. New hairs appear first on the back and sides, but shedding of old hairs is not observed. Possibly, at this time occurs only the undergrowth of supplementary hair cover (Shubin, 1972).

157

Sex and age composition of population. In Kazakhstan females usually predominate in all age groups of Pallas's pikas (Ismagilov*, 1962; Sludskii et al., 1980). However, in 1959 with high numbers, males were predominant both among adults and young animals (Ismagilov, 1961). According to the observations in Gornoi Altai, where 6,929 pikas were caught during 1962-1968, females (60%) predominated among young, but with age the sex ratio became close to 1:1, apparently on account of high mortality of young females (Lazarov, 1968). In Tuva, predominance of males was observed among embryos. In young, having begun to lead an independent life, the sex ratio levelled, and among adults females begin to predominate (Tables 8.6 and 8.7). Predominance of females (up to 58.9%, n = 1,311) in populations of Pallas's pikas was recorded also in Khangai (Tarasov, 1950).

In spring during the period of reproduction, the population of Pallas's pikas consists predominantly of overwintered animals, two- and three-year-old animals constitute the smallest proportion. According to the data of Y.A. Ustyzhin (1971), in Tuva until the start of reproduction, pikas 7-10 months of age constituted 25.5% of the population, one-year-olds—48.5, 2-year-olds and older—26%. In the western part of Tuva the current year's individuals begin to be encountered only from June, and by August their proportion in the population increases to 60% (Table 8.7). According to the data of T.V. Krylova** (1973), in the Mugun-Taiginsk district of Tuva,

*Not in Lit. Cit.—General Editor.
**In Russian original, T.A. Krylova in Lit. Cit.—General Editor.

Table 8.6. Proportion of males in Tuva populations of Pallas's pikas
in different age groups (from Krylova, 1973, 1974)

Region of study	Proportion of males, %				
	Embryos	Current year's individuals	1 year old	2 years old	3 years old
Mugun-Taiginsk district (June-August, 1972)	–	54.0	42.0	36.0	14.0
Kara-Bel'dyr River (June, 1973)	53.0	50.0	40.0	34.0	36.0

Table 8.7. Sex and age composition of populations of Pallas's pikas
in the western part of Tuva in 1975 (Boyarkin, 1984)

Month	Total number caught	Adult				Underyearlings			
		Absolute	%	Males, %	Females, %	Absolute	%	Males, %	Females, %
May	28	28	100.0	43.5	56.5	–	–	–	–
June	99	71	71.8	40.3	59.7	28	28.2	57.0	43.0
July	47	29	61.7	37.8	62.2	11	30.3	53.5	46.5
August	20	8	40.0	37.5	62.5	12	60.0	50.0	50.0

the current year's individuals predominate from June. In Kazakhstan (vicinity of the Bosaga station) already in May, resident animals in the population of Pallas's pikas constituted 68%, by the end of June their proportion increased to 71.1%, and at the end of the reproductive season the current year's individuals constituted from 87 to 92% of the population (Shubin, 1966).

Mortality of the juveniles is not high in the first month of their life. Death of a large part of Pallas's pikas occurs by one year of age. Thus, during analysis of jaws of Pallas's pikas from pellets of eagle owls in Tuva, it was established that this predator preys mainly on year-old animals (46.4%); the proportion of 4-year-old animals in pellets is relatively high— 2.2% (Krylova, 1973).

Population dynamics. The population of Pallas's pikas in some parts of the range has been subjected to considerable fluctuations—it changes both year-to-year and season-to-season. In Kazakhstan in mid-summer their population may increase 10-12 times, reaching 40-60 individuals/ha. By spring the population of Pallas's pikas decreases sharply (Shubin, 1958). The level of fluctuation in populations of Pallas's pikas over a period of several years may reach 10 times and sometimes 100 times the value. Thus, for example, the number of Pallas's pikas in the vicinity of the Bosaga station in 1960 was 200-400 times lower than in 1959; in 1962 here pikas were abundant, but in 1963 their numbers again dropped sharply. In 1967, in this region in July not even one pika could be found, but in 1972 here they were again abundant. With a quite high density there may be up to 30-40 animals per ha (Bondar', 1956; Sludskii et al., 1980).

In the southeastern Altai seasonal fluctuations in populations of Pallas's pikas can attain 6-10 fold values, and large multi-year fluctuations in population were not observed here (Derevshchikov, 1975). For example, in May 1965, the density of occupied burrows of Pallas's pikas in the area of Bol'shoi Kochkar-Bas was 8.3/ha, and in middle fork of the B[olshaya] Shibet River—5.5. In May 1966, the density of burrows in these places was 9.3 and 5.0, respectively (Eshelkin and Purtov, 1971). In neighboring regions of the plague focus, during the time of the epizootic of 1965, the density of occupied burrows at the end of August was 3-8/ha, but by November it was reduced to 2 burrows/ha. In these regions a decrease in population of pikas was noticed in winter of 1966/67, which occurred as a result of strong frosts and compact snow cover; in subsequent years (1968-69), a decrease in population occurred because of an almost total absence of snow cover and lower winter temperatures (Eshelkin and Purtov, 1971). On the spurs of the Sailyugem Range in the winter of 1959-60, a maximum mortality of these animals was recorded in places where the snow cover exceeded 20 cm. By spring about 2% of living colonies survived here. Most of all pikas survived in those places where the depth of snow was, on average, 5 cm—here pikas survived in 90% of colonies.

In Tuva, from 20 years of observations (from 1951 to 1971) in the territory of the Mugun-Taiginsk district, a low population of Pallas's pikas was recorded in 1951, 1956, 1961, and 1968, and high—in 1953, 1959, 1966, and 1970. The rate of population growth, in comparison with the fall, was faster. The duration of the decline in population varied from two to three years. The peak population in 1953 had a steeper peak than in 1959, and maintained itself for three years. After a sharp fall in population in 1968 as a result of the epizootic, by 1970 the population was restored to a high level (Ustyuzhin, 1972). In southwest Tuva from 1976, a sharp fall was recorded in the population of Pallas's pikas. In uniform open landscapes at altitudes of 1,800-2,300 m above sea level until the beginning of 1970s, their population fluctuated from 10 to 30 individuals/ha. In the winter of 1980 the population of Pallas's pikas here dropped sharply to 0.3/ha, and at many places pikas disappeared altogether. At an altitude of 1,900-2,100 m in highly dissected and mossaic terrains, Pallas's pikas survived; here their density was 3 to 12 individuals/ha (Zonov and Okunev, 1991).

Enemies. All small and medium-sized predators inhabiting alongside them are enemies of Pallas's pikas. Pallas's pikas are most frequently devoured by foxes, corsacks and steppe polecats, and also by steppe eagles, pale harrier and eagle owls (Shubin, 1962a). In Kazakhstan the most dangerous for them are foxes and eagle owls. Pallas's pikas constitute 5.1-6.9% of the diet (in % of the number of excrements or pellets investigated) of steppe polecat, 1.5-2%—in corsacks, 13.4-20%—in foxes, 2.8-4.3%—in pale harrier, 8-20%—in steppe eagle, and 38.5-56.2%—in eagle owls (Shubin, 1962a). In Tuva in July 118 remnants of Pallas's pikas were removed in 10 days from the nest of a black kite, of which 107 were juveniles (Ustyuzhin, 1971).

Competitors. In different parts of the range different species of rodents live in the nests of Pallas's pikas. Thus in the zone of montane steppes of the Mongolian Altai in 1965, long-tailed Siberian sousliks, clawed jirds, Eversmann's hamster, Siberian jerboas, thick-tailed three-toed jerboas, northern three-toed jerboas, Altai high-mountain voles lived in the burrows of Pallas's pikas. Pallas's pikas themselves often occupy burrows of marmots (Chugunov, 1961). In the northern Lake Balkhash area, gray voles live alongside Pallas's pikas, and in the Altai—flat-skulled voles.

In the Altai, in Mongolia and Tuva, the range of Pallas's pikas overlaps with the range of the Daurian pika. However, in years with a low population, their colonies are confined to different habitats—Pallas's pikas prefer drier habitats and their colonies are most often situated on mountain slopes, and Daurian pikas live at the foot of the mountains and in depressions overgrown with shrub thickets. In years of their high population, Pallas's pikas and Daurian pikas can form mixed colonies (Derviz and Proskurina, 1983). In such colonies aggressive interspecific contacts are observed; most often Pallas's pikas are the initiators of fights. Interspecific aggression of Pallas's pikas is manifest in the form of a chase with pushes, bites and "boxing." With this, noticeable changes do not occur in the distribution of territories of Pallas's and Daurian pikas—Daurian pikas continue to visit the territory, from which they are regularly chased away by Pallas's pikas, and they, in turn, continue to run over territories defended by Daurian pikas.

In Kazakhstan (the northern Lake Balkhash area), little pikas live alongside Pallas's pikas. On Mt. Bek-tau-Ata, for example, both species live at the very same altitude, but Pallas's pikas colonize crevices in rocks or dig burrows under individual stones, and little pikas live exclusively in thickets of pea shrub and other shrubs (E.Yu. Ivanitskaya). In the vicinity of the Bosaga station, little pikas penetrate slopes of hills and limits of dry ravines, which are usually colonized by Pallas's pikas, only in areas with wet ground and shrub thickets. In winter more intimate contacts between these species are observed in places, where more snow accumulates. Acute competitive interactions of Pallas's pikas with little pikas appear only in years of high population. But in Pallas's pikas, given their greater aggressiveness, apparently the process of dispersal beyond places of permanent residence is more successful (Smirnov, 1974).

Diseases and Parasites. In the Kazakh upland, larvae of warble flies—*Oestromyia leporina,* parasitize Pallas's pikas in large numbers (Grunin, 1962; Shubin, 1958).

In Pallas's pikas 7 species of helminths have been found: *Schizorchis altaica, Diuterinotaenia spasskyi, Cephalurus andrejevi, Dermatoxys schumakovitsci, Labiostomum vesicularis, Nematodirus aspinosus, Trichocephalus* sp. (Gvozdev, 1962). In addition, 4 species of coccids are found in them: *Eimeria kriygsmanni, E. erschovi, E. musculi, E.* sp. (Svanbaev, 1958) and blood parasites belonging to genus *Grahamella.*

In Pallas's pikas ixodid mites, larve and nymphs of *Dermacentor marginatus, Rhipicephalus rossicus, R. pumilio* are found in large numbers; occasionally on them are found *R. schulzei, Haemophysalis warburtoni* (Ushakova and Busalaeva, 1962; Sludskii et al., 1980). One species of gammasid mite—*Allodermanyssus sanguineus* has been found; also reported are 14 species of fleas, of which most numerous and specific are: *Amphalius runatus, Ceratophyllus desertus* and *Ctenophyllus bondari* (Mikulin, 1956). In Tuva 14 species of fleas have been found on Pallas's pikas (Ustyuzhina and Ustyuzhin, 1971). Of these, most numerous are: *Ctenophyllus hirticrus, Amphalius runatus* and *Paramonopsyllus scalonae.* According to other data (Letov and Letova, 1971), the number of species and subspecies of fleas parasitizing Pallas's pikas in Tuva reaches 26: *Ctenophysslus scaloni, Echidnophaga oschanini, Paradoxopsyllus dashidorzhii, Rhadinopsylla daurica sila, Fron topsylla elatoides longa, Ceratophyllus hirticrus, Amphalius runatus, Frontopsylla primaris, Neopsylla mana, Ceratophyllus tesquorum altaicus, Cilaeviceps ellobii, Paradoxipsyllus scorodumovi, Phadiropsylla li transbaicalica, R. altaica, R. rotschildi, Pestinoctenus pavlovskii, Oropslylla asiatica, Shfetopsylla* sp. and others.

Pallas's pikas are vectors of plague (Demina et al., 1961; Vashenok, 1962; and others). Infectivity of Pallas's pikas with plague has been confirmed for the Mongolian part of the range (Nekipelov, 1959a). Despite high susceptibility of Pallas's pikas to plague, extensive epizootics are rare in them. Investigations conducted in the territory of Gorntoi Altai point to the possibility of a prolonged harboring of the plague microbe by populations of Pallas's pikas (Bondarenko et al., 1971).

In Kazakhstan, Pallas's pikas are vectors of trypanosomiasis specific to them, caused by *Trypanosoma ochotona* and *T.* sp. (Galuzo and Novinskaya, 1961). Probably, these pikas can be vectors of paratyphus and other dangerous diseases. A sharp drop in the number of Pallas's pikas in individual years, as for example in 1968, in Tuva (Ustyuzhin, 1972), may happen as a consequence of epizootic diseases.

Practical Significance

Pallas's pikas with high population may exert adverse effect on the plant cover, destroying it almost completely. However, continuous dense colonies of Pallas's pikas are found rarely, and these pikas may not cause a serious damage to pastures. Occasionally Pallas's pikas may damage vegetable gardens.

In years of increasing populations Pallas's pikas may be the source of focal infections, in particular, of plague.

FAMILY OF HARES AND RABBITS
FAMILIA LEPORIDAE FISCHER, 1811

Dimensions quite large, body length varies from 300 to 600 mm. Hind legs more than 20% longer than forelegs. Tail short, usually quite distinct externally. Ears long, their length always exceeding half the length of the head; the tip of the ear is cuneate in form. Body color variable—in summer from monotone light gray to black-brown with a varying degree of expression of mottling; in winter often lighter, sometimes white. The lower side of wrists and feet with dense hair cover, forming tufts, hence digital pads are always concealed. Hairs on head, ears and ends of legs shorter and more densely appressed to the skin than on other body parts.

The size and shape of the skull varies widely. The most characteristic feature of the skull of hares is the reticulate structure of the lateral sides of the maxillary bones and the presence of supraorbital processes on the frontal bones. The process of the zygomatic bone is short. The bony palate is relatively narrow, and the length of the part of the palatine bone comprising it is less than the width of palatine processes of maxillary bone. Tympanic bullae are relatively small; their walls are not spongy. The zygomatic process of the squamosal bone with a shallow articular fossa. The interparietal bone is retained in the adult state only in members of the tribe Oryctolagini.

The dental formula is i 2/1, c 0/0, p 3/2, m 3/3 = 28. Upper molars (except the second) have a deep fold entering from the inner side. Third and fourth upper premolar teeth are roughly the same size. The third upper molar is very small. Lower molar teeth (except the third) have a deep fold entering from the outer side, and this fold borders the anterior and posterior segments; moreover, posterior segments have a variable height and differ in form. The third lower molar consists of two segments of different sizes, but of the same height, and are joined by cement. The anterior upper incisors have a V-shaped groove on the anterior surface, which in some species is filled with cement. The cutting edge of these incisors lacks a

161

deep groove, and on the posterior side there are stepped projections.

Thoracic vertebrae with short transverse and high spinal processes. Centrum of the lumber vertebrae is elongated, quite massive, and its transverse processes have a different form and length; the spinal [neural spine] process is short and comb-shaped.

Radius is as broad as the ulna, or 2 to 2.5 times as broad in which case the wrist articulates mainly with ulna.

The relative length of the intestines in hares is somewhat less than in pikas. The caecum is relatively shorter and broader, and the spiral fold forms fewer loops than in pikas. In the region joining the ileum and caecum, there is only one lymphoid diverticulum; a supplementary vermiform appendix, unlike in pikas, is absent. The ampulla is relatively shorter than in pikas (Naumov, 1981).

The os pennis is absent in hares.

There are insignificant interspecific differences in the number and morphology of chromosomes in the family—the diploid complement of a majority of species has 48 chromosomes.

Hares inhabit diverse landscapes in Europe, Asia, Africa, North and South America (to Patagonia*). As a result of economic activity of man, the range of hares has been expanded to include Australia and New Zealand, where rabbits and European hares have been successfully introduced. For species adapted to an open landscape, and also inhabiting temperate latitudes, extensive ranges are characteristic, whereas the majority of species of tropical and subtropical zones have restricted ranges.

A majority of species of the family (excluding rabbits) do not dig [burrows]. Burrowing species live in colonies. All hares lead a nocturnal or crepuscular mode of life; distant migrations are not characteristic of them. Hares feed on different plant types: herbaceous, shrubby and woody. Their characteristic feature is the ability to change over to feeding on woody and shrubby vegetation in unfavorable seasons.

The reproductive period of hares is extended—most often it lasts from February to September. For the reproductive season, depending on the climatic conditions, 1 to 4 litters (often 2-3) occur. A single litter may have 1-8 young. Rabbits, living in burrows, are born blind and helpless, others can see and can lead an independent life soon after birth.

The most ancient members of the hare family (subfamily Mytonolaginae) are known from the Upper Eocene of Asia and North America. In the Early Oligocene, when members of the Mytonalaginae still existed in the territory of Asia, and in North America, they had become already extinct, the more

---

*Editor: Introduced forms.

specialized forms of the subfamily Agispelaginae (Asia) and Megalaginae (North America) were widespread in both continents. Members of the extant subfamily Leporinae are known from the Miocene on the territory of Europe, Asia and North America. Hares penetrated Africa apparently not earlier than in the Pliocene, and in South America in the Pleistocene (Gureev, 1964).

The Family Leporidae includes three extinct and one extant (Leporinae) subfamilies. A.A. Gureev (1964) combined the extant and extinct members of Leporinae in 5 tribes, of which two comprise 7 extinct genera. In the tribe Pentalagini is included one extinct genus and three extant genera: *Pentalagus, Nesolagus,* and *Caprologus.* The tribe Oryctolagini consists of extant genera: *Romerolagus, Sylvilagus, Microlagus, Pronolagus,* and *Oryctolagus.* In the tribe Laporini is included the extant genus *Lepus.*

Hares have quite high economic significance. All species living on the territory of Eurasia are commercial. Their skins are used in the fur, wool and felt industry. Rabbit husbandry occupies a special place in the national economy. Some species of hares, particularly those in years of their high population, can cause damage to agriculture in winter by devouring bark of fruit trees and young saplings in nurseries. Being the host of many parasites which are carriers/vectors of diseases of man and animals, hares also can cause a definite damage.

162

Key for Identification of Genera of the Family Leporidae of the Fauna of the Former USSR

1(2). Length of ear shorter than length of head. Width of choana less than length of bony palate. Interparietal bone in adults separated and its order distinct .... Rabbit, *Oryctolagus.*

2(1). Length of ear equal to or exceeding length of head, width of choana equal to or exceeding length of bony palate. Interparietal bone in adults fused with neighboring bones, and its border indistinct .... Hares, *Lepus.*

Genus of Rabbits
Genus *Oryctolagus* Lilljeborg, 1871

1790. *Cuniculus* Mayer. Mag. Theirg., 1, 1: 52. *Cuniculus campestris* Meyer (= *Oryctolagus cuniculus*). Nom. praeocc., non Gronovius, 1763, Non Brisson, 1762.

1871. (?) *Oryctolagus* Lilljeborg. Sverig. og Norg. Ryggrodsjur, 1: 417. *Lepus cuniculus* Linnaeus. (While citing the above work different dates of publication are mentioned—from 1871 to 1874).

Dimensions small within the limits of the family, body length from 310 to 450 mm; overall length of skull from 62 to 84 mm (average 78.6 mm). Length of the ear from 60 to 79 mm (tip of the ear in bent condition does not reach the tip of the nose). Hair cover soft and dense. Hairs quite short and straight. Color of the upper side of the body brownish-gray or brownish. Light and dark ends of hairs form a ripple. Belly whitish. Tail above black-brownish or of the same color as on the back; white underneath.

Radius 1.25-1.5 times thinner than the ulna, and the proportion of participation in articulation of the radius and ulna with the wrist is roughly equal. The cranium and tympanic bullae are strongly bulged. The nasal section of the skull is narrowed, and the width of the nasal bones in the anterior part is 20-25% less than the width of the frontal bones in the postorbital region. Orbits round. Zygomatic arch thin, with a weakly developed crest of zygomatic processes of maxillary bones. Supraorbital processes narrow. Interparietal bone not fused with neighboring bones. The width of the choana is less than the length of the bony palate. Between the anterior and posterior segments of the lower molar teet, there are no enamel folds. The folds, entering from the inner side of upper molar teeth (p^3 - m^2), have strongly crimped lateral margins. The third lower premolar with one deep fold entering from the outer side.

Rabbits are distributed in the central and southern parts of western Europe and north Africa; introduced in South Africa, Australia, New Zealand and in several islands of the Pacific and Atlantic oceans. In the territory of southeastern Europe, introduced rabbit populations exist in the Ukraine.

The genus has one extant species, *Oryctologus cuniculus* (Linnaeus, 1758).

Wild Rabbit
Oryctolagus cuniculus (Linnaeus, 1758)

1758. *Lepus cuniculus* Linnaeus. Syst. Nat., 10 ed., 1: 58, "Germany."

1837. *Lepus vermicularis* Thompson. The Athenaeum: 468.

1843. *Lepus vermicula* Gray. List spec. Mamm. British Mus.: 128 (nomen nudum).

1867. *Culiculus fodiens* Gray. Ann. Mag. Nat. Hist., 20, 3: 225.

1903 (1904). *Oryctolagus cuniculus* Gray. Smithsonian Misc. Coll.*, 45: 402-406.

Species diagnosis agrees with the diagnosis of the genus.

---

*In Russian original, Koll.—General Editor.

Description

Body length from 310 to 450 mm (average 375.7 mm); length of foot 80-105 mm (average 98.5 mm); length of ear 60-79 mm (average 71 mm); body weight 1,300-2,200 g (average 1.612 g). Total length of skull 62-84 mm (average 78.6 mm); length of upper tooth row 10-13 mm (average 12.7 mm); zygomatic width 35-41 mm (average 37.8 mm).

The general tone of color varies from light gray to brownish-gray with a rusty tinge, on the upper part of the body—with dark brown ripple; head and ears have the same color as the back; tips of ears usually have a dark ring, and behind the ears—a rusty-ochreous spot. The ends of the legs are slightly lighter in color than the back. Belly white or grayish-white. Along the sides passes a diffused light stripe, which on the thighs merges into a more or less distinctly marked light spot. The upper side of the tail is of the same color as the back or darker, sometimes with an admixture of black hairs; the lower side is often white.

Not only geographic variation of color of wild rabbits, but also intra-populational polymorphism in respect of this character, have been recorded. Thus, in Scotland in areas with a predominance of heather and common gorse, a population of rabbits with the usual color (agouti) contains isolated individuals of white and black rabbits, and 3% of the population consisted of gray rabbits (Boag, 1986). From 3 to 5% of wild rabbits inhabiting southern Ukraine also have aberrant (black, light gray or ash) color; white rabbits are the rarest. In 1976, in the Black Sea region in the period of peak population, a sharp increase was recorded in the number of rabbits (up to 42%) with mottled color; later with the decrease in population density, mottled rabbits disappeared from the population (Shevchenko, 1986).

The form of the skull of wild rabbits is typical for the family. Nasal bones are long, their length exceeds the width by 2.5-3 times. Often a crest runs along the median suture. The anterior nasal opening is small; its length is less than or equal to the length of the anterior upper incisors. The crests of the zygomatic processes of the maxillary bones are weakly developed. The interpteregoid space is narrow; its length exceeds the width by 3.5-4 times. The interparietal bone does not fuse with adjoining bones in the skull. Tympanic bullae are relatively large. The articulated process of the mandible is considerably bent backward, particularly in the upper part (Fig. 40).

The diploid complement has 44 chromosomes: 11 pairs of meta-submetacentrics, 16 pairs of subtelocentrics; 4 pairs of acrocentrics (NF—76), X-chromosome—submetacentric, Y-chromosome—subtelocentric (Hsu and Benirschke, 1967).

163

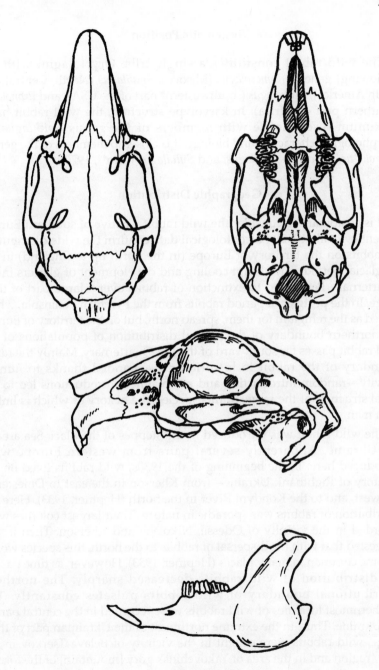

Fig. 40. Skull of the wild rabbit *Oryctolagus cuniculus*.

226

Systematic Position

The wild rabbit constitutes a single tribe Oryctolagini with the [following] genera: *Romerolagus* (Mexico), *Sylvilagus* (North, Central and South America), *Microlagus* (southwestern* part of the USA), and *Pronolagus* (southern part of Africa). In karyotype structure, the wild rabbit has a maximum resemblance with members of the genus *Sylvilagus*. In morphological characters and biology, it is closer to members of the genera *Caprolagus, Nesolagus, Poelagus,* and *Sylvilagus* (Luo**, 1970).

Geographic Distribution

It is customary to consider the wild rabbit a native of southern Europe (Terent'ev et al., 1952). Paleontological data confirm the wide distribution of rabbits on the territory of Europe (in the north up to England) in the preglacial period. The abrupt cooling and development of glaciers in the Quarternary Period led to extinction of rabbits over a large part of their range. In the postglacial period rabbits from the Pyrenian Peninsula, which served as the refugium for them, spread north, but on the territory of Europe the northern boundary of the natural distribution of populations of the wild rabbit passes far southward of the Pre-Quarternary. Mainly the range boundary of the rabbit in historic time expanded thanks to human activity—random introduction and pre-planned introductions led to the usual structure of the range of this species, the history of which is linked with man.

165

The wild rabbit was introduced in the steppes of the Black Sea area of the Ukraine. Apparently several pairs from western Europe were introduced here. In the beginning of the 1930s, wild rabbits lived on the territory of Rightbank Ukraine—from Kherson in the east to Dniester in the west, and to the Kondym River in the north (Heptner, 1933). Here the distribution of rabbits was sporadic in nature. Their largest colonies were recorded in the vicinity of Odessa, Nikolaev and Kherson. Then it was suggested that with the dispersal of rabbits to the north, this species would become common in many places (Heptner, 1933). However, as time passed the distribution of wild rabbits decreased sharply. The northern distributional boundary of wild rabbits pulsates constantly. The northernmost localities of wild rabbits were recorded in the central part of the Ukraine. Thus, in the extreme northwest of the Ukrainian part of their range, wild rabbits were caught in the vicinity of Belaya Tserkov in the Kiew region and in the area of Yakovchinsk gory [mountain] of the former

*Should be: northwestern; currently known as *Brachylagus*—Editor.
**Not in Lit. Cit.—General Editor.

227

Poltava district (collections of the Institute of Zoology, Kiev and ZM MGU; collections of 1951 and 1933). S.I. Ognev (1940) considered that colonies of wild rabbits on the territory of Rightbank Ukraine are not found north of the Kondyma River at the latitude of Balta (48° N. Lat.).

The western distributional boundary of wild rabbits in southern Ukraine enters the interfluve of the Dniester and Bug rivers along the Black Sea lowland (Fig. 41). The northwesternmost colonies of rabbits are found at present on the territory of the Voznesensk game farm of the Nikolaev region, where rabbits were released in 1974 (Arkhipchuk and Gruzdev, 1986). Farther south the boundary passes along the western districts of the Odessa region through Berezovka in the valley of the Tiligul' River, Kuyal'nitskii and Khadzhitbeisk liman* to Odessa (Ognev, 1940; collections of UkrZIN and KGU, collections of 1946 and 1977). From here wild rabbits dispersed east along the Black Sea coast, their colonies have been recorded along Tiligulsk liman, in the vicinity of Nikolaev, in the Snegirevsk district of the Nikolaevsk region, on Verevochnyi Ridge near the village of Chernobaevka of the Kherson region and in the vicinity of Cape Rakhovka (Ognev, 1940; Arkhipchuk and Gruzdev, 1986; collections, Kiev). According to the reports of S.I. Ognev (1940), in the east wild rabbits reached Aleksandrovsk (now Zaporozh'e).

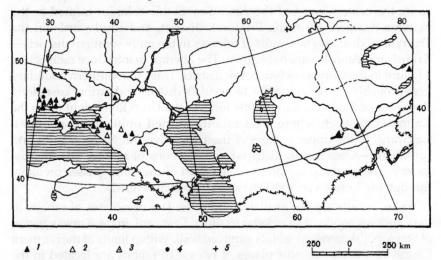

Fig. 41. Places of introduction of wild rabbits. 1—places of release of wild rabbits, where stable populations have formed; 2—places of unsatisfactory acclimatization of wild rabbits; 3—places of release of rabbits, where their further fate was not followed; 4—regions of dispersal of wild rabbits; 5—localities of escaped domesticated rabbits.

*Inundated river valley—Translator.

On the Crimean Peninsula, colonies of wild rabbits are confined mainly to steppe areas of the central part of the peninsula. West of the peninsula wild rabbits were caught during 1973-1977 in the vicinity of Tarkhankut and in Evpatoria; in the east, in the vicinity of Sudak and Kerch (Archipchuk and Gruzdev, 1986; collections of UkrZIN, Kiev). Colonies of wild rabbits, formed as a result of a release of over 1,000 rabbits in 1983 at the border of the Simferopol' and Belogorsk districts in the Simferopol' region, were liquidated by the spring of 1986 (Arkhipchuk and Gruzdev, 1986). By 1981 colonies of wild rabbits formed on the territory of the Nizhne-Kudryuchensk game farm, where in 1973 about 600 rabbits were released (Pavlov et al., 1984).

From 1959 to 1982 attempts were made to scatter wild rabbits in Kabardino-Balkaria, in Krasnodar territory, the Rostov region and in Stavropol'sk territory, in Uzbekistan (Dzhizaksk and Tashkent regions), in Kazakhstan (eastern part of Trans-Ili Alatau), and in Azerbaidzhan on the Apsheron Peninsula (Pavlov et al., 1984; collections of UkrZIN, Kiev).

In the northern Caucasus wild rabbits were released in 20 localities in the Krasnodar territory and at two localities in Stavropol'sk territory. Quite stable populations formed in the Krasnodar territory along the valley of the Urup River. The western boundary in this part of the range passes along the eastern coast of the Azov Sea. Stable colonies formed in the extreme northwest—in the Primorsk district, and southward—in the Temryuksk district. The results of releases in the more southern districts—Krimsk and Abinsk—are not known. The attempt to introduce rabbits was a failure in the Goryacheklyuchevsk district. Thus, the southern boundary here probably passes along the Temryuksk district and farther east along the valley of the lower and middle forks of the Urup River right up to the Kalininsk district, where wild rabbits formed quite stable colonies. Eastward, in the upper reaches of the Urup River (in Novokubansk and southward—Otradnensk districts) introductions of rabbits was unsatisfactory. The northern boundary in this part of the range passes probably through the Korenovsk and Timashevsk districts (Fig. 41).

So far, results are not available about the fate of releases of rabbits in the Caucasus region of Krasnodar territory. Eastward, in the Kirovsk region of Stavropol'sk territory, rabbits survived well. Within limits of the northern Caucasus, the easternmost places of release of rabbits are located in the valley of the Argudan River (Kabardino-Balkaria). However, rabbits did not survive here (Pavlov, 1984).

The origin of populations of rabbits inhabiting the Caspian coast in the area of the Apsheron Peninsula and on islands of the Baku Gulf (Peschanyi, Glinyanyi, Oblivnyi) is associated with repeated (starting from the end of the 19th century) releases of domesticated rabbits in these places. The affinity of domesticated rabbits along the Caspian coast is confirmed by morphometric analysis of skulls and fur color (Shevchenko, 1986).

In 1979, a batch of rabbits was released on the slope of the Chatkal Range, and two batches of rabbits were released in the southeastern part of Kyzylkums on the coast of Lake Aidarkul' (Uzbekistan). By 1983 the area of colonization of rabbits in the Aldarkul' area expanded to 15 ha (Pavlov et al., 1984).

Throughout Europe wild rabbits are found in Spain, Portugal, France, Poland, Czechoslovakia, Germany and Great Britain, where their colonies are of sporadic nature. Wild rabbits were introduced in South America (Chile and Argentina) and scattered in many states and provinces of North America. Particularly successful have been the translocation of rabbits to Australia and New Zealand, where wild rabbits multiplied and began to inflict appreciable damage to agriculture.

Geographic Variation

Many subspecies of wild rabbits have been described. However, human intervention in the process of dispersal of this species makes it difficult to study the evolutionary and genetic links between individual populations. In the territory of the Ukraine, the type form of wild rabbit—*O. c. cuniculus* L.—apparently occurs.

Biology

Habitat. Wild rabbits prefer to occupy open areas with dissected relief, for example, in ravines, gullies, abandoned open-cut mines and quarries. They may also colonize small areas of forests, forest-steppes and open fields, making their shelters in stacks of straw or hay. Wild rabbits do not avoid proximity of human dwellings. As a rule, colonies of rabbits are found on lands not used for agriculture, and with intensive exploitation of the territory by man, the number of wild rabbits decreases significantly down to their total disappearance (Arkhipchuk and Gruzdov, 1986).

In the vicinity of Kherson, wild rabbits live in caves and old mines of the Verovochnaya ravine, the upper part of which consists of the exposed layers of loose limestone. As a result of industrial processing of shell rock, a multitude of caves have developed here, which have been occupied by wild rabbits. A still larger number of rabbits was observed in the lower part of the ravine, where large shell rocks form a complex labyrinth of underground tunnels with a series of vertical mines (Ognev, 1940).

In the Odessa region near the town of Petrovka, until 1982 wild rabbits in large numbers inhabited the forest on the driest belt of the slope of the

Balaya River valley, an area covered with dense herbage. Here the soil is sandyloam with an abundance of stones. There are groves of low trees and shrubs. In many places this zone is dissected by narrow shallow gullies overgrown with hawthorn. In the vicinity of Evpatoria, wild rabbits inhabit hummocky localities on the coast of Lake Sasyk, where on the rubbly substrate with stone outcrops the vegetation is quite poor and represented mainly by harmel [*Peganum*].

In Slovakia wild rabbits inhabit altitudes of 460 m above sea level, though data are available about the prolonged existence of a small colony of introduced rabbits near the Moravian Dam at an altitude of 600 m above sea level. However, these conditions for wild rabbits are extreme. The most suitable types of habitats for this species, apparently, are the areas with light soils (sandy, clayey-sandy, sand-clay), where rabbits can readily dig burrows. Very often colonies of wild rabbits are found in the foothills at the boundary of burozems* and grayish-brown soils, and also in places with chernozems**. Colonies of rabbits have been recorded in pine forests, and in western Slovakia—in deciduous forests. Here rabbits dig their burrows in shrub thickets on the forest edges. In cultivated plains rabbits prefer to colonize burozem soils. Also, the affinity of wild rabbit colonies has been noticed for specific agricultural crops; they frequently colonize corn and beet fields and less often, potato fields. The important factor for the existence of rabbit colonies in agricultural regions is the presence of small forest blocks and uncultivated areas of land with weeds and tree-shrub vegetation. In places with heavy soils, wild rabbit colony mounds occur in gullies and other areas with dissected relief, often anthropogenic in origin. Wild rabbits avoid wet habitats with heavy soil, because here the conditions are most unfavorable for digging burrows. Moreover, in wet habitats rabbits are more susceptible to parasitic diseases. Rabbits are practically not found on substrates with bedrocks (Hell and Bakoš, 1970).

In Slovakia wild rabbits are frequently encountered in areas where the maximum precipitation is 650-700 mm, but in the vegetative period the quantity of precipitation should not exceed 450 mm. Excessive rainfall facilitates the spread of coccidosis, which adversely affects survival of the young. The most important factor in the dispersal of wild rabbits is the duration and depth of the snow cover. The most favorable areas for habitations of rabbits are where the snow cover does not persist longer than 50-60 days, although individual colonies are encountered in places with the duration of snow cover being 60-80 days. The optimal depth of snow cover for successful wintering is 30-40 cm; with deeper snow, food procurement becomes difficult (Hell and Bakoš, 1970).

168

*Brown soils—Translator.
**Black soils—Translator.

In rabbits massive and distant migrations are not observed. They change the place of their habitation only when their burrows are destroyed or the plant cover around the burrows changes sharply. Then rabbits search for new, more convenient habitats; they can disperse gradually and to quite far off distances.

Population. In the Ukraine, the population of wild rabbits was insignificant in all years. In 1929, in the Ukraine, about 5,000 skins of wild rabbits were processed. In the early 1930s processing did not exceed 10,000 skins (Heptner, 1933). By 1971 the total population of wild rabbits in the Ukraine was about 8,000 (Gizenko and Shevchenko, 1973).

Shelters. Rabbits make burrows in light soils, on slopes of hills and mounds, under roots of old trees in dense forests or sparse forest, abandoned vineyards, gardens, live fences, or abandoned cellars. Sometimes rabbits enter large cities. It is known, for instance, that a colony of wild rabbits lived in Berlin near the Brandenburg gate. In populated areas rabbits build shelters in warehouses, delapidated houses, cemeteries and in railroad dumps. They mainly occupy dry parts, and areas where the soil water is close to the surface, select the highest places.

According to the data of B.A. Kuznetsov (cited from Ognev, 1940), there are two types of burrows in wild rabbits. The first type is distinguished by simple construction—the central chamber is situated at a depth of 30-60 cm and most often 2 tunnels connect it to the surface (sometimes 1, very rarely 3 tunnels). The dimensions of the central chamber vary, but most often its width is 40-60 cm and the height is 25-40 cm. Bedding is usually absent in the central chamber. Burrows of the second type have 4-8 (and sometimes more) exits; not all the exits are used always. Burrows of the second type, apparently, are permanent and serve several generations of rabbits.

In the Kherson region wild rabbits dig deep burrows with a large number of exits. On elevated and dry places, burrows reach a depth of 1.5-2 m and more, and the length of underground tunnels may reach 10-15 m. Burrow entrances are broad, their diameter is 35-40 cm, and the form is funnel-shaped. Nesting chambers (25 × 20 × 30 cm) are situated at a depth of 1.5-2 m. Before births the female plucks downy hair from its stomach to serve as bedding for newborns. For the rest of the time, rabbits do away with bedding (Gizenko and Shevchenko, 1973).

Mainly males dig burrows. The dominant female in the harem of the male owning the territory lives in the male's burrow and before parturition digs a brood chamber in the side tunnel. In this manner a male, dominant female and her brood live in such a burrow. Remaining females living in the territory of the said male give birth to young in individual burrows. The burrow of a female has a length of 100-150 cm and terminates in a broad nesting chamber, the diameter of which is about 30 cm. The female burrow usually lies at a depth of 30-50 cm. Four weeks after parturition the female leaves its preceding brood and in the same burrow digs a new side tunnel, at the end of which it builds a new nesting chamber for the next brood (Hell and Bakoš, 1970).

169 The old occupied burrows of wild rabbits are a branched network of tunnels with numerous chambers and branch tunnels. The length of corridors sometimes exceeds 40 m. Besides the usual oblique tunnels, there are vertical tunnels, and also stepped corridors, leading down below the level of the system of main tunnels. Tunnels at times are so narrow that rabbits pass through them crawling. When digging tunnels a rabbit digs the earth with its fore paws, collects it under its belly, and throws it back with hind paws and pushes the hind part of its body outward through the tunnel.

In dense shrub thickets wild rabbits can live on the land surface; here they dig tunnels and make nests, where females feed their young (Hell and Bakoš, 1970).

Feeding. Wild rabbits feed on plant food. Because wild rabbits lead quite a sedentary mode of life, the range of plants devoured by them is determined mainly by the plants growing around their habitations; rabbits do not have any special selectivity in food. Winter feeding of rabbits differs considerably from that in summer. In winter rabbits feed predominantly on dry grass, seeds, and roots of different plants. In forest habitats a large percentage of their winter diet consists of bark and young shoots of trees and shrubs, including branches of willow, leaves of black locust, Siberian pea, tree branches of European aspen, ash, honey locust, filbert and oak. According to the data of T.M. Homolka (1988), the winter diet of rabbits living at the boundary of fields contains 30% winter wheat and 15% lucerne; 17% of the total volume is made up of woody parts of plants, 13% leaves of *Armeniaca vulgaris* and 15%—sugar beet.

Green parts of plants predominate in the summer diet of wild rabbits, which may constitute up to 98% of the entire volume of food (Homolka, 1988). In the Kherson region the diet of rabbits registered about 150 species of plants (Gizenko and Shevchenko, 1973). According to observations in the Netherlands, wild rabbits change their places of foraging in summer and may translocate to a distance of up to 200 m. And even those rabbits, the burrows of which are located in forest, prefer to feed at night on agricultural crops, heathers or on improved meadows (Wallage-Drees, 1983). In one attempt a wild rabbit eats 40-60 g of green food (Gizenko, 1968).

Sometimes rabbits may eat food wastes of other animals. In Dresden park rabbits were found to feed on remnants of meat and other wastes. In captivity rabbits also eat meat. It is also suggested that rabbits may eat insects in different stages of their development. For rabbits, as also for other hares, coprophagy is characteristic. The excretion from the caecum in rabbits occurs in the day, when they are sleeping. During this time a rabbit licks its anal opening roughly 10-14 times. A fourth of the excretion of the caecum consists of proteins. Moreover, bacteria contained in these excretions facilitate digestion of cellulose. Thus, coprophagy has great importance in the feeding of rabbits (Hell and Bakoš, 1970).

Activity and behavior. Daily activity of wild rabbits depends on the type of their habitat. In anthropogenic habitats they lead a predominantly nocturnal mode of life, and in areas where nobody threatens them, a diurnal mode. Thus, in Verevochnaya ravine in the vicinity of Kherson, in summer wild rabbits are active from dawn to 10-11 hr in the morning and from 15 hr to the onset of dark dusk (Gizenko and Shevchenko, 1973).

Rabbits do not venture far from their burrow, they feed most frequently in a radius of not more than 100 m from the shelter (Gizenko, 1960). Their movement is most frequently by leaps, which are quiet and fast. During leaping the front paws lie alongside and not one behind the other as in hares. The length of a leap in rabbits is less than in hares and does not exceed 2 m. Rabbits may develop high speed of run only for a short distance. Rabbits do not leap high, except for leaps during play. Hence they most frequently avoid large obstacles. In dense shrub thickets rabbits move by crawling (Hell and Bakoš, 1970).

Wild rabbits lead a family-group mode of life, and in their colonies a definite hierarchic structure exists. The relationship between females within 170 a group is usually more stressed than that between other members of the group, because they compete for places to nest. Occasionally there are aggressive fights between adult and young females, while males are quite friendly to young individuals, not only to their own but also to those of a stronger brood. Instances have been noticed when the aggressive behavior of females within the group was put to rest by the dominant male, who intervened in their fight (Cowan, 1987). Males, lower in rank, usually do not have an area within the nest and are all the time forced to remain on the surface. The dominant male has access to the territory of subordinate males in the said group. When the place of dominant male in the group is open, the remaining males struggle for the place, often winning it in fights.

Chemocommunication is quite well-developed in wild rabbits. Animals living in the group recognize each other by smell. Specific glands attain the maximum development in the breeding period. The activity of discharge of secretion from glands depends on the time of the year, weather conditions (in unfavorable conditions exogenous glands work intensively), and the social status of an individual. There also exist sex and age-related differences in the secretory activity of glands. The presence of females results in more frequent marking of the territory by males, whereas the presence of males does not produce an effect on this type of behavior in females. For individual recognition the maximum role is played by the secretion of inguinal glands, with which rabbits mark their progeny. There are instances when a young rabbit incidently receiving the scent marking of another group is chased away from the nest by its own parents (Soares and Diamond, 1982). Marking of the territory is done with the help of secretion of the anal and buccal glands, although 15% of the cases of markings with secretion of buccal glands are not aimed at marking the territory, but individuals of one's own type. Females in estrus more often leave marking of secretions from buccal glands than pregnant ones. The marking with secretion of buccal glands is occasionally accompanied by release of urine and feces, which also serve as territorial marking. During aggressive fights rabbits raise the hind part of their body and move their tail; during pauses between clashes they release marked excrements. Wild rabbits exhibit three types of urination: 1) normal; 2) urinating while tending; and 3) spraying urine (Bell, 1980; Soares and Diamond, 1982).

In wild rabbits there is a unique vocal signal in the event of danger; they stamp the ground with their hind paws. This signal is well heard to a distance of 30-50 m, and all rabbits in the radius of its audibility quickly hide in the burrow or in shrub thickets. Moreover, during enemy attack or when injured, rabbits produce a plantive chirp, which differs from the squeaking of hares by being softer.

Area of habitation. During a study of the social organization of wild rabbits in southern England it was shown, that on a territory with an area of 10 ha lived 11 to 14 breeding groups, which consisted of 1-8 males and 1-12 females. Most often such a group comprised 8-10 individuals. Only 11% of males and 4.2% of females lived in groups which did not comprise individuals of the same sex. The area of the territory of a family group varied from 0.1 to 2 ha. Each group had common underground structures. The area of habitation of groups overlapped somewhat. The territory of a group is defended against alien threat. The number of rabbits in a group (particularly with reference to females) depends on the number of places convenient for building nests. In groups occupying small burrows, there is a lesser number of young rabbits in a litter. Other conditions being equal, all indices characterizing the success of breeding were higher in groups with two or more females. In groups with one female, the mean longevity of females was higher; and the survival of males was higher in groups with one male (Cowan*, 1987a).

Breeding. In Europe a strict seasonality of breeding is characteristic of wild rabbits.

---

*Only one citation in Lit. Cit.—General Editor.

171   Start of breeding in wild rabbits in Slovakia occurs in February, and the first brood appears in March (almost simultaneously in all rabbits) (Hell and Bakoš, 1970). Similar periods of the start of reproduction are observed also in rabbits from Holland, where in the spring of 1981 about 50% of females were pregnant already in the first half of March. It was shown that an increase in the proportion of herbs in the winter diet hastens the onset of breeding (Wallage-Drees, 1983). Weather conditions in spring also effect the start of breeding periods. The breeding season terminates most often in August, but sometimes with favorable conditions females bear the last litter in September or even in October. In rabbits from southern Ukraine there is a partial cessation of breeding in summer, which is associated with summer draughts, and the reproductive period extends from January to November (Gizenko and Shevchenko, 1973).

On the whole, in the European part of the range of wild rabbits, the duration of the vegetative period, which is the main factor determining breeding periods, is not less than 200 days, and often surpasses 250 days. In New Zealand and Australia lower limits of temperature generally do not restrict the vegetative period, which provides wild rabbits with a potential possibility of uninterrupted breeding. However, in this part of the range also there exist periods when the intensity of breeding of rabbits falls. In southeastern Australia and New Zealand, irrespective of the temperature regime, the cyclic functioning of gonads has been observed, which also determines reproductive periods. In South Australia, breeding is interrupted in the middle of the vegetative period because of the spread of draught and absence of full-value foods. This period usually extends up to 200 days (Naumov and Shatalova, 1974).

In males, during the annual sexual cycle, considerable changes occur in the size of sex glands, which is regulated, apparently, by the length of the daylight period. Thus, in England, in captive rearing of wild rabbits with prolonged illumination there were no signs whatsoever of a change in weight of testes. However, in natural conditions the weight of testes attains the maximum in May-June, and their minimum size is recorded in the period between October and December. A decrease in the weight of testes starts after the summer solstice and increases after the winter solstice (Boyd, 1985). In Slovakia and Holland the maximum weight of testes (5 g) in wild rabbits was observed in April and March, respectively; at the end of the breeding period (June-August) testes lose 50% of their weight and become soft and spongy. In late fall the mass of testes begins to increase again (Vandewalle*, 1989). In the current year's individuals the increase in the weight of testes occurs up to October, and in January-February young rabbits do not differ from adults in the weight of testes. In females,

*Not in Lit. Cit.—General Editor.

sexual maturity sets by the age of five months (Boifani and Lieckfeld, 1989).

Ovulation in wild rabbits, as also in hares, is induced and sets in 10-11 hr after mating. Each subsequent pregnancy in females may occur soon after parturition (repeat mating occurs in the first two hours after parturition), and usually rabbits bear offspring every 5-8 weeks. Consequently, during the breeding season each female may bear up to 5 broods, although some available data indicate that about 7 litters in a year are possible (Boijani and Lieckfeld, 1989). In those cases, when rabbits are not fertilized during the first two weeks after parturition, new mating only occurs 56-60 days after births. Hence, in some females more or less prolonged intervals are possible between individual broods.

The number of young in a brood varies from 2 to 12 (average 6); in the first and last broods, as a rule, the number of offspring is less. Young of the first brood, particularly females, may participate in breeding by the end of summer. According to some data, rabbits attain sexual maturity in 5-8 months after birth; according to other data, after 4-5 months (Hell and Bakoš, 1970; Gizenko and Shevchenko, 1973).

In wild rabbits pregnancy lasts for 28 to 32 days, and most often for 31 days. Usually births occur in the night or in the morning and last from 5 to 60 min. (Terent'ev et al. 1952). Mothers lick newborns, place them in the nest, and cover them with the down. Up until the 17-20th day, young feed exclusively on mother's milk. Each female rabbit produces from 50 to 200 g of milk per day. Milk secretion starts slightly before parturition, reaches its maximum on the 20th day after parturition, and decreases on the 25-30th day, after which it remains almost at the same level, decreasing gradually toward end of lactation. Milk of rabbits is marked by high caloric values and high content of nutrients. It contains from 10.4-22% fat, 10.5-15.5% protein and from 1.8-2.1% lactose. Its water content is only 69.5% (Terent'ev et al., 1952).

172

Females feed young rabbits with milk in the nest for about three weeks, roughly twice a day and always at night. Each feeding lasts from 5 to 10 min, after which the female rabbit leaves young in the nest, carefully closes the entrance to the burrow from outside with earth collected near the entrance (Terent'ev et al., 1952).

For the breeding season a female European wild rabbit bears from 10 to 30 offspring, and if the progeny of its daughters are considered, then, on the whole, the progeny of one female for one year may comprise from 36 to 42 offspring. However, the actual annual increment of a rabbit family is much lower than the theoretical, since usually 3 of the 5-6 newborns die, and the current year's female in the year of their birth bear broods irregularly. Thus, the actual increment of young from one female does not exceed 15 young. Under unfavorable weather conditions and exceptional population density, resorption of embryos occurs. On average 15% of embryos are resorbed. However, embryonic mortality depends on the status of the female in the system of hierarchical structure—the lower the status, the higher is the percentage of mortality of its embryos (Hell and Bakoš, 1970).

Growth and development. Wild rabbits are born naked and blind, but by the end of the first day rudiments of hairs of the primary hair coat on their head begin to appear. The body of 3-day old rabbits is covered with short, not more than 1 mm long, guard and directed hairs. At the age of five days, hairs reach a length of 5-6 mm. By this time the rudiments of downy hairs are formed. A 10-day old individual possesses an appreciably grown primary hair coat 14-16 mm high; however the down is still weakly developed. Complete development of the primary hair coat is attained on the 20th-25th day. Young rabbits begin to see on the 11th day after birth, and at the age of 12-13 days their eyes are completely open. At the age of 25 days, young rabbits begin to lead an independent life. Four weeks after parturition the female begins to build a new nest, and young are left behind in the old burrow (Terent'ev et al., 1952; Gizenko and Shevchenko, 1973).

Newborn rabbits weigh from 37 to 54.3 g (average 46.1 g), but already after two days their weight increases roughly by one-third. On the 6th day after birth, the weight of newborns usually doubles, and on the 10th day exceeds the weight at birth by 2.9-3.0 times. By the end of the second week their weight is roughly 3.3 times the weight at birth, by the end of the third week—3.5-4 times, and by the end of the fourth—7 to 10 times. At the age of 4-5 months the weight of rabbits is 30-50 times their weight at birth (Terent'ev et al., 1952).

In rabbits the growth of different body parts after birth occurs unevenly. Ears grow the fastest of all. The width of the head, the distance between the outer angles of the iliac bone, and the length of the hind paws increase slowly. The dimension of the chest and body are the slowest to increase (Terent'ev et al., 1952).

Newborn rabbits have 16 milk teeth, which fall very early, and the front upper milk incisors fall down even before birth. The change of milk teeth starts from the 18th day and terminates by the 28th day after birth (Terent'ev et al., 1952).

Molt. In domesticated rabbits the change of juvenile hair coat to the adult occurs gradually, starting roughly from the age of one and one-half months (Terent'ev et al., 1952). During this time the fur becomes dull and sparser. Soon it is possible to notice a darkening of individual parts of skin along the spine and on the sides, since new secondary hairs start growing there. Hairs of its secondary coat begin to grow after two weeks, and juvenile hairs fall simultaneously. The process of molting gradually spreads over

173

the entire spinal part and sides and is completely terminated by the 4th-5th month, that is, by the time of sexual maturity.

In adults of both sexes, spring molting starts simultaneously in March. Winter fur becomes dull and starts becoming sparse. However, in males molting proceeds more slowly, and already in April the winter fur is retained on them. Females molt faster, and by the end of April the fur becomes sparse on them. By the time of the birth of first litters females pluck out hairs from the belly, such that it becomes almost naked. Such character of the hair coat is retained in females during the entire breeding period. In males the slow replacement of winter fur by summer fur extends throughout the spring and summer (Terent'ev at al., 1952).

The fall molt in adult females begins vigorously and continues for 1.5-2 months. Winter fur appears in patches—first on the nape and in the anterior part of the back, and later sequentially on the shoulder blades, posterior part of the back, on rumps, sides and belly. In males the change of summer hair coat to winter starts by the end of summer and lasts 2-2.5 months (Terentev et al., 1952). In southern Ukraine the fall molt in adult rabbits starts from the middle of October and ends in the 2nd-3rd decades of November (Gizenko and Shevchenko, 1973).

Sex and age structure of population. The sex ratio in a population of wild rabbits is close to 1:1; however, usually there are slightly more females (Hell and Bakoš, 1970).

The maximum population size of wild rabbits is found at the end of summer; later the number decreases, first slowly, then more intensively. The maximum number in a population is recorded at the end of winter-beginning of spring. In Australia, where the breeding season of wild rabbits begins with the first winter rains (March-April), the number in a population attains a peak in October-December. Here every year during 10 months the proportion of the current year's young constitutes more than 50% of the population, and the tempo of replenishment of the population is very high (King and Wheller*, 1985).

The mortality of wild rabbits is considerable not only in young animals, but also in adults. Only a small proportion of rabbits survives up to two years. In the first year of life about 80% of young perish, a large part of which constitutes the young of first broods, of which not more than 20% survive to the next spring. On average, rabbits live for 18 months; only 2% of the population survives up to 3 years. However, in better conditions rabbits may attain an age of 8 years. In rabbits, as compared to hares, the replenishment cycle of the population is shorter.

---

*Not in Lit. Cit.—General Editor.

Population dynamics. The number of wild rabbits is subject to considerable fluctuations. With a quite high fecundity, rabbits often perish as a result of epizootics, poor weather conditions and absence of a sufficient quantity of food. Anthropogenic factors exert a suppressive effect on population numbers of wild rabbits, especially the development of land for agriculture and the use of toxic chemicals, and also immoderate hunting (Hell and Bakoš, 1970).

In the Crimea, from 1966 to 1970, the total number of wild rabbits remained at the level of 4,000-5,000; in 1975 it was 2,000; in 1975-1979 a rise of numbers was noticed up to 20,000-23,000 (Arkhipchuk and Gruzdev, 1986). The most successful place for acclimatization of wild rabbits in the Crimea is apparently the Tarkhankut Peninsula (Black Sea area, "Bolshoi Kostel" ravines). Here, in the beginning of the 1960s, 50 rabbits were released, which by the beginning of the 1970s multiplied to 3,000 (Gizenko and Shevchenko, 1973). In 1982 the number of rabbits in the majority of regions of their habitation for various reasons (mainly because of myxomatosis attacks) decreased so much, that it was possible to restore it just partially only in the Crimean region.

174   The oldest and the most numerous population of wild rabbits in the Ukraine is in the Kherson region in the Verevochanaya ravines near the village of Chernobaevka. Here in individual years the number of rabbits reached 7,500. However, presently this population is in a depressed state because of rapid development of residential construction in this region. In 1985, this population was estimated at only 200 individuals (Gizenko and Shevchenko, 1973).

In the middle of May 1986 one of the northernmost places of release of rabbits in the Ukraine was investigated—the vicinity of the Malaya Ivanovka town in the Kommunarsk district of the Lugansk region, where in 1978, 200 wild rabbits were introduced. In 1986, at this place, represented by a host of ravines and outcrops of stone blocks with shrub thickets along the streams, only five occupied and an equal number of old burrows of rabbits were found. Thus during eight years the number of rabbits dropped to 20-30 individuals. Apparently, this population of wild rabbits is doomed for extinction (V.V. Gruzdev, unpublished data).

Considerable fluctuations have been reported in population numbers of wild rabbits in the Voznesensk district of the Nikolaevsk region. Here in 1974, 100 rabbits were released. Their number rose 10-fold in 1977, and from 1978 to 1982 the head count of rabbits here remained stable—from 1,500 to 2,000 individuals. In 1982 after the epidemic of myxomatosis, the number in this population dropped to 70 individuals. In the same year many rabbits died of myxomatosis in the vicinity of Evpatoria. Here, rabbits were introduced in 1973, and after a year their number increased from 30 to 2,000. From 1974 to 1981 with great complexity 7,649 rabbits were removed (shot and caught for dispersal in other areas). In 1983, after the epizootic the number of rabbits dropped to 800, and by July 1983 rose to 1,500 (V.V. Gruzdev, unpublished data).

On the whole, in the territory of the Ukraine the number of wild rabbits rose from 4,000-5,000 in 1970-1975 to 20,000-25,000 in 1975-1979; and by 1982 decreased sharply. The economic activity of man, epizootics, poaching, abundance of predators and irrational management of commerce of this species should be considered as the main reasons for the decrease in the number of wild rabbits in this part of the range (Arkhipchuk and Gruzdev, 1986).

Enemies. Mustelids hunt wild rabbits, attacking them directly in burrows. In the Ukraine the following have been reported among enemies of rabbits: steppe and forest polecats, stone martens, weasels and ermines. Brown rats may attack young in burrows. Outside burrows, rabbits are hunted by foxes, stray dogs, and wild and homeless wandering cats. Among birds the enemies of wild rabbits in the Ukraine are the short-eared owl, eagle-owl, reed-harrier and hen-harrier, tetreonid-sparrowhawk, white-tailed eagle, and hooded crow. Hooded crows, more than any other species of birds, destroy the young (Gizenko and Shevchenko, 1973).

Competitors. Data are not available about competitive interrelationships of wild rabbits with other mammals in the territory of the Ukraine.

Diseases and parasites. The family-group mode of life of wild rabbits in their burrow system favors the spread and high degree of infectivity of this species both with ectoparasites and endoparasites. On the whole, for wild rabbits the same ectoparasites are characteristic as for hares. Specific to wild rabbits are many species of mites and fleas. Thus *Demodex cuniculi* parasitizes the roots of hairs and skin glands of rabbits, which causes the appearance of bald patches, purulent nodes, excrescence and enlargement of glands. Scabies and cat mites *Sarcoptes scabici* forma *cuniculi* and *Notoedres cati* parasitize head parts with little fur, making tunnels under the skin surface and causing its ulceration with high infestation of these parasites. Rabbits may die of these infections. Other species of mites (*Psoroptes equi* forma *cuniculus* and *Chorioptes ovis* forma *cuniculus*) parasitize the skin surface of rabbits. Most often they infest the pinna in the outer auditory meatus.

175   Of the specific fleas, *Spylopsyllus cuniculi* parasitizes wild rabbits. The reproductive cycle of this species of fleas is linked with the breeding period of rabbits—mating of fleas occurs only on newborn rabbits. The rabbit flea is one of the vectors of myxomatosis—a dangerous disease of rabbits.

Among endoparasites the cestode *Ctenotalnia stenoides* has been identified in wild rabbits. Its level of infectivity increases with an increase in population density (Gizenko and Shevchenko, 1943).

Wild as well as domestic rabbits are subject to acute infectious disease—myxomatosis. The pathogen of the disease is a DNA-containing virus, which is transmitted from rabbit to rabbit by the air-drop route through the mucous secretions, and is also spread by blood-sucking insects—midges, fleas, and mosquitoes. In the salivary glands of mosquitoes the virus may persist up to 6-7 months. In the dried skins of rabbit, the virus remains viable for 10 months, and in moist medium at temperatures of +10° to + 12°C for 3 months. This epizootic may erupt any time of the year, but occurs frequently in late spring and early summer, since at this time there is the maximum number of blood-sucking insects. Mortality due to myxomatosis reaches 40-100%. The recovered rabbits acquire prolonged immunity (Veterinary Encyclopedia*, 1973).

Practical Significance

Skins of wild rabbits are quite beautiful and are used in the natural state. In the beginning of the 1930s in Russia about 10,000 skins were processed each year. In addition, the meat of wild rabbits has valuable gustatory and nutritive properties and in quality does not differ from the meat of domestic rabbits.

On mating wild rabbits with mongrel domestic rabbits, it has been established that in the first and second generation hybrid progeny, as compared to the control groups, the number of young increases. The first and second generation hybrid individuals grow and develop well, and by the age of 7-8 months already attain body weight of 4.5-5 kg, whereas the original forms weighed not more than 2.5 kg. Thus, in the first generation hybrids there is a phenomenon of heterosis: fecundity increases by 25-30% and body vigor increases. The first generation hybrids have a color of wild rabbits (Gizenko, 1968).

Genus of Hares
Genus *Lepus* Linnaeus, 1758

1758.   *Lepus* Linnaeus. Syst. Nat., 10 ed., 1: 57. *Lepus timidus* Linnaeus.

1828.   *Lagos* Brookes. Cat. Anat. Zool. Mus., 1: 54. *Lepus arcticus* Ross (Palmer, 1904: 361).

1829.   *Chionobates* Kaup. Entw.-Geshch. Nat. Syst. Europ. Thierwelt, 1: 170. Based on *Lepus variabilis* Pallas, *Lepus borealis* Pallas.

---

*Not in Lit. Cit.—General Editor.

1867. *Eulagos* Gray. Ann. Mag. Nat. Hist. 20: 22, *Lepus mediterraneus* Wagner (Ellerman and Morrison-Scott, 1951: 429) (= *Lepus capensis*).

1899. *Eulepus*. Acloque. Faune de la France, Mamm.: 52. *Lepus timidus*. Linnaeus (G. Allen, 1039: 272).

1929. *Allolagus* Ognev. Zool. Anz. 84: 72. *Lepus mandschuricus* Radde.

1940. *Eulagus* Ognev. Zvery SSSR i prolezhashchikh stran [Animals of the USSR and Adjoining Countries]. 4: 109.

Description

Dimensions small to medium in the family, body length from 390 to 680 mm. Hair coat soft and dense. Color of the upper side of the body gray, ochreous-gray, sand gray, ochreous-brown or brown (may be various combinations of these colors). In the majority of species on the back and the upper part of head there is a small or large mottley outline, formed by a combination of light and dark tips of guard hairs in different proportions or the presence of entirely dark guard or directed hairs. In summer the tail above is dark reddish-brown, black-brown or gray.

176 Radius thicker than the ulna (1.5-3 times). The forearm is articulated 2/3-4/5 with the wrist on account of the radius.

Skull simple in construction. The cranium is bulged; tympanic bullae are small to quite large. The width of the frontal bones between the bases of the supraorbital processes is less than the width of the nasal bones in their anterior part. Supraorbital processes are narrow and elongated or broad and triangular in form, and not attached to frontal bones. The interparietal bone is fused with the occipital bone, and its boundaries are inconspicuous.

Members of the genus are known from the Pliocene, and in their origin they are linked with the ancestors of the genus *Alilepus* Dice.

The main direction of specializations is adaptation to faster locomotion through leaps, and to feeding on diverse plants with a sharp change in the composition of winter and summer diet.

Hares occupy diverse landscapes in Europe, Asia, North America and Africa.

The majority of species of the genus are valuable game animals. Some species may cause damage to agriculture and forestry. Being hosts of many ectoparasites, they may be vectors of diseases of domestic animals and man. Apparently 20 species (4 subgenera) are included in the circumscription of the genus.

Key for Identification of the Species of Hares in the Fauna of the Former USSR

1(2). Length of ear less than length of head. Hair coat coarse. Lower side of tail brown or ochreous-brown. Width of choana not exceeding length of bony palate ....
.... *Lepus (Allolagus) mandshuricus*—Manchurian hare.

2(1). Length of ear more than length of head. Hair coat coarse. Lower side of tail brown or ochreous-brown. Width of choana not exceeding length of bony palate .... 3.

3(4). Ear with white stripe on outer margin. Tail rounded in form, in summer with gray spot on upper side, in winter entirely white. Winter fur white, excluding black tips of ears. Articular process of mandible high and massive, steeply rising upward. Grooves on anterior surface of upper incisors located closer to their inner margins .... *Lepus (Lepus) timidus*—Blue hare.

4(3). Ear without white stripe on outer margin. Tail cuneate in form, with black stripe on upper side both in winter and summer. Winter color never white on entire body. Articular process of mandible not very high, bent backward. Grooves on anterior surface of upper incisors located closer to their middle .... 5.

5(6). Broad black stripe passing along margin of tip of ear. Summer color ochreous, brown or ochreous-brown, relatively bright, with large mottley patches. Summer underfur and guard hair at base white or silver-gray. In winter over large part of range, excluding Crimea and Caucasus, sides of head and body becoming white. Base of upper anterior incisor not reaching region of suture connecting intermaxillary and maxillary bones. Zygomatic arches wide apart, their width in anterior section, usually not less than 40.5 mm .... *Lepus (Lepus) europaeus*—European hare.

6(5). Narrow dark stripe passing along margin of tip of ear or absent. Summer color gray, ochreous-gray, yellowish-gray or brownish with small streaked pattern, formed by dark and light tips of guard hairs. Summer underfur and guard hairs brown or gray-brown at base. Winter color similar to summer, and in montane regions and in north of the range becoming lighter in the region of sacrum and mottley on back. Base of anterior upper incisor reaching region of suture, connecting intermaxillary and maxillary bones. Zygomatic arches weakly apart, their width in anterior section usually not exceeding 39.5 mm ....
.... *Lepus (Lepus) tolai*—Tolai hare.

Blue Hare
Lepus (Lepus) timidus Linnaeus, 1758

1758. *Lepus timidus* Linnaeus. Syst. Nat., 10 ed., 1: 57, Sweden, Uppsala.
1778. *Lepus variabilis* Pallas. Nov. Spec. Quadr. Glir. Ord.: 2; Nom. nov. pro *Lepus timidus* Linnaeus.
1795. *Lepus septentrionalis* Link. Beitr. Naturg., 1, 2: 73, Nom. prov. pro *Lepus variabilis* Pallas, 1820, *Lepus borealis* Nilsson. Scand. Fauna, 1: 21, Nom. nov. pro *Lepus variabilis* Pallas.
1883. *Lepus timidus tschuktschorum* Nordquist. Vega Exped., 2: 84. Magadan region, Chukotsk A[utonomous] R[egion], Pitlekai.
1899. *Lepus lugubris* Katschenko. Izv. Tomsk. Un-ta: 57. Altai territory, Gorno-Altaisk A[utonomous] R[egion]. Biya River, Ongudai.
1890*. *Lepus timidus altaicus* Barret-Hamilton. Proc. Zool. Soc. London: 90. Altai territory, Gorno-Altaisk A[utonomous] R[egion]. ("Altai mountains").
1903. *Lepus gichiganus* J. Allen. Bull. Amer. Mus. Nat. Hist., 19: 155. Magadan region, Gizhiga.
1922. *Lepus timidus kolymensis* Ognev. Biol. Izv. 1: 106; Yakutsk ASSR, Kolyma River, 80 km north of Nizhnekolymsk, Pokhodskoe.
1922. *Lepus kamtschaticus* Dybowski. Arch. Tovar. Naukow Lwowe, 1: 354, Kamchatka region, Kamchatka Peninsula. Nom. nudem.
1923. *Lepus timidus sibiricorum* Johansen. Izv. Tornsk. Un-ta, 72: 59. Tomsk region, Chulym River (basin of Ob' River), Novokuskovo.
1928. *Lepus timidus orii* Kuroda. J. Mammal. N.** 9: 223. Sakhalin region, Sakhalin Island, tomari ("Tomarioro Noyoro") .
1929. *Lepus timidus kozhevnikovi* Ognev. Zool. Anz., 84: 79. Moscow region, Noginsk district ("Bogorodsk district "). Chernaya.
1929. *Lepus timidus transbaicalensis* Ognev. ibid., 84: 81. Buryatia, Barguzinsk district, Sosnovka.
1931. *Lepus timidus saghalinensis* Abl. J. Sci. Hiroshima Univ., ser. B, 1: 49. Sakhalin region, Sakhalin Island, "Ogomari."
1933. *Lepus timidus mordeni* Goodwin. Amer. Mus. Novit., 681: 15, Khabarovsk territory, Amur River, Troitskoe, Manoma River.
1935. *Lepus gichiganus robustus**** Urita. Karafuto Dobuts. Kansuru Bunkan: 16, Sakhalin region, Sakhalin Island, Nom. nudem.

---

*In Russian original, 1990—General Editor.
**In Russian original, m—General Editor.
***In Russian original, *rubustus*—General Editor.

1936.   *Lepus timidus begitschevi* Kolyushev. Tr. Biol. In-ta, Tomsk, 2: 304. Krasnoyarsk Territory, Taimyrsk A[utonomous] R[egion], Pyasinsk Gulf, western coast.

1938.   *Lepus timidus abei* Kuroda. List Japan. Mamm.: 42. Sakhalin region, Iturup Island ("Yetorofu, Toshimoi").

Diagnosis

Dimensions large. Body color in summer rusty-gray, gray or slate-gray. Tail rounded in form with gray spot on upper side, white below. Winter color of body white, all over, excluding black tips of ears.

Skull with relatively short facial section and broad postorbital space: ratio of width of frontal bones behind supraorbital processes to width of base of rostrum is from 49 to 69% (average—59%). Length of bony palate 1.25-1.5 times less than width of choana. Alveolar length of upper tooth row 17.5-19.9 mm (average 18.7 mm). Zygomatic width in anterior part 40.4-45.2 mm (average 42.8 mm). Supraorbital processes broad, almost triangular in form. Base of anterior upper incisor reaching suture, connecting inter-maxillary and maxillary bones. Length of lower diastema in relation to length of lower tooth row more than 80%. Articular process of mandible steeply raised above.

Description

Body length from 440 to 740 mm; length of tail from 49 to 108 mm; length of foot from 132 to 190 mm. Length of ear from 75 to 100 mm.

Ear weakly broadened and relatively short—appressed to head, reaching just the end of face or slightly jutting forward (Fig. 42).

178   Summer color of hairs on back rusty gray, gray or slate-gray, sometimes with dark brown ripple. Head is colored the same as on the back or slightly brighter. Dark and white spots are absent on sides of the head. The anterior part of the snout and cheeks are of ochreous-chestnut or ginger-rusty color. Ears are dark brown with a white stripe on the outer margin and with a black tip. Around the eyes there is a narrow whitish or whitish-ochreous ring. Lips, chin and the upper part of the head are white or grayish-white. Sides of the neck are grayish or rusty gray with an ochreous ripple. Occiput and upper side of the neck are ochreous-rusty or grayish-brown. The paws are light gray with an ochreous or ochre tinge. Belly and inner side of the legs are white.

The skull is massive and broad; occipital part is broad and flattened. The sides of the frontal bones are depressed anterior to the supraorbital processes. Zygomatic arches are massive and wide apart; the anterior part

246

Fig. 42. Young blue hare *Lepus timidus* (Photo by B.S. Yudin).

slopes down. Crests of the zygomatic processes of the maxillary bones are well-developed (Fig. 43). The mandible is massive with the articular process steeply raised upward. Bases of the lower incisors almost reach the level of the alveole of the first premolar tooth.

The diploid complement has 48 chromosomes: 8 pairs of meta-submetacentrics, 15 pairs of subtelocentrics and acrocentrics (NF = 88), X-chromosome—submetacentric, Y-chromosome—small acrocentric (Gustavsson, 1971).

Systematic Position

In the supraspecific systems of S.I. Ognev (1940) and A.A. Gureev (1964), the blue hare is identified as a separate subgenus *Lepus* L. 1758. In many morphological characters the blue hare is close to both European and tolai hares, which offers the basis to relate these three species to the subgenus *Lepus*.

180   S.I. Ognev (1940) pointed to the morphological similarity of the North American species *L. arcticus* Ross, 1819 and *L. othus* Merriam, 1901; with *L. timidus*, and in the system of A.A. Gureev (1964) *L. arcticus* (along with

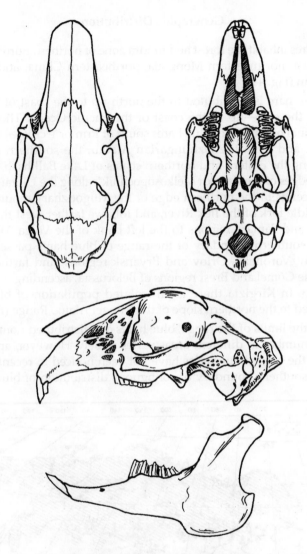

Fig. 43. Skull of the blue hare *Lepus timidus*.

L. othus) is considered a subspecies of *L. timidus*. B. Rausch (1963), while analyzing the history of amphi-Beringian links of mammals, concludes about the Holarctic distribution of *L. timidus*, recognizing thereby the conspecificity of *L. othus* (and possibly, of *L. arcticus*) with the European hare. The majority of American systematists recognize the species independence of *L. arcticus* and *L. othus* (Banfield, 1974; Hall, 1981, and others).

Geographic Distribution

Blue hares inhabit the forest and tundra zone of northern Europe, Siberia, the Far East, northwestern Mongolia, northeastern China, and northern Kazakhstan (Fig. 44).

The blue hare is distributed in the north up to the coast of the Arctic Ocean, in the east—up to the coast of the Pacific Ocean. The southern distributional boundary of blue hares south of Primor'e crosses the border of Russia, and from the Dzhungarian Alatau, the southern boundary passes along the eastern and northern coasts of Lake Balkhash, along the southern edges of the Kazakh Melkosopochnik*, along the Ulutai mountain massif, ascends to the northern edge of the Mugodzhary mountains and to the middle fork of the Ilek River, and follows farther west through the Orenburg and Samara regions to the left bank of the Volga. West of the Volga the southern boundary of the range of blue hares passes through the Saratov, Voronezh, Orlov and Bryansk regions, and farther west— through the Gomel and Brest regions of Belorussia, ascending in the north to Grodno. In Kirgizia there is an isolated population of blue hares, acclimatized to the northern slope of the Kungei Alatoo Range (Fig. 45).

Within the limits of the range, blue hares are distributed nonuniformly and their number in different areas varies widely. However, appreciable change in the range boundaries have not been noticed in recent decades. Thus, the southern boundary of contiguous distribution of blue hares in

180

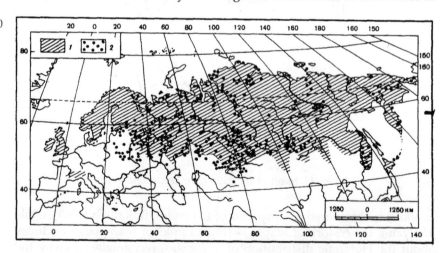

Fig. 44. Range of the blue hare across Eurasia.
1—schematic range according to Corbet (1978); 2—our data.

*Area of low rounded isolated hillocks—Translator.

eastern Europe has almost remained unchanged since 1940. Blue hares disappeared or became fewer only in places, where intensive shooting was undertaken and where adequate satisfactory conditions for breeding were lacking (Gruzdev and Osmolovskaya, 1969).

182   Blue hares are found throughout the Kola Peninsula except in the tundra zone in the extreme north of the peninsula. Blue hares are common along the coast of the White Sea (except in the northern areas of Kanin tundra). Farther east the distributional boundary of blue hares passes along the Timanskaya and Bolshazemelskaya tundra to the coast of the Barents Sea. According to the data of S.I. Ognev (1940), blue hares are absent in the extreme north of the Yugorsk Peninsula and the Kolguev and Vaigach islands. However, at the present time blue hares are met with throughout the Yogorsk Peninsula (collections of ZM MGU), and on the Kolguev and Vaigach islands blue hares are encountered quite rarely (Laptev, 1958). On the Yamal Peninsula blue hares almost do not enter the Arctic tundra, and the Lake Neito region (about 70°-71° N. Lat.) is apparently the extreme northern distributional limit of blue hares in this part of the range.

The distributional boundary of blue hares extends along the coasts of the Ob' and Tazov gulfs. On Gydansk Peninsula the distributional boundary of blue hares coincides with the boundary of tundra, that is, it extends to 71° N. Lat., and northward this species is found extremely rarely. For example, I.P. Laptev (1958) reported the localities of blue hares in the valley of the Yubilei River and in the vicinity of Cape Khanarasalya (71°40'). Northwest of the Taimyr Peninsula blue hares are common in the vicinity of Dikson, and in winter cases of their strayings in Dikson Island have been reported (Ognev, 1940). On the whole, in Taimyr the distributional boundary of blue hares coincides with the coastline (except, apparently, in the central part of the peninsula). Thus, blue hares were caught or their traces encountered at different times 80-170 km east of Dikson Island, along the western coast of the Pyasinsk Gulf, on the coast of the Laptev Sea— from Cape Chelyuskin to the islands of Andrei, Samuell, Fram and Faddei. Evidently, in years of intensive breeding and migrations, blue hares stray onto the islands adjacent to the coast (2-12 km), passing over the ice (Ognev, 1940). Thus, in the region of Taimy, the northernmost localities of blue hares are situated at 71°40' N. latitude (cf. Fig. 45).

Farther east the northern boundary of the range of blue hares passes along the coast of the Laptev Sea—through the mouths of the Olenek, Lena and Yana rivers and along the coast of the East Siberian Sea—through the mouths of Khorma and Kolyma rivers (Ognev, 1940). Small populations of blue hares were recorded on the Novosibirsk islands and on Begichev Island (Tavrovskii et al., 1971). Later the boundary passes along the tundra of eastern Siberia to the northern coast of Chukotka. The extreme

181

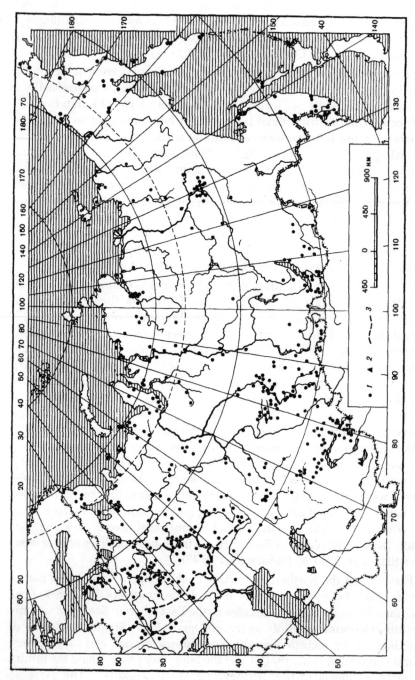

Fig. 45. See next page for caption.

northeastern locality of blue hares is the village of Uelen (collections of ZM MGU).

From Uelen the distributional boundary of blue hares turns south and passes along the coasts of the Bering and Okhotsk seas, including Sakhalin and some other islands. In Primorsk* territory blue hares are encountered in the basins of the Sitsa and Sankhobe' rivers, and southward in the Sudzukhinsk Preserve and surrounding Tigrovaya station.

In Russia the eastern distributional boundary of blue hares goes as far south as Vladivostok (collections of ZM MGU), although the number of this species is quite low here. In Sakhalin blue hares were caught in the valley of the Rukutama River, and in the vicinity of the village of Gastello (Nayoro), and also on the islands of the southern Kuril chain—Kunashir and Iturup (Voronov, 1974).

The southern distributional boundary of blue hares passes the northeastern regions of China and Mongolia. Along the state border blue hares were recorded in the Primorsk territory—in the vicinity of Khanka Island (Ognev, 1940) and along the Khor River (collections of ZM MGU); in the eastern Chita region—from the lower reaches of the Argun' River to the towns of Kailastui and Kuti (Nekipelov, 1961), and westward—in the vicinity of Chindant station, along the valley of the Arei River and the Malakhansk Range. South of Buryatia blue hares were captured in the vicinity of Troitskosavsk (now Kyakhta), and in the valley of the Khamnei River and in the vicinity of the village of Mondy (collections of IGU and ZM MGU). In southern Tuva blue hares were found along the valley of the Karga and E'rzin rivers (collections of BIN).

183   The distributional boundary of blue hares goes northward through the Kulumys Range and the upper reaches of the Bolshoi Arbat River, turns south and passes through the upper reaches of the Bolshoi to the villages of Tashanta and Chagan-Uzun (Firstov, 1957; collections of BIN). From Chagan-Uzun the boundary ascends to the north and proceeds along the Il'gumen' River to the village of Ongudai and Seminskii Pass (Ognev, 1940; Koneva, 1983). Farther west blue hares are encountered along the southern spurs of the Kalbinsk Altai. South of the Semipalatinsk region blue hares are reported in the vicinity of Ust'-Kamenogorsk, village of Kainda, and along the valley of the Urdzhar River (Sludskii et al., 1980; collections of 2M MGU).

---

*The Pacific coastal area—General Editor.

Fig. 45. Distribution of blue hares. 1—localities of blue hares (mainly based on the data of collectors' collections); 2—places of artificial dispersal; 3—southern boundary of contiguous distribution of blue hares according to the data from the winter census of 1964 (Gruzdev and Osmolovskaya, 1969).

In Tarbagatai blue hares, although rare, are found on both southern and northern slopes (excluding southern spurs of the Arkaly, Dzhaitobe and Siirektau). In the Alakol' basin blue hares are absent, but along the valleys of the Urdzhar and Tentek rivers they may reach the villages of Uch-Aral and Maiskoe. On eastern spurs of the Dzhungarian Alatau, blue hares are common in the upper reaches of the Tentek River and the basin of the Usek River, they were also recorded on the Altynemel' Range. Blue hares were not found south of the Dzhungarian Alatau—on the left bank of the Ili, and on the Ketmen' and Trans-Ili Alatau ranges (Sludskii et al., 1980).

From the Dzhungarian Alatau the southern boundary of the range of blue hares turns north and passes through the desert zone along river valleys up to the southeastern edges of Lake Balkhash. This part of the distributional boundary of blue hares is the most arbitrary, since blue hares do not regularly stray in the desert and that too only in winter. In this part, according to the data of S.I. Ognev (1940), blue hares are not found west of the village of Abakumovskii (Abakumovka). From the eastern edge of Lake Balkhash, the boundary proceeds north to the village of Ayaguz and the Chingistau mountain system, and farther west and southwest—along the Bakanas River and its left tributaries to the village of Madaniet and the Kalmakemel' mountains. The distributional boundary of blue hares is not precisely established in the region from Kalmake'mel' to the Tokrau River. Northwest of the Kalmake'mel' blue hares are reported in the lower reaches of the Espe River (50-60 km west of the Kotane'mel' mountains), in the region of confluence of the Karshigali and Kusak rivers and farther— at the eastern edge of the Arkarly mountains. From here the boundary passes south and reaches apparently up to the Bektauta mountains. Habitations of blue hares eastward and southward of Bektauta (valley of Tokrau River) need confirmation (Sludskii et al., 1980).

From Bektauta the southern boundary proceeds northwest along the southern edges of the Kazakh Melkosopochnik through the Zhanet mountain massif, the valley of Espemeirman River, the valley of the Zhamshi River to the vicinity of the village of Akchatau and the railroad station of Kiik. Descending south, the boundary reaches apparently up to the Shunak mountains, and then along the valleys of the Sarybulak and Shazhagoi rivers up to the Zhel'dytau and Munly (Mungly) mountains. Along the valley of the Atasu River, the boundary turns north and reaches up to the Taldymonak River (left tributary of the Sarysu). From Taskaraala the boundary proceeds west up to the Ulutau mountains, south of which blue hares are apparently not encountered (Sludskii et al. 1980) .

The southern boundary of the range of blue hares proceeds west along the Kazakh upland, up to the southern districts of the Kustanai region. According to the data of A.O. Solomatin (1975), blue hares earlier inhabited environs of the town of Turgai, but now forest groves suitable for blue hares have remained only on the slope of the Turgai trough north of the Moilda River. The present-day southern distributional boundary of blue hares on the Turgai Plateau passes along the line of Tersek-Karagai pine forest—Naurzum-Karagai pine forest—to the upper reaches of the Damda River (Solomatin, 1975). South of the Turgai trough, blue hares have been reported in the vicinity of the village of Irgiz. From Irgiz the boundary passes northwest to the middle fork of the Ilek River, passing through the northern edge of the Mugodzhary mountains (Ognev, 1940; Sludskii et al., 1980).

In the plains part of the Uralsk region, blue hares are extremely low in number. Probably the boundary of permanent habitations of blue hares in western Asian and eastern European parts of the range passes along the line from Aktyubinsk to Orenburg south of the Samara region. West of the Uralsk region, blue hares are distributed in the south up to the villages of Kirsanovo and Burli (Ognev, 1940; Sludskii et al:, 1980). S.I. Ognev (1940) presumed that from the southern edges of Obshchyi Syrt*, the boundary abruptly ascends north to the Samara bend, bending around the Trans-Volga steppes. According to more recent data (Gruzdev and Osmolovskaya, 1963; Tomilova, 1975), the present-day southern boundary of contiguous distribution of blue hares passes south of the Orenburg and Samara regions. Reports are available about encounters of blue hares in the Buzuluksk pine forest, environs of the town of Bolshaya Glushitsa and in the Samara preserve (Ognev, 1940; collections of ZM MGU and Kharkov University). The southern distributional boundary of blue hares in the Volga River area has not been established precisely. Apparently, at present it does not reach Kamyshin (cf. Fig. 45).

184

According to the data of V.V. Gruzdev and V.I. Osmolovskaya (1969), the southern boundary of the range of blue hares passes south of the Saratov region, farther north to Atkarsk, and later to the west-south of Voronezh, Kursk and south of the Bryansk region. The boundary, drawn by T.P. Tomilova (1975), passes north of Saratov and the central part of the Voronezh regions and then turns north and proceeds along the eastern and northern districts of the Orlov region and farther west—south of the Bryansk region (cf. Fig. 45).

According to S.I. Ognev (1940), the southern distributional boundary of blue hares east of Europe passes between Volsk and Saratov, then turns north to Atkarsk (Saratov region) and Serdobsk (south of the Penza region). Later the boundary proceeds northwest, passing between Morshansk and Tombov, in the vicinity of Ranenburg (now Chaplygin, Lipetsk region) to the northern districts of the Tula region. Thus, the southern distributional

*Common watershed upland—Translator.

boundary of blue hares, according to the data of S.I. Ognev, passes farther north than the boundary drawn at the present time. Thus the localities of blue hares in the vicinity of Borisoglebsk, Pavlovsk, and Voronezh is viewed by S.I. Ognev as disjunct from the main range of the insular part, while V.V. Gruzdev and V.I. Osmolovskaya (1969) and T.P. Tomilova (1975) include this part in the contiguous range of blue hares (Fig. 44). According to the data of S.I. Ognev (1940), from Tula the boundary turns southwest and goes to the city of Odoev (Tula region), the city of Bolkhov, and the Khotynets station (Orlov region), covering the basin of Desna with its tributaries the Navlei and Nerusoi, and proceeds west through Novgorod-Severskii (Chernigov region of the Ukraine) and southward of the city of Novozybkovo (Bryansk region).

The distributional boundary of blue hares passes along the territory of Belorussia. According to data from 1964-1969 (Gaiduk, 1970), it passes along the line Buda—Koshelevo—Rechitsa—Kalinkovichi—Peterikov—David-gorodok—Pinsk—Drogichin—Grodno. From Grodno the boundary of the range of blue hares moves west over the territory of Poland, and in the north reaches the coast of the Barents Sea.

Besides the main boundary in the southwest of the range, S.I. Ognev (1940) cites data about isolated areas of habitation of blue hares in the vicinity of Suma, in the individual districts of Poltava, Kharkov, Cherkassk, Khmelnitskii and Zhitomir regions. All these data refer to the early part of the present century (to 1930), and at present blue hares are apparently absent in these districts.

Many cases are known about attempts to introduce blue hares. Thus, in 1962, 138 blue hares were released in the Chili district of the Alma-Ata region at the lower limit of forest. Hares ascended in the forest zone, but their further fate is unknown (Sludskii and Afanas'ev, 1964). In April 1965, blue hares were introduced in the Bolshoi Kemin (northern slope of the Kungai-Alatoo Range). Here, blue hares survived well on the mountain slopes along the lower limit of spruce forest. In dense forest and shrubbly flood plain of the Chon-Kemin River, blue hares did not stray, although the distance to it is not great. According to observations in 1969 and 1977, blue hares were quite numerous in this region (Aizin, 1979).

Geographic Variation

Geographic variation of color of fur and dimensional characters are well expressed in blue hares, on the basis of which S.I. Ognev (1940) identified 10 subspecies for the territory of the former USSR: *L. t. timidus* L., 1758 (Scandinavia, Karelia, Murmansk, Vologodsk, Kostromsk, Kirovsk and Perm regions, vicinity of Tobolsk, Surgut and Narym); *L. t. kozhevnikovi*

Ognev, 1929 (from the southern boundary of the species to the southern boundary of the type form); *L. t. sibiricorum* Johans, 1923 (plains part of western Siberia, in the south—up to the Abakan, Semipalatinsk, Kalbinsk Range, Zaisan basin, northern part of Semirech'e, Kustanai region and Turgai); *L. t. transbaikalicus* Ognev, 1929 (Trans-Baikal, Bolshoi [Great] Khingan), *L. t. gichiganus* J. Allen, 1903 (upper and middle reaches of the Anadyr River, basin of the Penzhina River, the Koryaksk upland, Kamchatka, northern and western coasts of the Sea of Okhotsk); *L. t. mordeni* Goodwin, 1933 (valley of the Ussuri, possibly, lower and middle forks of the Amur River, Primor'e); *L. t. tjrii* Kuroda, 1928 (Sakhalin), *L. t. kolymensis* Ognev, 1922 (basins of the Kolyma, Yana, and Indigirka rivers); *L. t. begitschevi* Koljushev, 1936 (Gydansk Peninsula, Taimyr, lower reaches of the Yenisei, in the east—to the Khatanga Gulf); *L. t. tschuktschorum* Norquist, 1823 (Chukchi and Anadyr lowland).

According to the system of A.A. Gureev (1964), blue hares inhabiting the territory of the former USSR, belong to the type subspecies, and beyond the borders of Russia the existence of three more subspecies is recognized: *L. t. hibernicus** Bell, 1827 (Ireland), *L. t. arcticus* Ross, 1819 (extreme north of North America) and *L. t. groenlandicus* Rhoads, 1896 (north and northwest Greenland).

In the works of R. Angermann (1967) and O.L. Rossolimo (1979), devoted to the analysis of geographic variation of metric characters of the skull, clinal variations have been demonstrated. The results of our studies (multifactorial analysis of 18 metric characters of the skull of 200 blue hares) almost completely agree with the scheme of clinal variation mentioned by Rossolimo (1979), which provides the basis to doubt the advisability of identifying subspecies in *L. timidus* (E.Yu. Ivanitskaya).

Biology

Habitat. The most typical habitats occupied by blue hares are forests and shrub thickets. Blue hares avoid deeper areas of taiga. The maximum number of blue hares is usually recorded in forests with a large number of glades. Such ruggedness of the territory presents blue hares their defensive properties, since it helps combine shelters with open space, necessary for maneuvering in the moment of danger.

Blue hares may colonize areas in direct proximity to human habitations, and their number sometimes is even higher in the vicinity of villages. In such places blue hares feed in winter on winter crops and in hay fields and where hay is gathered near roads. In strips of forest adjoining fields,

hiberinicus in original—Editor.

blue hares are considerably more numerous than deep inside the forest, where they are often altogether absent (Aspisov, 1936). In western Siberia the maximum population density of blue hares is confined to birch forest-steppe; even here blue hares are adapted to conditions that change as a result of human activity. In the forest zone places with the maximum density of blue hares coincide with territories occupied by man (Bedak, 1940; Laptev, 1958).

In winter blue hares concentrate near remnants of fellings, particularly in those places where European aspen has been felled. The presence of people and technology in forest clearings does not frighten blue hares, and they settle for daytime naps often in direct proximity of felled trees. In summer, escaping from dampness and mosquitoes, blue hares often come to logging trails; here in dry weather they indulge in "dust baths". Log-earth embankments, formed during the construction of logging roads, may serve as places for breeding young. Near such roads, in the beginning of winter, there is highest density of blue hares (Agafonov, 1975).

Blue hares are found throughout the Kola Peninsula, including also in mountain tundra, but nowhere do they reach high numbers. Most often, particularly in winter, blue hares may be found in spruce forests and along banks of streams. Here, in spruce forests there is a relatively large number of deciduous species, serving as winter food for blue hares, and the depth of snow cover is not as great as in other places. In summer spruce forests are rich in green foods (berry shrubs and semi-shrubs); and in damp places, herbaceous vegetation. Considerable litter and dense underbrush, typical of spruce forests of this area, serve as good places for shelters. In the snowless period blue hares are quite common in mountain tundra (Vladimirskaya, 1958).

In Karelia in the beginning of winter, blue hares are most often found in mixed spruce-birch forests along banks of water bodies, along stream and dense green spruce forests, and in broad-leaved sparse forests. In January-February hares disperse to forest fellings and edges of sedge swamps. At the end of winter the number of blue hares increases appreciably in the swampy herbaceous and stream-bed spruce forests, and blue hares are particularly numerous in anthropogenic habitats—along groves and cultivated lands, in fields with shrubs and leafy young growth along boundaries. The habitats of blue hares when their numbers are high are much more diverse than with low numbers (Ivanter,1969). In the Leningrad [St. Petersburg] region, 46% of encounters of blue hares were in broad-leaved forests and shrubs with fellings, 40%—in mixed forests with fellings, 7%—in spruce forests, 3%—in pine forests, and 1%—in mossy swamps. In winter blue hares prefer to settle for rest in spruce forests (Novikov et al., 1970).

In mixed forests of the Russian plains (Moscow, Vladimir and Ivanovsk regions) the character of distribution of blue hares is determined mainly by the height of the woodstand and to a lesser extent by species compostion of the woody vegetation. Moreover, the biological affinity of blue hares changes during winter. In December the maximum proportion (to 65-69%) of trails and fattening of blue hares was recorded in middle-aged and old forests and the lowest (9-11%) in young forests with tree heights of 1-3 m. On open lands at this time, hares do not venture out at all for feeding. In February-March in young forests with tree heights of 1-3 m, the number of trails and fattening increases to 24%, and in middle-aged and old forests decreases to 17%. The food resources of young forests 4-7 m high and open lands (to 68 and 33% fattening, respectively) are utilized most intensively in this period. Thus in the beginning of winter at any density of blue hares, the main feeding bases are middle-aged and old forests. In the first half of December, blue hares with any numbers are absent in young forests 1-3 m high. In February-March the main feeding activity of blue hares shifts to young forests with tree heights of 4-7 m and with high numbers—also in young forests 1-3 m high. Young forests with heights of 1-3 m are not used by blue hares intensively at low and medium densities—4-18% fattening, as compared to 24% with high density (Mamaev, 1979).

According to the data of V.A. Agafonov (1975), in the Kirov region the dispersal of blue hares is also determined mainly by the defensive properties of lands. The number of blue hares here is higher than in places where there are more spruce groves and edges. With continuous concentrated fellings of forests, for blue hares unfelled conifers along the stream banks and ravines play a favorable role. Southward (in Tataria) blue hares at any time of the year prefer mature forests with a predominance of European aspen and broad-leaved young forests, and in the second half of summer and in the beginning of winter—purple osier willow thickets in the flood plains of rivers (V.A. Popov, 1960).

In Kazakh melkosopochnik blue hares live in woody plantations, in dense thickets of chee grass, in thickers of pea shrub along hill slopes, and low hill valleys (Beme, 1952). In insular and winding pine forests (Karkaralinsk, Borovoe, Kyzylrai, etc.), blue hares remain in young pine or birch-pine forests with an underbrush of rose, pea shrub and other shrubs, and also in mixed dense forest with osier beds. In the vicinity of Karkaralinsk, blue hares were found in flood plain tall pine forest, and in Zerendinsk pine forest—on lowlands near swamps or in birch thickets (Sludskii et al., 1980). In the Ulutau and Bektau-Ata mountains (northern Lake Balkhash area), blue hares inhabit birch-European aspen groves, in juniper stands on mountain slopes, along valleys and flood plains of steppe rivulets with banks, overgrown with shrubs or European aspen thickets,

258

and sometimes they are found on rocky slopes of hills, poor in shrubs (Afanas'ev and Sludskii, 1947; Sludskii, 1953).

In the Alakulsk basin blue hares are found along stream thickets of osier and also dense herbage with isolated bushes of spirea, winter fat, and wormwood (Sludskii, 1953). In the Zaisan basin blue hares are found very rarely. Here, they remain in sandy depressions with birch groves (Afanas'ev and Bazhanov*, 1948).

187   Blue hares occur throughout the Altai and colonize all altitudinal zones from the foothills and montane forests, including the bald peak zone, but prefer to remain in river valleys with thickets of birch, European aspen and shrubs, forest edges and meadows. Blue hares are also found in fir forests, if there are broad-leaved species there (Sludskii et al., 1980). According to the data of B.E. Yudin, in the subgoltsovoi [sub-bald mountain and goltsovoi [bald mountain] zones of the northeastern Altai, blue hares are also numerous, as well as along river valleys (Yudin et al., 1979).

South of Western Siberia blue hares are most frequently found along valleys of large rivers, where they colonize willow thickets and fellings with birch and European aspen stands. In pine forests, blue hares remain on forest edges. In the zone of forest steppes, blue hares are most numerous along lowlands overgrown with birch groves, and in steppe—in hummocky marshes and places overgrown with grass. The numbers of blue hares are high in places where fields alternate with groves of broad-leaved trees and thickets of osier and also among shrubs on the edges of swamps (Bedak, 1940).

In the western Sayan blue hares live in forest-steppe and taiga zones, where they prefer mixed forests with swampy herbage and shrubs. Along valleys of small rivers and streams, blue hares ascend to high montane tundra at an altitude up to 2,200 m above sea level. Continuous coniferous forest devoid of underbrush is avoided by blue hares, but in taiga they may be found only in places reclaimed by man (Dul'keit, 1964; Yudin et al., 1979). In the forest steppe and forest zones of Khakasia, blue hares colonize osier beds along river valleys, thin forests, forest edges and in the vicinity of human habitations. In foothills, besides the valleys of rivers and streams, the favorite places of colonization by blue hares are ravines overgrown with rose and birch (Kokhanovskii, 1962).

In the area of the Baikal basin, blue hares colonize all altitudinal zones, but prefer larch forests with an admixture of birch, European aspen or shrubs. In sub-goltsovoi [sub-bald mountain] zone they live in thickets of osier and ground birch, and sometimes-dwarf arctic birch. Large stone

*Not in Lit. Cit.—General Editor.

mounds of the sub-goltsovoi [sub-bald mountain and goltsovoi [bald mountain] zones are used by blue hares as shelters, and for foraging they visit nival and lacustrine meadow patches, and willow and birch groves (Shvetsov et al.* 1984). On the Barguzin Range the number of blue hares in the sub-bald mountain zone is even higher than in forests (Zharov*, 1978). In the eastern Cis-Baikal area, blue hares are common in forest-steppes, on slopes with thinned-out forests and riverside shrubs.

In the central part of Yakutia in early spring, blue hares are most frequently found on rejuvenating burned-out forests, in coastal willow groves and larch nurseries. In the beginning of summer (to June), their favorite places are open areas warmed by the sun, where greens appear earlier than at other places. According to the data of Yu.V. Labutin (1988), at this time up to 80% of hares forage on open mounds lying closer to the old burned-out forests or dense groves of young larch. In young forests, despite better protective conditions, blue hares are absent at this time, since the snow remains here for a long time and there is almost no herbaceous food (M.V. Popov, 1960). From the second decade of June, blue hares are met with in all habitats for which leafy vegetation is characteristic to some extent. Most of all they are found in rejuvenating burned-out forests, large groves with varying degree of development of underbrush and nursery, and in young forests. All these habitats have good protective conditions and a large stock of food, which is very important during the reproductive period and the feeding of young. At this time females remain mainly in streambed larch forest habitat with a dense underbrush and rejuvenating burned-out forests. Males often occupy dense larch forest massives on slopes of ridges. In pine forests without underbrush, swamps and plowed fields, hares are not found at this time of the year (Labutin, 1968).

In the second half of July to the beginning of August, blue hares may leave their places of daytime resting to a distance of up to 300 m in search of succulent foods. At this time they are most frequently seen in humid glades and meadow patches, forming quite large groups. With the onset of cold they begin to move to lands richer in ramal foods—in ground birch forest and coastal thickets; moreover the first to disperse are young, the winter molting in which is prolonged, and early balding adults remain longer in larch forests with dense underbrush. During this period blue hares are generally distributed uniformly, occupying those places which were avoided by them in summer. In mid-winter blue hares are most numerous in swamps, alases (interdepressions with shrub and meadow vegetation) and in burned-out forests, and also in floodplain pine forests with undergrowth (Lablutin, 1989).

188

---

*Not. in Lit. Cit.—General Editor.

In southern Yakutia the most favorable places for blue hares are the valleys of rivers with a broad flood-plain, insular and high flood-plain terraces. In such habitats usually there are a large number of shelters in the form of deposition of driftwood and felled forests, and the rich and diverse vegetation serves blue hares as a good fodder base. Along valleys with montane streams, blue hares ascend to the level of water divides; however, their number here is not high. In southern Yakutia, unlike other regions, blue hares are not numerous in burned-out forests (Revin, 1989).

In the northern Far East blue hares are found from the plains tundra and valleys of mountain taiga streams to peaks of mountain ranges. However here, and also in other parts of the range, blue hares have a definite biological selectivity, determined primarily by the nature of the landscape. In mountain-taiga landscape blue hares prefer to settle in riverside willow groves, and in mountain terrains along valley bottoms and mountain slopes with large groves of cedar elfin woods, and in taiga terrains—in rejuvenated burned-out forests and fellings with an undergrowth of larch and developed herb and shrub cover. In coastal tundra blue hares settle in flood plain shrub thickets and piles of driftwood, and in mountain tundra—on undulating parts of slopes with large stone mounds and rich herbaceous vegetation (Yudin et al., 1976).

In the northern part of their range, massive migrations of blue hares have been reported. Large migrating groups of blue hares were observed in the beginning of April 1955 on the coasts of the Barents Sea and Kara Sea. Such groups comprised from 100 to 500 hares, and on April 7, 1955 on the Silova-Yakha River, a "herd" of blue hares was observed, occupying an area of 0.5 ha; after 11 days this "herd" was still at a distance of 12-15 km from the place where it was first encountered (Makridin, 1956).

In Yamal fall migrations of blue hares have been observed to take a northwestern direction, and during spring migration—in the opposite direction. At the end of May 1954 in the vicinity of the village of Yara of the Priuralsk area of the Yamalo-Nenets autonomous district, two groups of blue hares with 1,500-2,000 hares in each were recorded, which moved in the direction from west to east. And in fall, blue hares in this region moved in a northwestern direction. The daily translocation of hares may constitute up to 15 km (Pavlinin, 1971).

In the territory of Yakutia, massive migrations of blue hares were observed in years of high numbers. Thus in fall of 1942 on the right bank of the Lena near the mouth of Vilyui, in the course of two nights a large group of blue hares was observed migrating north. In the Verkhoyansk district in fall 1955, numerous traces and well-beaten tracks of blue hares were observed along the banks of the Yana before complete freezing (Labutin and Popov, 1960). In March 1958, in the second year after a peak population in Verkhoyane,* partial and irregular migrations of blue hares were observed, which occurred against the backdrop of an acute shortage of main winter foods. During such migrations hares can form groups of 5 to 15 individuals (Labutin, 1988).

However, massive migrations of blue hares, while a quite rare phenomenon, occur in years with high numbers in territories with extreme conditions. More common and regular for blue hares are seasonal strayings from one habitat to another, which are linked with the disposition of shelters and foraging groups. In western Siberia seasonal strayings of blue hares are determined, in addition, by the quantity of precipitation. In rainy years, when lowlands with groves are flooded with water, blue hares move to elevated places, while in dry periods they prefer to remain in thickets, which are more reliable shelters. In winter blue hares move from those places where snow is high and loose, for instance, from willow groves. In the northern part of their range and in the foothills, where arboreal vegetation is absent, blue hares stray from lowlands to higher places (Bedok, 1940; Pavlinin, 1971).

Seasonal migrations of blue hares in the north of the range depend on blood-sucking flies, which drive hares into open places. The maximum concentration of hares in the period of massive attacks by blood-sucking insects is observed on the banks of rivers, sandy deposits, and in wind-swept places with low vegetation. After the onset of the first frosts, blue hares abandon open places (Pavlinin, 1971; Labutin, 1988).

Population. According to the data of the first All-Russian route census of game animals, conducted in 1964, the total number of blue hares in the territory of Russia constituted not less than 7 million; the share of the European part and the Urals was more than 2.5 million. On the whole, the number of blue hares in the year of the census was at a higher level. The maximum population density of blue hares in European Russia was reported in Moscow, Kostroma, Yaroslavl' and Ivanov regions (Novikov and Timofeeva, 1965; Priklonskii and Teplova, 1965). Moreover, according to the data of this census, a link was established between natural conditions of individual territories and population density of blue hares (Gruzdev and Osmolovskaya, 1969).

The territory with high numbers of blue hares in the European part in general coincides with the zone of mixed forests of the Russian plain, where birch and European aspen forests predominate. Here blue hares prefer small, disjunct parts of broad-leaved sparse forests alternating with spruce groves (less often with pine groves), plowed lands and meadows, etc. In such places forests are usually strongly altered by fires and fellings, and

189

---

*Upper Yana River area—General Editor.

agricultural holdings occupy 40-50% of the area. The territory with a moderate population of blue hares is situated eastward, where birch and European aspen forests are considerably fewer. The northern boundary of the territory with moderate and high population of blue hares almost coincides with the northern boundary of the southern taiga and passes along regions where agricultural holdings occupy 15-25% of the territory.

In the eastern part of the range, the same pattern is observed—a decrease in population of blue hares in the north and south of the optimal belt, covering mainly the forest-steppe zone. Moving north, that is, with an increase in areas covered by continuous forests, the population of blue hares decreases. In central districts of the Novosibirsk region (left bank of the Ob' river) and in the rightbank areas, agricultural holdings occupy an almost similar area (52.6-58.3%). However, the population density of blue hares in these areas differs quite considerably. The highest population of blue hares is observed in the rightbank territory.

Here blue hares inhabit foothill taiga, mixed and flood-plain forests, and in forest-steppe—numerous valleys and ravines with willow groves and birch forests (Polyakov*, 1969). Here, higher numbers of blue hares exist because of the dissected relief.

In Barabinsk flat-plain leftbank forest-steppe, the population density of blue hares is not high because of fewer places suitable for shelters and high loose snow in winter. Earlier the forest-steppe population of blue hares in Barabinsk was higher, since the forest was not trampled by cattle and more herbs remained in fields after harvest. Now only in the vicinity of Lake Chan shrub-thickets are still present and hares are appreciably more abundant there, than in other leftbank areas.

In the northern Kazakhstan region during 1965-1970, the population density of blue hares fluctuated in different habitats from 11 individuals/1,000 ha in grassy birch forests to 62-69 individuals/1,000 ha in flood-plain thickets and birch forests with sedge (Utinov, 1979). In the Ermitau mountains (Tselinograd region), in different habitats and with different methods of counting, in 1970 the density of blue hares fluctuated from 20 to 25/1,000 ha. In summer 1967, in valley forests of birch-European aspen and pine of the Kent mountains (Karaganda region) the density of blue hares was 27 individuals/1,000 ha, and roughly the same density (30 individuals/1,000 ha) was noted in fall of 1966 in the vicinity of Borodulikha (Semipalatinsk region). A very low population of blue hares (2/1,000 ha) was reported in summer 1958 in the Karkaralinsk mountains, and a very high density (200 individuals/1,000 ha) was observed in summer 1962 at the Begenevsk forest farm of the Semipalatinsk region (Afanas'ev and Grachev, 1970; Sludskii et al., 1980).

190

*Not in Lit. Cit.—General Editor.

In the eastern part of the Dzhungarian Alatau, where blue hares are fewer throughout, their polpulation density in different habitats varies considerably. Thus, in 1967 in the Tentek basin in spruce groves, 3 hares/ 1,000 ha were recorded, and in birch groves—52; in other habitats blue hares were altogether absent (Sludskii et al., 1980).

In Tuva blue hares are most numerous in shrub thickets of the subalpine zone and in river flood plains (4.0 and 3.3 individual for a 10 km route, respectively) and quite rare (1.4 for 10 km) in dark coniferous taiga (Boyarkin, 1984).

In the territory of Yakutia, the highest population of blue hares was reported in the basin of the upper stream of the Vilyui River in the western part of the Lena-Vilyui interfluve, in the southwestern part of the Aldan basin, in the lowland part of the Indigirka basin and in the Kolyma basin (10-30 individuals/1,000 ha), and also in the mountain-taiga habitats of the Verkhoyansk and Chersk ranges, where commerce of blue hare may reach 50 skins/1,000 ha. In Yakutia the zone of high population is in its central part (basin of lower and middle forks of the Vilyui, Lena-Vilyui interfluve and the northeastern part of the Lena-Aldan interfluve), where, in years of peak population of blue hares, up to 100 skins were obtained from 1,000 ha. The inter-range depression of the Yana basin is the part most densely occupied by blue hares, and up to 190 skins were collected from 1,000 ha. In the lower reaches of the Lena with an overall low population of blue hares, there are regions, where under favorable conditions (mainly in the absence of high floods), the density of blue hares may reach several hundred individuals per 1,000 ha. On islands regularly inundated with water, blue hares are not found; they are very rare in the majority of places of terraced flood plains and supra-flood plain terraces (Vol'pert et al., 1988).

According to the data of M.V. Popov (1960), for 18 trading seasons (1938-1956) 85% of skins of blue hares were processed in the Central and Vilyui group of regions of Yakutia, which constitute just 14% of the area of the republic. The proportion of agricultural holdings in the given case has no significance. In the Central and Vilyui group of the region it constitutes, as also in southern group of regions, about 3%, but in southern regions, which occupy 13% of the total area of the republic, only 0.6% of the total number of skins of blue hares are processed. In Yakutia the places with a maximum density of blue hares, to a still greater degree than in western Siberia, are determined by the degree of dissection of relief and forest canopy. It is in the Verkhoyan'e area, were the population of blue hares is highest, mottley relief is associated with a large number of lacustrine, and other depressions are manifested most strongly.

191   *Shelters.* The nature of shelters of blue hares depends on their habitats

and time of the year. Usually, blue hares have three types of shelter: open, semi-open and closed. To open shelters belong resting places of blue hares, in which hares are exposed to the action of precipitation. Semi-open shelters usually are situated under individual standing firs or under branches of shrubs, under snow-bent branches of young broad-leaved trees, and also upper trunks or roots of fallen trees. To closed shelters belong snow burrows or such shelters, which protect hares from the effect of sharp fluctuations of air temperature, from wind and precipitation (in stands of dense spruce underbrush, in piles of fallen trees, under dense shrubs). In winter, closed shelters, as a rule, have an extensive snow cover and the entrance to such a shelter is oval or arcuate in form with a diameter of 15-25 cm, sometimes—to 50 cm (Mamaev, 1979).

The proportion of closed shelters in different parts of the range of blue hares varies from 17 to 69%. Closed shelters are most often used by blue hares in the northern part of their range. In the very same locality, with the lowering of air temperature and an increase in the quantity of precipitation, the occurrence of open shelters of blue hares is reduced, and with an increase in the quantity of precipitation the number of instances of open shelters of blue hares decreases and the number of semi-open shelters increases. With strong rain and wet snow, and also in the period of strong frosts, blue hares prefer shelters of a closed type or burrows. Toward the end of winter blue hares use open shelters more often than in the beginning (Mamaev, 1979).

For blue hares the best protective cover is offered by middle-aged and old forests, where there is an excess of the most frequently used shelters. It is in these habitats that there is a maximum proportion of blue hare shelters (Table 9.1). The height of the underbrush and growth of seedlings (particularly of spruce) also play an important role in the choice of places to rest. With height of spruce from 1 to 3-4 m, the development of lower branches is optimal for resting places of blue hares.

Table 9.1. Disposition of resting places of blue hares in the center of the European part of Russia (in % of 270 counted resting places (from Mamaev, 1979)

Population level	Land-use type				
	young forest, 1-3 m	young forest, 4-7 m	middle-aged forest	old forest	open habitats
Low	0.0	10.0	46.7	36.7	6.6
Medium	7.0	11.6	28.7	50.4	2.3
High	5.4	9.0	39.6	38.7	7.2
Average	5.6	10.4	35.2	44.1	4.8

In selecting any type of resting places, blue hares prefer places with brushwood and fallen trees. In constructing shelters of closed and semi-open types, preference is shown for groves of spruce with seedling growth. Since shelters serve hares not so much as protection from unfavorable weather conditions, as much for masking resting places from enemies, brushwood and spruce undergrowth should be considered ideal shelters for blue hares.

In winter in fresh fellings, blue hares use hollows under snow, formed among the remnants of striped trees, for shelter. Moreover, usually for hares in such places there are two resting places—one open on the snow surface, and another—in a burrow under the snow. At dusk hares use the upper "scouting" resting place, and in midday and in bad weather—the burrow. In instances when the danger of detection is very late, hares do not run away, and enter the burrow. If there are many hollows under the snow, hares have the possibility to move under the snow. Such dependable shelters are used by blue hares as more or less permanent shelters; some hares lying in such places do not perform the ritual swelps* (Agafonov, 1982).

In Karelia, in the north, blue hares predominantly make their burrows under snow, while in southern regions open shelters predominate. Most often (in 65% of cases) shelters are located under low young spruce, less often—willow thickets, under juniper, in young birch seedling growth or in brushwood (Ivanter, 1969).

In the Yamal and Polar Urals in winter, blue hares always use snow burrows as shelters, which run horizontal in snowbanks, and in level places—with a down slope of up to 40-45°. Burrows are usually straight and are 50-60 cm long (rarely 100 cm). Hollows under the snow may be used as shelters, which are formed in dense osier thickets. In snow burrows hares not only build resting places, but, possibly, also use them for feeding. Resting places outside of burrows are extremely rare and, as a rule, are used with low snow cover and on warm winter days. Hares, scared off from the burrow, do not build a second burrow but lie down directly on 192 the snow (Pavlinin, 1971). In summer, to ward off blood-sucking flies, blue hares may dig hollows, the depth of which may reach 70-80 cm. Such burrows in the period of massive flights of blood-sucking insects, are used by hares repeatedly, and they periodically restor them (Labutin, 1988).

In northern Kazakhstan blue hares select resting places usually in open areas, where the field of view and the possibility of an unobstructed run in the event of a sudden danger are good. In the absence of causes of

---

*Sudden sideward long hops—General.Editor.

anxiety, their resting places are relatively permanent. In summer in resting places, hares only partake of grass; in winter they may dig undersnow burrows which do not extend 1 m. However, more often, for places of winter resting, blue hares just dig snow, making a shallow (from 8 to 12 cm deep) pit, the length of which is 40-44 cm, and width—18-22 cm. In the Kazakh upland, resting places of blue hares are located most often on the edges of osier thickets or thin forest. In warm wather resting places are located at the base of trunks of grown dense osier beds in well aerated places or on crests of ridges—among small stone blocks, in the shade of rocky ledges or in individual thickets of juniper grove. To avoid mosquitoes in summer, in river valleys blue hares build shelters in open places—in depressions between large stones or under fallen trees. In dry, hot hills in the southern Kazakh upland, blue hares prefer northern slopes for shelters, where they settle down in the shade of overhanging rocks (Sludskii et al., 1980).

In western Siberia, in the first powdery fresh snow, blue hares prefer open dry mounds for their daytime resting. Even in the case when a swamp is bordered with dense osier beds, in a majority of cases hares prefer lying down on open mounds. In fall, in open fields and dry meadows, blue hares lie down among weeds along old plowed fields and in solitary groves of saw thistle; and in flood plains, in small osier beds. In dissected places blue hares prefer to lie down in depressions with hummocks. Deep in winter when choppings are covered with snow, blue hares rest in osier beds or in open fields, where they dig deep burrows in snow banks (Ponmarev, 1934).

Feeding. Diet of blue hares has a clearly expressed seasonal character. In the summer diet herbaceous plants predominate; in winter, ramal food. Transition from a winter to summer diet occurs gradually and depends on the periods of thawing of snow and beginning of vegetative growth of herbaceous plants. Usually the bulk of the diet of blue hares is made up of background species of plants, although selectivity of food in these hares is manifested quite clearly.

According to the data of M.I. Vladimirskaya (1955), the bulk of food in the winter diet of blue hares of the Kola Peninsula is made up of thin
193 branches of birch (31.4% of the total consumption). However, here the preferred food of blue hares is the rarely occurring willow (22.3%). Mountain ash (18.3%) and stones of berry shrubs (16%) play a considerable role in the winter diet; blue hares actively search mountain ash and dig up branches of berry shrubs from under the snow. Since European aspen occurs rarely in the north, it does not have great significance (2.9%) in the diet of blue hares of the Kola Peninsula. In Finland, where, as in the Kola Peninsula, blue hares in winter most avidly consume willow, given the choice, they prefer to nibble at shoots from the tips of mature plants, and not the young shoots. Directed by their sense of smell, hares choose plants

or their parts with lesser concentrations of phenolic glycosides (Helle et al.,* 1986).

In Karelia the preferred foods of blue hares are European aspen, mountain ash and willow—up to 30-40% of trees of these species are usually nibbled at by blue hares, while for birch, being the background species, this proportion is 8.7% (Ivanter, 1969). These very tree species (96% of all consumption) constitute the bulk of winter diet of blue hares in the Leningrad [St. Petersburg] region (Novikov and Timofeeva, 1965). In the Vologoda region 42% of consumption by blue hares consists of willow; 25.7%, birch; 17%, European aspen; the rest shared by juniper, cowberry, blueberry, and herbaceous plants. In the Moscow region, European aspen constitutes 30% of consumption, various willow species—25.4%, birch—7.5%, filbert—22.2%, and oak—10.2% (Naumov, 1947).

The dietary regime of blue hares has its own peculiarities in different periods of winter. According to the data collected in the Vologoda region in October-November (with height of snow cover being 10-20 cm), some features of the summer dietary regime are retained in the feeding of blue hares, when the bulk of the diet comprises herbaceous plants, and also shoots and leaves of berry shrubs (50.5%). In December-January with a snow height of 30-45 cm, woody and shrub species (90%) play a major role in the diet. In the second half of winter (February-March), berries are excluded from the diet, and the significance of woody species and shrubs increases still more (Naumov, 1947).

It is reported that the spectrum of winter diet of blue hares may differ even within limits of a relatively small territory. Thus, in the Volga-Kama territory in mixed forests, consumption of birch and European aspen (39.4 and 24.9%), in oak forests—oak and filbert (20.5 and 20.5%) and, in river flood plains—willow and filbert (61%and 19.9%) play a predominant role in their feeding, that is, the food composition is determined to a large extent by the predominance of any one species in the underbrush and seedling growth (Naumov, 1947). According to the data of V.A. Popov (1960), in the territory of Tataria, the fall-winter diet of blue hares inhabiting broad-leaved forests is most varied. However, in November, blue hares consume dry grass, willow, European aspen, birch, filbert and oak roughly equal to their extent (from 10-20% of consumption). Only in February-March does European aspen begin to predominate appreciably in the diet. The fall-winter diet in the flood plain of the Volga and Kama consists of predominantly willow and dry grass (particuarly in November), and in dry-bottom forests—birch (Aspisov, 1936; V. Popov, 1960).

In northern Kazakhstan during winter, the dietary regime of blue hares not only changes the ratio of herbaceous and ramal food, but also the

---

*Not in Lit. Cit.—General Editor.

268

fraction of any species of woody trees. Thus in the flood-plain holdings rich in osier beds, the food species predominates throughout winter, and in birch forests most often blue hares feed on birch shoots (Table 9.2). However, the winter diet of blue hares on the Ermentau mountains is not predominated by the European alder, which grows there in abundance, but rather the shoots of willows, European aspen, birch, spirea, black berried cotoneaster and European bird cherry (Sludskii et al., 1980). In the Kazakh melkosopochnik [low-hilly area] in winter, trees and shrubs of artificial plantations—poplar, *Elaeagnus*, Siberian cherry and box elder— are damaged to a great extent by blue hares (Beme, 1952).

Table 9.2. Plants found in the winter diet of blue hares in flood plains and birch groves of northern Kazakhstan (in %) (Sludskii et al., 1980)

Food species	Flood plain holdings			Birch groves		
	November n = 543	January n = 1013	March n = 580	November n = 277	January n = 913	March n = 482
Willow	43.0	68.0	73.0	4.0	14.0	16.0
Aspen	4.5	12.0	15.0	16.0	24.0	22.0
Poplar	–	6.0	4.0	–	–	–
Birch	2.0	8.0	6.0	44.0	54.0	56.0
Rose	12.0	2.0	0.5	9.0	2.0	1.0
Black cherry	9.0	0.5	–	–	–	–
Bird cherry	4.0	0.3	–	–	–	–
Laurel willow	–	–	–	–	2.0	3.0
Ground cherry	2.0	–	–	4.0	0.5	–
Spirea	3.0	0.2	–	2.0	–	–
Herbs	20.5	3.0	1.5	21.0	3.5	2.0

The complement of winter foods of blue hares in the Altai and in western Siberia, as also in other parts of their range, depends to a considerable extent on their habitats. In forests blue hares feed exclusively on small branches of birch, European aspen and osier beds. In forestless areas they dig up herbaceous plants from under the snow, and on wastelands they eat seeds of various weeds (Bedak, 1940).

In the Angara River area, in almost all habitats hares most intensively eat goat willow, and in those habitats, where the willow is rare, preference is given to European aspen and birch. In the forest-steppe part of central Siberia, hares prefer shoots of European aspen and hawthorn, often damaging plantations of poplar and fruit trees; here they eat birch and poplar as the last choice. In Evenkia in the zone of forest-tundra and young larch plantations, the winter diet of hares shows a predominance of bilberry and in leafy young plantations of birch and European aspen (Shishikin*, 1988).

The winter diet of blue hares in Yakutia, consisting as also in other parts of the range mainly of woody-shrub vegetation, has some preculiarities. First, here blue hares almost do not utilize European aspen in their diet, which in other parts of the range together with willows and birch constitutes the bulk of the winter diet of blue hares. Second, in Yakutia one of the main foods of hares is Daurian larch, found here more often than other woody trees (cf. Table 9.3). However, given the choice, blue hares eat larch species as first priority. Thus in experimental conditions with an abundance and rich choice of foods, blue hares preferred branches of birch (41% of total intake of food) and willow (26%),while European aspen and larch constituted only 9% and 5%, respectively. In that case, when blue hares were offered only one food species, in a day they consumed 783 g of birch, 689 g of willows, or 315 g of larch; with an exclusive feeding on larch, blue hares lost weight (M. Popov, 1960). Thus, Yakutian blue hares are forced to feed on larch only because of its abundance in this region.

Table 9.3. Winter diet of blue hares in Yakutia, based on multiyear observations (from Tavrovskii et al., 1971)

Plants	Frequency of consumption, % of total number of cases (n = 9839)	Composition of arboreal species in places of occurence of blue hares, %
Birch	24.1	27
Willow	23.9	28
Larch	20.3	38
Rose	3.8	3.8
Spirea	1.6	2.1
Northern red currant	0.3	1.0
Alder	0.2	–
Poplar	0.21	–
Pine	0.2	–
Hawthorn	0.1	–
European aspen	0.1	0.1
Herbaceous plants	16.8	
Berries	8.24	
Hay	0.1	
Ledum	0.04	
Fungi	0.01	

*In Russian original, spelled so throughout, but Shishkin in Lit. Cit.—General Editor.

In the case of larch, blue hares in more than 50% of cases nibble only thin (to 2 mm in diameter) shoots, whereas in birch and willow they eat shoots thicker than 2 mm (6% and 86%, respectively). Moreover, while feeding on larch, blue hares usually nibble only isolated shoots, but while feeding on birch and willow most often massive nibbling of shoots is observed (Tavrovskii et al., 1971).

In the northern Far East in valleys of tundra rivulets, blue hares in the winter period, because of thick snow cover, feed mainly on apical shoots of willows and considerably less often—alder groves. In elevated places and cliffs, where thick snow cover is absent in the beginning of winter, blue hares dig out green shoots and berries of cowberry, cones and alpine ptarmigan berry. In the taiga zone of this region, the diet of blue hares is more varied and besides willow shrubs includes branches of other shrubs (Middendroff birch, and *Betula exilis*, mountain ash, rose, spirea, northern red currant, etc.), and also seedling growth of larch (Yudin et al., 1976). In the Koryaksk upland blue hares eat branches of willow and *Chosenia*, and also bark of poplar. On mountain slopes in the zone of cedar groves, blue hares may dig out cones from under the snow and feed on nuts (Portenko et al., 1963).

In Kamchatka willows and Erman's birch predominate in the winter diet of blue hares. Blue hares rarely consume alder and poplar. In this region horsetail, rose, dry stems of spirea and some shrubs are the second-order foods (Chernikin, 1965).

A comparative analysis of winter diet of blue hares in different parts of the range shows that in mixed forests broad-leaved species predominate in their diet, while in the taiga zone the proportion of coniferous species increases considerably (Table 9.4). Over a large part of the range twigs of willows, birch and European aspen constitute about 50% of the winter diet of blue hares. In the zone of continuous forests, hares can destroy a large part of seedling growth of willow and European aspen in winter, which is usually situated along forest edges and glades (Naumov, 1947).

According to observations in captivity, the daily food requirement of blue hares in January in the territory of Yakutia is more than 35% of the body weight of the animal; while in March-April, not more than 49%. During the period of daytime resting, hares exhibited coprophagy—in 13-16 intakes they ate from 70 to 167 g of soft feces containing from 24 to 36% crude protein (Pshennikov et al., 1988). A positive correlation has been established between the quantity of food and the content of crude protein in feces of hares. Protein content in the feces of hares below 7.5% confirms the critical state of the quality of foods, at which hares are in a condition to maintain normal body weight (Sinclair et al.,* 1983).

---

*Not in Lit. Cit.—General Editor.

Table 9.4. Principal composition of the winter diet of blue hares in different parts of the range, in % of total number of consumptions of trees and shrubs (Novikov and Timofeeva, 1965)

Species	Taiga forests						Mixed forests		Oak forests
	Yakutia	Murmansk region	Arkhangelsk region	Vologodsk region	Leningrad region	Southern Finland	Moscow region	Tataria	of Chuvashia
Willows	22.3	25.4	6.0	37.5	54.1	58.4	25.0	12.5	–
Birches	29.3	35.8	9.0	28.5	15.9	22.9	4.6	39.4	17.0
European aspen	0.2	3.2	–	16.4	18.1	16.7	26.6	24.9	12.5
Raspberry	–	–	–	3.4	10.3	0.8	1.9	–	–
Mountain ash	–	22.8	49.0	0.1	0.3	0.4	1.1	0.1	–
Alder	9.5	3.8	–	0.6	1.1	–	–	–	–
Rose	2.6	–	–	–	0.1	–	0.4	0.1	–
Broad-leaved	–	–	–	–	–	–	37.1	15.6	70.5
Conifers	33.7	7.1	36.0	4.1	0.1	0.4	0.1	2.7	–

272

Summer diet of blue hares differs from that in winter primarily by the predominance of herbaceous plants. Moreover, summer diet of blue hares in different parts of the range has its peculiarities, and to a greater extent than in winter, characterized by the diversity and individual variability. In the summer diet of blue hares inhabiting the northeast of the range, as in winter, willows play a great role (Chernyavskii, 1984), and in the northwest of the range shoots of blueberry make up the bulk of summer diet (Novikov et al., 1970). Over a large part of the range, in the summer diet of blue hares herbaceous plants predominate, the species composition of which changes with the development of the grass stand.

Periods of transition from winter to summer feeding are determined by the time of emergence of new growth of herbaceous plants. In the southern taiga zone of Central Siberia, the transition to summer diet is completed by the beginning of May. At this time the stomachs of blue hares contain not more than 1-2% remnants of wood. From the third decade of May, when leaves of trees and shrubs open up and grass attains a height of 10-30 cm, blue hares stop utilizing ramal food in their diet (Shishikin*, 1988). The most frequent foraging places of blue hares are the clearings and glades, characterized by low but species-rich herbaceous vegetation. In selection of foraging places, an important role is played by the proximity of convenient shelters. Coarse herb and grass associations are not used by blue hares as foraging places, since reconnoitering in high and dense grass is restricted and movement is obstructed. According to the observations of A.S. Shishikin (1988), blue hares actively feed in clearings up until the height of grass does not surpass 50 cm. In such cases, first of all, all shoots of willow weed and leafy euphorbia, half the leaves of garden burnet, white clover, and northern bedstraw are eaten, and grasses starting vegetative growth are left untouched. After mowing, blue hares again actively visit these areas but give preference to leaves of legumes.

The intensity of eating any grass species depends on the abundance of that species, and also its combinations with other species. The plants eaten in foraging areas usually constitute 10-15% of the species composition. At places with poor species composition of herbaceous vegetation, often such plants are eaten, which are ignored by blue hares in other areas. During the snowless period in the very same holdings, the intensity of consumption of some species changes (Shishikin*, 1988). In the Leningrad [St. Petersburg] region summer diet of blue hares is quite uniform and often shoots of blueberry constitutes 60% of the diet (Novikov et al., 1970).

In the Altai sedge, nettle, reed grass, and also wild and cultivated grasses predominate in the summer diet of blue hares (Yudin et al., 1979).

---

*See footnote on p. 269.—General Editor.

In the taiga zone of western Siberia, blue hares most avidly consume sedge, vetch, clover, timothy, vizil*, wheat grass, blue grass, cultivated grasses, and also strawberry and klubnik** (Bedak, 1940). In the basin of Lake Baikal, of main importance in the summer diet of blue hares are the legumes, some composites (dandelion, hawk's beard), and also grasses, horse-tails and sedges (Shvetsov et al., 1984). According to the data of F.B. Chernyavskii (1984) in the middle fork of the Omolon River, the summer diet of blue hares, besides herbaceous plants, includes several species of willow; the majority of these is absent in the diet of blue hares in spring and fall. The leaves of willows (together with chosenia, crowberry, bog bilberry, willow weed, reed grass, goat's beard, and cloud berry) are present in the summer diet of blue hares also in the Koryaksk upland (Portenko et al., 1963).

According to observations on the diet of blue hares in captivity, it has been shown that of the 28 most frequently occurring herbaceous plants in the "Kivach" preserve (Karelia), blue hares most avidly consumed clover, sedge, dandelion, yarrow, cultivated oat grass, pea vine, reed grass, bear fistula, shoots and berries of blueberry and branches of European aspen; willows were not nibbled at all in summer (Ivanter, 1969). In the Volga-Kama territory, blue hares in captivity avidly consumed nodding [or mountain] melik, white bedstraw, chee reed grass, Greek valerian, brown-scale centaurea, hawkweed oxtongue and rattleweed. Besides grasses growing in the territory of cages, blue hares fed on other species, among which they preferred: barley, dandelion, lady's mantle, bitter pea vine, bird vetch, clover, common globe flower, and white bedstraw (Aspisov, 1936). In the northern Kazakhstan region blue hares living in cages during summer were offered 45 species of herbaceous plants, of which they avidly consumed common oat, bird vetch, glossy saltbush [*Atriplex nitens*], common asparagus, common dandelion, garden burnet and bastard lupin (Sludskii et al., 1980). In Yakutia, of the 60 species of herbaceous plants, blue hares preferred purple and many-stemmed vetches, garden pea, onidium, common oat, rough dandelion, field saw thistle, bird vetch and ground pea vine, and among woody-shrubby plants—needles of larch (Tavrovskii et al., 1971).

197

According to the data of A.S. Shishikin (1988), during cage rearing of blue hares, particularly young, get used to new type of food quite fast. Thus, winter cress introduced in the ration of young blue hares was consumed as avidly on the fourth day as the habitual species of grasses, and on the introduction of agricultural crops in the ration, the proportion of the earlier preferred legumes in the daily diet dropped from 92% to 42%. If hares were

*Local name of herb—General Editor.
**Same as strawberry—General Editor.

offered uniform diet consisting only of the preferred food, then in the first two days the volume of food requirement dropped by 30-40%, but was restored on the fifth day. On transition of hares to a uniform diet, the proportion of leftover food increases from 6 to 35%. In those cases, when the diet was varied, the unconsumed parts of plants predominantly contained inflorescences of clover, white cress, leaves of willow weed, pods of pea vine, awns of barley and oat grains, while succulent leaves and stems were eaten up completely (Shishikin, 1988).

In summer Yakutian blue hares eat, on average, 614 g of food per day, and females—704 g (Tavrovskii et al., 1971). Roughly similar data were obtained in a study of blue hares feeding in northern Kazakhstan (Sludskii et al., 1980). Here in cage maintenance, blue hares ate in a day from 340 to 980 g (average—735 g) of plants. Young hares consumed in a day, on average, from 266.6 to 604.4 g of green food (Shishikin, 1988).

The summer feeding period terminates with the first frosts. Until the formation of snow cover, the main foraging places of blue hares are the damp glades, where green leaves still remain. At the end of September in the zone of southern taiga, ramal food appears in the diet of blue hares, although in habitats with poor grass the transition to winter diet may begin even earlier (Shishikin, 1988). The composition of fall diet of blue hares depends on the type of habitat, and the time of transition to winter feeding is determined by the periods of formation of snow cover in the given part of the range and in any habitat. In summer and fall, Hart's truffle occupies a significant place in the diet of blue hares. Blue hares dig out these fungi not only from the moss cover and forest litter, but also from soil from depths of 4-6 cm. In winter until the height of snow cover does not exceed 15 cm, blue hares dig out Hart's truffle from under the snow (Naumov, 1947).

Blue hares periodically supplement the deficit of mineral salts in their body. They find salt inclusions in soil even in small concentrations. According to the data of A.S. Shishikin (1988), in all blue hares shot in summer, sand or small stones were present in the stomach, the weight of which was up to 18 g. The composition of mineral nutrition of blue hares is usually determined by the nature of the underlying parent rock or surface layer of the nearby roads. Blue hares particularly willingly visit places where salt of natural or anthropogenic origin is present in the soil. Adult females most often approach salt-licks during pregnancy and lactation. Appearance of young at salt-licks coincides with the period of their changeover to plant food. If salt-licks are present, blue hares visit them also in winter, though not so frequently as in spring and summer. Hares, like mouselike rodents, may use shed moose horns as a mineral supplement (Agafonov, 1982).

Activity and behavior. The character of daily activity of blue hares is determined mainly by a combination of conditions, primarily ensuring safety of food acquisition and raising young. On intensification of safety factors, activity of blue hares decreases and hiding becomes the dominant element of behavior. In summer and fall blue hares leave for foraging at dusk, and in mid-summer because of short nights they feed even after dawn. The start of activity of blue hares is most intimately connected with the time of the day, which practically always begins at sunset, whereas the end of activity does not always coincide with the dawn. The duration of the active period changes during the year and depends on the duration of the twilight phase. During the year the rhythm of daily activity also changes. Moreover the nature of activity of blue hares depends on age. Young hares to the age of 1 month are active mainly late at night. In fall young, irrespective of the time of birth, proceed to forage earlier than adults (Shishikin, 1988).

Physiological processes occurring in the large intestine, having strict periodicity, play a definite role in the formation of the rhythm of daily activity of blue hares. At night and in early morning, when animals are active, only hard feces are formed in the intestine; but during the resting time, soft. It is during daytime resting that the act of coprophagy is observed in blue hares (Pshennikov et al., 1988).

In summer blood-sucking insects exert great influence on the activity and behavior of blue hares. An abundance of mosquitoes forces hares to change their hours of foraging; they feed in the days when the insect activity is low (Pavlinin, 1971).

In Yakutia in winter, during captivity the daytime resting in blue hares takes from 25 to 33% of the daytime; it becomes more prolonged with the rise in temperature. Daytime resting extends from 7.00-8.30 hr in the morning to 12.30-15.55 hr (often 13-14 hr), after which hares proceed to forage. In January with a mean daily temperature of -39 to -41.5°C, hares spend from 65 to 67% of the daytime foraging, and intervals between feeding constitute 7-10%. In March-April, with a mean daily temperature of -13 to -16°C, less time is spent for feeding—from 28 to 29%, and the frequency and duration of intervals between feedings increase, which, in general, constitutes 38-44% of the daytime. Individual and sex-related differences have been noticed in the activity of blue hares. Thus, the very same male was the first to rise from daytime resting, and three other males started feeding with a delay of 1-5 to 10-20 min. Females appeared 44-135 min later for foraging as compared to the first male (Pshennikov et al., 1988).

As observations on blue hares in the lower reaches of the Lena showed, their maximum activity in summer occurs in the evening and night hours—from 21:00-04:00 hr in the night (Vol'pert et al., 1988). In Karelia blue hares are most active in evening and morning hours, but forage often also at night. Thus in spring hares feed throughout night, sometimes with a small interval of 1-2 hr, during which time they lie close to the place of foraging. During the day, blue hares usually lie down in shelters, but some animals also feed during the day (Ivanter, 1969).

Hearing is very well developed in blue hares, but vision is somewhat poor. They become frightened if they observe any change or movement along the path from that with which they are accustomed. Such changes force blue hares to abandon their customary path. Blue hares conduct themselves very cautiously at reappearing objects. The fright response in blue hares is very high. In the event of danger, fright overtakes the animal so much, that it stops being cautious and attempts only one thing—to hide. In midwinter and at the height of summer blue hares rest less soundly, not allowing anything closer than 15-20 m. During the molt they are less frightened and permit considerable closeness, sometimes almost touching them (Ivanter, 1969).

The stereotypic peculiarity of behavior of blue hares is, that, having described a large irregular circle, they lie down not far from the entrance of the tunnel. Here, the trace to circle is confused by a sufficient number of double backs (passing here and there at the same place) and sweeping or "hopping" (leaping far to the side of the path). At the head of this circle the hare moves at different angles to the wind and then lies with its nose to the wind and to his track (Ponomarev, 1934). The daily tracks of Karelian blue hares usually are an ellipse in shape (often open); less often they circumscribe or zig-zag. The distance from the resting place to foraging in a straight line is usually less than half the daily straying of a blue hare, which is 0.5-0.6 km (Ivanter, 1969).

199 In Kamchatka with its exceptionally snowy winters, all major traverses are made by blue hares along their beaten paths in snow. These paths are used also by sables and foxes. After abundant snowfall, moving on these paths, blue hares become tired very soon. In the event of danger they attempt to hide, burying themselves as deep as possible in the snow (Chernikin, 1965).

During the period of the rut, blue hares become less frightened and more mobile. According to observations in captivity, it has been established, that in the course of 3-4 hours, blue hares mate many times with a male discharge interval of 15-20 minutes shifting the attention of females.

Area of habitation. The territorial behavior of blue hares has been little studied so far. However, it is known that with normal numbers and in the absence of extreme external conditions, blue hares lead a settled mode of life and each hare is attached to the area which it knows best. The area of individual territory of blue hares depends on the time of the year, quality of habitat and the number of animals. Thus in the Angara River area, the individual summer territories of blue hares, depending on the type of habitat, varies from 3 to 30 ha. The largest individual territories have been recorded in broad-leaved young forests and in cowberry-pine groves, where blue hares are found very rarely. In the same-age young forests, the area of individual territory of blue hares increases depending on their sparseness. In dark coniferous forests the size of an individual territory depends on the degree of rubbish and presence of coniferous undergrowth. The total area of an annual territory increases with the seasonal change of habitats; for example, in light coniferous forests, where such changes do not occur, the area of an annual territory does not exceed 50 ha, and in the event of seasonal differentiation of foraging places, the annual territory increases 2-3 times (Shishikin 1988).

According to the data of V.A. Agafonov (1982), the size of individual winter territories of blue hares, depending on weather conditions, varies from 0.01 to 49.2 ha and, on average, constitutes 5.2 ha. In heather associations of southern Scotland, the area of daily territory, on average, constitutes 10-15 ha; in females during the breeding season and with young, the area of individual territory is less—4 and 2 ha, respectively (Flux, 1970).

Reproduction. In blue hares the rut begins earlier in the southern and western parts of the range. Thus, in Belorussia in blue hare, the first rut starts in the first half of February (Serzhanin, 1955), but in Kazakhstan, usually in the middle of February (Sludskii et al., 1980). In the central and western belt of the European part of Russia, in Volga-Kama territory, in southwestern Siberia, in the Altai-Sayan mountain system and in the Baikal basin rut begins sometimes at the end of February, but often in the first half of March (Barabash-Nikiforov, 1957; Savinov and Lobanov, 1958; V.A. Popov, 1960; Novikov et al., 1970; Yudin et al., 1978; Shvetsov et al., 1984; Shishikin, 1988). In northeastern Europe, northwestern Siberia, southern Yakutia and Sakhalin, the rut begins at the end of March to the first days of April (Naumov, 1947; Pavlinin, 1971; Voronov, 1974; Revin, 1989), and in central Yakutia, the lower reaches of the Kolyma and the tundra of western Chukotka, by the end of April-beginning of May (Tavrovskii et al., 1971; Chernyavskii, 1984). Moreover, the beginning of the rut and its duration depend on weather conditions. Thus in Karelia in prolonged and cold springs, the rut is extended, but in mild springs it occurs in condensed periods. In northern Karelia the rut begins two weeks later than in the south (Ivanter, 1969).

A second rut (in those parts of the range, where more than one brood is produced) starts immediately after the birth of baby hares of the first litter and usually continues to about 20 days. A third rut occurs only in blue hares from the southern parts of the range and begins usually in the second half of June. In regions where there are only two broods, for example in western Tuva, the reproductive period terminates in June (Boyarkin, 1984).

Not all adult females produce a spring brood. In Belorussia, according to the data of V.E. Gaiduk (1973), only 43% of females were involved in 200 spring reproduction during 1964-1970. In the Vologoda and Kirov regions in 1938-1940, 13% of females did not participate in spring reproduction. In Tataria in the 1930s, 3% of females did not bear a spring brood, and in the forest-steppe part of northern Kazakhstan in 1964-1971, 13.4% of females did not reproduce in spring (Naumov, 1947; Sludskii et al., 1980).

In the north of the range—in forest tundra of western Siberia, the Cis-Polar Urals, Yakutia, northeast Siberia and also in Sakhalin and along the Okhotsk coast, females of blue hares most often bear one brood each (Pavlinin, 1971; Tavrovskii et al., 1971; Voronov, 1974; Chernyavskii, 1984).

In the middle belt of western Siberia and in western Tuva, blue hares, as a rule, bear two broods in a breeding season. Three broods happen to 201 occur in blue hares inhabiting Karelia, the Vologda region of the Volga-Kama territory, in Kazakhstan, and also south of the Leningrad region (Novikov et al., 1970), in the southern taiga zone of western Siberia, in Khakassia (Kokhanovskii, 1962) and in the forest steppe part of the Angara River area (Shishikin, 1988). In these regions in summer, as a rule, females participate in reproduction. In Belorussia up to 60% of females have a fourth brood (Gaiduk, 1973).

Fecundity of blue hares varies in different parts of the range (Table 9.5). The highest indices of the number of embryos were recorded in blue hares of Belorussia—on average 8.6 per female (Gaiduk, 1973; Table 9.6). In summer broods baby hares are usually more in number, and the size of the second brood (according to the observations in northern Kazakhstan) is more in females which began reproducing late (Sludskii et al., 1980). In northern Leningrad [St. Petersburg] and the Pskovsk regions, where blue hares most frequently have two broods in a year, the spring broods are also smaller than those in summer; here, on average, in spring there are 3.1 embryos per female, and in summer—4.3 (Kogteva and Morozov*, 1972). In the Verkhoyan'e [Upper Yana River area], where only about 10% of females reproduce twice, spring broods are larger, which is explained by large embryo mortality in repeatedly pregnant females (Revin et al., 1988). Thus, the size of the brood in blue hares depends on the time of year. Moreover, fecundity of females is linked with their age; young and old females bear, as a rule, smaller progeny, than females in the age of two-three years.

Long-term observations on fecundity of blue hares in different parts of the range confirm that, on average, the size of broods in the very same region varies insignificantly in different years (Table 9.5**). However, the size of broods in different parts of the range can differ significantly. On

*Not in Lit. Cit.—General Editor.
**Table 10.5 in original—Editor.

Table 9.5. Yearwise fecundity of blue hares in different parts of the range (M—mean numbers of embryos per female)

Vologda Region (Savinov and Lobanov, 1958)		Northern Kazakhstan (Sludskii et al., 1980)		Central Yakutia (Tavrovskii et al., 1971)		Verkhoyan'e (Tavrovskii et al., 1971)		Karelia (Ivanter, 1969)	
Year	M	Year	M	Year	M	Year	M	Year	M
1935	4.7	1941	5.1	1952	5.7	1954	6.5	1957	4.3
1936	5.2	1946	4.4	1953	6.7	1955	6.7	1958	3.8
1937	4.4	1947	4.6	1954	5.9	1956	6.2	1959	4.2
1938	3.9	1948	4.6	1955	5.9	1957	6.3	1960	4.9
1939	3.5	1949	4.5			1958	6.7	1961	4.6
1940	3.5	1951	4.7			1959	6.5	1962	4.1
1941	3.9							1963	4.0
1942	3.8								
1944	4.1								

Table 9.6. Size of broods of blue hares in different parts of the range
(from number of embryos and placental spots)

Regions	Average per female	Limits	Authors
Belorussia	8.6	4-11	Gaiduk, 1973
Vologoda region	4.1	2-6	Smirnov and Lobanov, 1958
Karelia	4.3	1-9	Ivanter, 1969
Leningrad region	2.6	1-3	Novikov et al., 1970
Polar Urals	5.5	–	Pavlinin, 1971
Yamal	5.4	4-7	Pavlinin, 1971
Northern Urals	5.2	–	Pavlinin, 1971
Evenkia	5.7	4-7	Shishikin, 1988
Middle Lena valley	6.7	1-11	Solomonov, 1973
Cis-Verkhoyan'e	7.1	5-10	Revin et al., 1968
Verkhoyan'e	6.5	1-11	Tavrovskii et al. 1971
Valley of middle fork of the Kolyma	6.0	3-10	Labutin et al., 1974
Western Chukotka	6.9	4-10	Chernyavskii, 1984
Lena-Amgun interfluve	8.2	4-11	Solomonov, 1973
Angara River area	5.0	4-8	Shishikin, 1988
Tuva	5.9	1-9	Boyakin, 1984
	5.6	4-6	Ochirov and Bashanov, 1975
South of western Siberia	6.5	2-11	Bedak, 1940
Northern Kazakhstan	5.8	3-10	Sludskii et al., 1980
Volga-Kama territory	4.0	–	Aspisov, 1986

average, one brood of females from the northern populations produces more baby hares (Table 9.6*), but then the southern population females, as a rule, bear more** than one brood in a year, which leads to a leveling of general fecundity of blue hares living in different natural zones, and sometimes also to a more intensive reproducibility of southern populations.

The highest reproductive potential has been observed in blue hares of the Volga-Kama territory (6.5), somewhat less (5.2) in Karelian blue hares, and the lowest (4.7 and 4.65) in blue hares from Yakutia and Vologda regions (Solomonov, 1973).

Fecundity of blue hares depends on the level of embryo mortality. As a

*Table 10.6 in original—Editor.
**"not more" in original, an obvious error—Editor.

rule, in young females the level of embryo resorption is higher, which is possibly linked with high susceptibility of young individuals to helminth invasions. Embryos of more fecund females are often susceptible to resorption. Thus, according to the data of E.V. Ivanter (1969), in Karelia embryo resorption in females having 1-4 embryos constituted 9.4%, and in females with 5-8 embryos—14.7%. Quite high embryo mortality (from 4 to 33%) was recorded in the central Yakutian populations; here embryo resorption occurs often in females pregnant for a second time (Revin et al., 1988). Embryo mortality in blue hares increases in years of peak population.

Thus, for blue hares throughout the range, quite condensed periods of reproduction, almost total participation in reproduction of over-wintered individuals, relatively large broods, and high fecundity are characteristic.

Growth and development. In blue hares pregnancy lasts for about 50 days (Grigor'ev, 1940; Shishikin, 1988). Embryos develop particularly intensively in the last 9-10 days, when their body weight increases almost three times (Sludskii et al., 1980).

Baby blue hares are born with sight and are entirely capable of independent locomotion. Their body is covered with a dense downy fur. Newborns weigh from 130 to 160 g. Their body weight depends on the size of the brood—with more siblings, the lesser the size of individual baby hares. Baby hares grow fast. At the age of 20-25 days they weigh, on average, 573 g, and at the age of 45-50 days—1,950 g (Sludskii et al., 1980). In northern Kazakhstan by the end of August, the weight of baby hares of the first brood reaches, on average, 2,840 g. According to some data (Sludskii et al., 1980), baby hares of second and third broods grow faster than baby hares of the first brood, according to another (Shatalova, 1970) baby hares of first broods grow faster.

According to the data of N.D. Grigor'ev (1940), in captive conditions baby hares feed on mother's milk to the age of one month. However, in nature, already 9 days after birth, baby hares begin to taste grass, and a complete changeover to plant food occurs, apparently, at the age of two weeks (V.A. Popov, 1960).

Sexual maturity in blue hares sets in the year after birth.

Molt. Blue hares molt twice a year. The periods and intensity of the molt depend on the air temperatures and the daylight regime. A change in the duration of daylight triggers the process of molting, and temperature determines the rate of its progress. Each area of the body of blue hares molts and changes color at a definite temperature range (Gaiduk, 1975). Readiness for molting is also subject to a biological rhythm, which is clearly manifest in adult individuals. Hence the fall molt in warm weather begins earlier in adult blue hares.

The spring molt starts gradually, but later proceeds very fast. Depending on the weather conditions, the periods of start of molting could shift to any side by 10 days. The onset of peak molting and change of color of the hair coat falls in the period of gathering of snow. Usually males are the first to start molting (M.V. Popov, 1960; Gaiduk, 1975).

In the forest steppe zone of northern Kazakhstan, the spring molt starts towards the beginning of March. The process of spring molt extends in this part of the range to 75-80 days and terminates by the middle of May, except in the high mountains of the Dzhungarian Alatau, where molting terminates by the beginning of June (Sludskii et al., 1980). The spring molt occurs roughly in the same periods in blue hares in the Vologoda and Leningrad [St. Petersburg] regions and in Belorussia (Serzhanin, 1955; Savinov and Lobanov, 1958). In the northeast of the range (Chukotka, Anadyr, Omolon, Koryak upland), spring molt starts later (in May) and occurs in more condensed periods—for one month (Portenko et al., 1963; Chernyavskii, 1984).

According to observations on blue hares in cages, growth of summer fur starts on the head (between the eyes) and simultaneously in the region of the sacrum. Summer fur appears after a week on the forehead and as individual islets all over the back. In the following week, winter hairs molt successively on the nose, cheeks, and over the entire loin, and by the end of this period summer hairs appear on the lower parts of the legs. Still later summer hairs appear on upper parts of the sides and around the eyes. Roughly a month after the start of the molt, the back and head molt completely, while sides and limbs—to half. The last areas to molt are the ears, the upper part of nape, rump, thighs and paws (Sludskii et al., 1980).

The fall molt in blue hares almost throughout their range begins at the end of August-September but, depending on weather conditions, a massive start of molting may shift within limits of 10-20 days. The end of the peak of the fall molt, as a rule, coincides with periods of formation of snow cover. However, it happens, that the end of molting and time of formation of snow cover do not coincide—in cold falls blue hares may completely change summer fur before the snowfall or, conversely, in warm falls molting may be extended and hares remain for sometimes not white with the already established snow cover (Gaiduk, 1975; Shishikin, 1988). Adult males are the first to molt, followed by adult females, and last to molt are the young hares. In diseased and weakened animals, the fall molt may extend to December. The duration of the fall molt is the same as the spring molt—78-80 days (Sludskii et al., 1980).

The development of winter hairs in blue hares occurs with the formation of new hair buds, and hence shedding of summer hairs is gradual and to some extent independent of this process. Especially, the first winter hairs appear on the back surface of the ears, in the region of tail and on the lower part of hind feet. In the first 30 days of molting, hind paws and the back surface of the ears become noticeably white and front paws begin to turn white. In the next 30 days, summer fur remains on the back and face (as individual spots) and on the sides, rump and front surface of the ears as only individual dark hairs. Last to molt are the end of the face and the area around the eyes (Larin, 1980).

Sex and population age structure. In blue hares the sex ratio changes during the year and depends on the state of the population. Differences have been noticed in the ratio of males and females in different age groups (Table 9.7). According to the data of V.A. Popov (1960), in spring and fall, a predominance of females (about 60%) is recorded among embryos of blue hares, while among embryos of summer broods males (55%) predominate. Quite considerable predominance of females among adult individuals is recorded in the southern Lake Baikal area (E'khirit-Bulagaisk, district)—here in 1984-85 females were twice as common as males, and in 1986-87—four times as common. However, among young, on the whole, males were predominant—1 : 0.7 (Dvoryadkin and Dzyuba, 1988).

Table 9.7. Proportion of males (in %) of blue hares in different age groups

Region	Embryos	Young	Adults
Karelia (Ivanter, 1969)	44.5	54.9	57.3
Verkhoyan'e (Tavrovskii et al., 1971)	57.0	57.0	–
Central Yakutia (Tavrovskii et al., 1971)	58.0	52.0	47.0
Southern Anqara area (Shishikin, 1988)	–	52.7	34.9
Southern Lake Baikal area (Dvoryadkin and Dzyuba, 1988)	–	59.0	31.3

Based on the analysis of histological sections of the lower jaws of blue hares from the territory of Karelia (n = 104), it was shown that at the time of decline in numbers in fall (1977-1978), hares of 1-2 years age (36.5%) predominated in the population, 13.5% of the population was comprised of 3-year-olds, 11.5%—4-year-olds and 1% each—5- and 6-year-old hares. With an increase in numbers of blue hares, the proportion of old individuals decreases in the population, and the average age of blue hares in 1979 decreased from 2.3 years to 1.7 years (Belkin, 1984). In Belorussia (according to the data of 1964-1970) fall populations of blue hares consisted of up to 82% underyearlings, and in winter the proportions of survivors decreased, on average, to 56.5% (Gaiduk, 1973).

In the southern Lake Baikal area in the initial phase of decline in numbers (1986-1987), underyearlings made up only 50-60% in catches. For every adult female there were 6.6 underyearlings; moreover 75% of them comprised individuals of the second brood (Dvoryadkin and Dzyuba, 1988).

In central Yakutia in the fall (in the absence of large-scale mortality of young animals), the population of blue hares comprised 75-88% underyearlings. In the course of winter, the numbers in a population (if the population was commercially being reduced) declined to 1/2-1/3 and the ratio of the young decreased to 65% by the spring. However, such a character of the age structure of population is not stable. Thus 1954, being the beginning of a decline in numbers, was accompanied by large-scale mortality of young, and by the end of summer the population had only 50% underyearlings, and in the next year still less (Tavrovskii et al., 1971). A similar picture of change of ratio of age groups was noticed in the Cis-Verkhoyan'e in the 1980s, when in September-October in the course of three years (1980-1982) young animals constituted from 61.2 to 64.3% of the population, and in fall of 1985 their proportion in the population fell to 34.5% (Revin et al., 1988).

Population dynamics. The fall in numbers of blue hares is linked mainly with epizootics, helminthoses, natural disasters, cold weather in spring, and activity of predators. As a rule, unfavorable weather conditions destroy the food base of hares, which leads to weight loss and decreased resistance of the organism. Predators often attack weakened animals, which are also more susceptible to helminth invasions and infectious diseases. Low spring temperatures particularly adversely affect numbers of blue hares, which leads to high mortality among young. In some places high floods adversely affect numbers of blue hares.

Development of cycles of numbers of blue hares from the lowest values to peaks occurs mainly on account of concentration of individuals of the summer generation in the population, and a relatively small rise; and fall in numbers between the extreme phases of the cycle is determined by the proportion of animals from the spring generations (Tomilova, 1975). In different parts of the range, because of the differences in reproductive potential, the duration of cyclic changes in numbers are different: in central regions of eastern Europe this cycle lasts on average, 5-6 years, in the European north—9.9, in western and eastern Siberia—9.6, and in Yakutia—12 years (S.P. Naumov, ed., 1960).

The sharpest fluctuations of numbers of blue hares are characteristic in Yakutia, where the maximum numbers exceed the minimum by 300 times and more (Tavrovskii et al., 1971), while in other regions of Siberia fluctuations in numbers of blue hares do not exceed 65 times (Naumov, 1960), and in forest zone of eastern Europe—30 times (Volodin, 1974). In Kazakhstan fluctuations in numbers of blue hares are estimated at 3-4 times (Sludskii et al., 1980).

In some parts of the range of blue hares, along with periodic fluctuations in numbers, there occurs a general gradual fall in numbers of this species.

The main reason for the shrinking of numbers of blue hares is the economic activity of man—felling of forests, plowing of lands, and creation of water reservoirs. In such places European hares begin to squeeze out blue hares. Thus in the territory of Tataria, destruction of a considerable part of forest areas and scrub fallows and also felling of inundated forests led to a situation, that for 30 years (from 1924 to 1954) the number of blue hares was reduced roughly three times, while the number of European hares increased more than twice. Thus, the southern boundary of distribution of blue hares gradually receded to the north, and even in such places as the Udmurtia and Kirov regions, there are signs of squeezing out of blue hares by European hares (V.A. Popov, 1960).

Enemies. Potential enemies of blue hares are many carnivorous mammals and birds living in the same places as hares. Blue hares become prey to carnivores much more often than European hares. Even in such regions as the Volga-Kama territory, where the number of European hares is higher than blue hares, the majority of predators attack blue hares twice as often (V.A. Popov, 1960).

Throughout their range the most constant enemy of blue hares is the lynx, in the diet of which blue hares may constitute up to 87.5% (Sludskii et al., 1980). Foxes may also inflict significant losses on blue hares. The proportion of blue hares is particularly high in the diet of foxes in the Volga-Kama territory (82.1%), in the basin of the Vilyui (86.5%), the Adychan uplands (89.6%), and in the Verkhoyan'e depression—67.4% in summer and 98% in winter (V.A. Popov, 1960; Egorov and Labutin, 1964; Egorov, 1965; Tavrovskii et al., 1971). In the diet of foxes, blue hares, in comparison with mouse-like rodents, usually constitute a smaller proportion, although in Yakutia the incidence of blue hares in the food of foxes may reach 94.4% (Tavrovskii et al., 1971). In northeastern Siberia polar foxes can cause quite high losses to young (Yudin et al., 1976).

Among mustelids, wolverines and sables most frequently attack blue hares. Thus, in eastern Kazakhstan and in the Altai, blue hares constitute up to 20-28% in the diet of sables (Afanas'ev, 1962; Sludskii et al., 1980). In Yakutia the proportion of blue hares in the diet of wolverines may reach 43.6% (Tavrovskii et al., 1971). In the Volga-Kama territory blue hares are relatively more often (13.8%) found in the diet of mink (V.A. Popov, 1960), and in the Primorsk territory kharza* hunt blue hares.

Over a large part of the range of the brown bear, blue hares do not play a significant role in its diet, although in the territory of Yakutia the proportion of blue hares is quite significant in the animal food of brown bears. Thus according to the analysis data of excrement in the area of the

*Polecats?—General Editor.

Adych' basin in 1956 and 1959, blue hares constituted 13.3 and 24.8%, respectively, of all animal food in the diet of brown bears (Tavrovskii et al.*, 1972).

Among birds such species as raven, eagle owl, snowy owl, marsh harrier, tetronid sparrow hawk, sea eagle, kite, golden eagle, spotted eagle, imperial eagle, steppe eagle, golden eagle**, bearded owl, and rough-legged buzzard, hunt blue hares (Tavrovskii et al., 1971; Yudin et al., 1976; Sludskii et al., 1980).

All the above listed species of carnivorous mammals and predatory birds (including lynx, eagle owl and bearded owl specialized to feeding on blue hares) do not play a significant role in the change in numbers of blue hares. However, carnivores may exert some influence on the state of populations of blue hares in years when their numbers are depressed. Moreover, intensification of predator stress on populations of blue hare occurs in years of decrease in numbers of small rodents, when the proportion of hares begins to increase in the diet of foxes and small mustelids. Mortality of blue hares due to predators increases in winter with deep loose snow cover, and also in the period of the rut, when hares are most active and least careful.

Competitors. The competition for food is most significant for blue hares, which is manifest most acutely in winter time and increases where there are a large number of species feeding on ramal food. The main competitors of blue hares are the roe deer and moose, especially when they begin to claim reforested areas earlier than hares. In these species, habitats, as well as an assortment of preferred foods, coincide with high numbers of these ungulates. The winter food base of blue hares may become significantly undermined; moreover, moose could destroy the undergrowth, thereby worsening the protective conditions for blue hares. However, the influence of moose on feeding of blue hares has not only an adverse effect, since in their foraging places appear additional foods for hares in the form of broken tips of trees.

To a much lesser extent than ungulates, some species of rodent pose competition for food to blue hares. Mouse-like rodents, in years of their high numbers, may destroy up to 5-10% of the volume of undergrowth, which may adversely affect the spring ration of blue hares. In those places where beavers live together with blue hares, a joint exploitation of food resources by these species has been noticed. Thus, in the southern Angara River area the majority of European aspen and birch dumped by beavers were consumed by blue hares, and only with high numbers of beavers their

---

*Not in Lit. Cit.—General Editor.
**Repeated in the original—General Editor.

activity may have adverse consequences on feeding conditions of blue hares (Shishikin, 1988).

In those regions where the area of distribution of blue hares coincides with the range of European hares, a competition between these species of hares is possible, and most often European hares force out blue hares. In spite of the fact that the number of blue hares is usually higher in territories reclaimed by man, excessive exploitation of holdings for cattle grazing leads to extinction of hares (Aspisov, 1936).

Diseases and parasites. Among ectoparasites 7 species of mites have been recorded in blue hares. In the taiga zone of the European part, Siberia and southern Far East, forest species of mites parasitize blue hares: *Ixodex persulcatus* and *I. ricinus*; in forest-steppe part of western Siberia—meadow mite *Dermacentor pictus* (Dunaeva, 1979; Shishikin, 1988). In the Far East the following have been detected on blue hares: *Haemaphysalis japonica, H. concinna* and *I. silvarum* (Dunaeva, 1979; Kolosov*, 1980). In northern Kazakhstan [the tick] *Rhipicephalus rossicus* is found on blue hares (Sludskii et al., 1980). The index of abundance of mites on blue hares is usually high, but its significance varies over the range and depends on the time of the year. Besides ticks and mites, fleas parasitize blue hares: *Pulex irritans, Cedopsylla simplex, Hoplopsylla affinis,* and lice—*Hoplopleura acanthopus* and *H. lyriocephalus* (Dunaeva, 1979).

Blue hares, more than any other species of hare, are susceptible to helminthoses. In all, 15 species of parasitic worms have been recorded in blue hares (Kontrimovichus, 1959). The most widely distributed are the lung nematodes of the genus *Protostrongylus* (*P. terminalis* and *P. kamenskyi*) and intestinal nematodes *Nematodirus aspinosus.* The nematodes—*Trichostrongylus colubiformis* and *P. retortaeformis* have been found in blue hares from the taiga zone of the European part of the range; another species, *Trichocephalus leporis*, also in hares from the Vilyui basin, has been found in blue hares from Yakutia and western Siberia.

Among cestodes the widely distributed species are: *Mosgovoyia pectinata, Taenia pisiformis, T. macrocystis,* and *Multiceps serialis.* The trematodes *Fascicola hepatica* and *Dicrocoelium lanceatum,* have been found in blue hares of the European north, and another species is found in hares from southern Yakutia. In northeastern Siberia, *Plagiorhis vespertilionis* has been found (Bedak, 1940; Naumov, 1947; Kontrimavichus, 1959; Tavrovskii et al., 1971; Shishikin, 1988).

The level of infection with some helminth species may reach 60%, and sometimes even higher. For blue hares of Yakutia, for instance, exceptionally high infectivity with lung nematodes is characteristic. Here, in

*Not in Lit. Cit.—General Editor.

specific seasons, the average intensity of infection of adult blue hares is up to 185 nematodes (Tavrovskii et al., 1971). Quite frequently blue hares are infected with several species of helminths.

The infectivity of blue hares with nematodes of the genus *Protostrongylus* in different age groups has its own peculiarities. Thus, according to observations in Yakutia, in young starting from July-August there is a gradual intensification of infectivity, which attains a peak by the end of winter. In adult individuals the intensity of infection in summer is higher than in young; but it decreases appreciably from July to September, and in the course of fall remains at quite low level, and by January, as in young, increases sharply. Among adult individuals, as a rule, the level of infectivity with nematodes is higher in females (Tavrovskii et al., 1971). Roughly similar is the character of dynamics of infectivity of blue hares with the nematodes *N. aspinosus*. One more widely distributed species of helminth, *M. pectinata*, infects usually only young hares, and the maximum infectivity with this cestode species occurs in July-August (Tavrovskii et al., 1971).

A comparison of the data of population density of blue hares in some regions of Yakutia with the level and intensity of helminth infection showed that the higher the population density of hares, the higher the intensity of their infectivity. With a decrease in numbers of hares, the level of their infectivity for sometime does not change, but the intensity (number of helminths per hare) begins to fall synchronously with the decrease in numbers of blue hares (Tavrovskii et al., 1971). However, according to other data, obtained from multiyear observations in the territory of the Pechora-Ilych preserve, outbreaks of helminth diseases in blue hares are not linked with their population dynamics (Teplov*, 1952).

Some peculiarities of habitats (for instance, soil composition) and weather conditions influence the degree of infectivity of hares with helminths. Thus, in the territory of Yakutia, blue hares living on sandy soils are infected with nematodes less intensively than hares living in places with loamy and sod-forest soil. On the whole, throughout their range weather conditions have the greatest influence on the spread and level of helminth infections. Dry summers and warm winters facilitate a decrease in helminth infection, and after severe and protracted winters and rainly cool summer, outbursts of helminth infections are observed (Teplov*, 1952; Tavrovskii et al., 1971; and others). Helminthoses very often are the direct cause of massive mortality of blue hares or, to a considerable extent, they weaken the organism, facilitating the spread of infectious diseases in populations.

207   Among protozoan diseases of blue hares, the most common and widely distributed is coccidosis, and the maximum level of infectivity is observed

---

*Not in Lit. Cit.—General Editor.

in adult individuals in May-June and in September. (Tavrovskii et al., 1971).

Among infectious diseases of blue hares, the most widespread and epidemiologically important is tularemia. Tularemia infection of hares occurs through ioxdid ticks and mosquitoes. Development of tularemic epizootics in hares may precede winter epizootics of this disease in voles (Tavrovskii et al., 1971). Moreover, blue hares, like European hares, may suffer from pseudotuberculosis and listeriosis. Blue hares are very sensitive to the pathogen of pseudotuberculosis; they suffer more often in the colder period of the year (November to March), the disease proceeds in a more acute form, leading to the death of hares. Blue hares are less sensitive to the pathogen of listeriosis and suffer from this disease lightly (Dunaeva, 1979).

Practical Significance

Blue hares in the area of high numbers are an important object of the fur trade, and practically everywhere they are the object of sport hunting.

Significant is the role of blue hares in feeding of such important fur-bearing animals as fox and lynx, the number of which to a great extent depends on the number of hares.

As massive consumers of woody plants, blue hares exert influence on the process of restoration of woody plantations. Constant damage to the tips of young trees by hares facilitates an appearance of tiered structure, undergrowth of birch in the upper tier, formation of birch groves in clearings, and delays in afforestation of burned out areas, clearings and felling areas. Blue hares may inflict considerable losses in gardens and truck gardens, and sometimes in grain crops.

In the territory of Siberia, numerous cases are known of human infection of tularemia from blue hares (Tavrovskii et al., 1971).

European Hare
Lepus (Lepus) europaeus Pallas, 1778

1778. *Lepus europaeus* Pallas. Nov. Spec. Quadr. Glir. Ord. 30. Poland, southwestern part (Ognev, 1940; 141—cited from Trouessart, 1910).

1811. *Lepus variabilis* var. *hybrida* Pallas. Zoogr. Rosso-Asiat: 147. Moscow region, Nom. nudum.

1822. *Lepus hybridus* Desmarest. Mammalogie, 2: 349.—First suitable name for *hybrida* Pallas.

1833. *Lepus caspicus* Ehrenberg. Symb. Phys., 2. Astrakhan region, vicinity of Astrakhan.

1842. *Lepus aquilonius* Blasius. Amtl. Bericht XIX Versam-ml. Naturf. u. Aeretze, Braunschweig: 89. Central regions of European part of Russia ("Central Russia").

1850. *Lepus timidus* var. *hyemalis* Tumac. Eversmann. Estestv. Istoriya Orenburg Kraya [Natural history of Orenburg territory), 2: 201, Nom. nudum.

1871. *Lepus campestris* Bogdanov. Ptitsy i zvery chernozemnoi polosy Povolz'ya [Birds and animals of chernozem belt of the Volga River area]: 175. Nom. praeocc., non Bachman, 1937.

1889. *Lepus timidus tumak* Tichomirov et Kortchagin. Izv. O-va Lyubit. Estestv. 56, 4: 31. First suitable name for *Tumac* Eversmann.

1901. *Lepus transsylvanicus* Matschie. Sitzungsb. Ges. Naturf. Er. Berlin: 236. Romania, Tasleu.

1905. *Lepus cyrensis* Satunin*. *Izv. Kavk. Muzeya*, 2: 60. Azerbaidzhan ("Elisavetopolsk governance, Dzhevanshirsk distr."), Barda.

1923. *Lepus europaeus tesquorum* Ognev. In book: Ognev and Vorob'ev. Fauna nazemnykh pozvonochnykh Voronezh. gub. [Fauna of terrestrial mammals of Voronezh governance): 115, Voronezh region, Bobrovsk distr., "Dokuchaev experimental station (Kamennaya Step)."

1929. *Lepus europaeus caucasicus* Ognev. *Zool. Anz.* 84: 75. North Ossetia. Vladikavkaz.

1929. *Lepus europaeus caucasicus ponticus* Ognev. ibid. Krasnodar territory, Gelendzhik distr. "Beta"—infrasubspecific name (unsuitable).

1940. *Lepus europaeus cyrensis lenocaranicus* Ognev. Zveri SSSR i prilezhashchikh stran [Animals of the USSR and adjoining countries]. 4: 158. Azerbaidzhan, Lenkoran distr., Kyzyl-Agach—infrasubspecific name (unsuitable).

1940. *Lepus europaeus caspicus turgaicus* Ognev. ibid. 4: 161. Kazakhstan, Kustanai region, Naurzumsk distr., Sypsyn: infrasubspecific name (unsuitable).

208 1944. *Lepus europaeus borealis* Kuznetsov. In book: Bobrinskii et al., Opredelitel' mlekopit. SSSR [Identification Keys to the Mammals of the USSR]: 271. Bashkiria (northern part). Nomen praeocc. non Pallas, 1788, non Nielsson, 1820.

1948. *Lepus europaeus biarmicus* Heptner. *Dokl. AN SSSR*, 60, 4: 709. Nom. nov. pro *Lepus europaeus borealis* Kuznetsov.

1956. *Lepus europaeus orientalis* Stroganov et Yudin. Tr. Tomsk. Un-ta, 142: 299. Novosibirsk region, Iskitimsk distr., Borodavkino.

---

*In Russian original, Satunun—General Editor.

Diagnosis

Hair coat soft, curly. Color of upper part of body varies from ochreous-gray to ochreous, ochreous-rusty, gray and olive brown; large dark iridescence always present on back. Tail cuneate in form, upper side of tail always with black stripe, tail white beneath. Winter color often similar to that in summer, but considerably lighter in area of rump; sometimes fur becoming whiter in winter, but not all over the body.

Skull with relatively long facial section, narrow postorbital space and a broad nasal part: ratio of width of frontal bones behind the supraorbital processes to width of the base of the rostrum constitutes from 41 to 61% (average 51%). Bony palate short, choanae broad—length of the bony palate 1.5-1.75 times less than the width of choanae. Alveolar length of upper tooth row 14.5-21.3 mm (average 17.9 mm). Zygomatic width in anterior part 39.4-46 mm (average 42.7 mm). Supraorbital processes elongate in form, not reaching the posterior margin of the frontal bones. Base of the upper anterior incisor not reaching the region of the suture connecting the

208

Fig. 46. European hare *Lepus europaeus* (Photo by B.S. Yudin).

intermaxillary and maxillary bones. Length of the lower diastema in relation to length of the lower tooth row is more than 80%. Articular process of the mandible strongly bent backward.

210
Description

Dimensions large—body length varies from 550 to 675 mm; tail length—from 75 to 140 mm; length of hind foot—from 124 to 167 mm; length of ear—from 100 to 118 mm.

Color of the summer fur varies widely from ochreous-gray to yellowish-gray with large dark iridescences, which are formed by the emerging dark underfur. Bases of underfur hairs are silver-gray or white, and the tips are black or black-brown.

The upper side of the head and the anterior surface of the ears are darker in color than on the back. The posterior surface of the ears is white or grayish-white with a black or black-brown spot at the tip. Around the eyes is a ring of white color. Below the anterior margin of the eye is a reddish-brown spot. Cheeks are ochreous-brownish or ochreous-gray.

The outer side of the limbs is of an ochreous brown tone, while the inner side of the limbs and belly are white. Tail above with a black stripe; white beneath.

Winter fur is dense and fluffy. In the northern part of the range the fur turns white over a large part of the body, such that only the anterior part of the back, head, and ears remain dark. In the southern part of the range, the winter color of the fur does not differ significantly from that in summer.

The skull is large and relatively broad in the nasal section (Fig. 47). Nasal bones are broad and rounded at the tip. The anterior nasal opening is broad. The interpteregoid space is broad and the bony palate short. The furrow on the anterior incisors is situated in the middle. Bases of the incisors do not reach the margin of the suture connecting the maxillary and intermaxillary bones. The mandible is of a light structure; its articular process is bent backward. The posterior margin of the lower incisor does not reach the base of the first molar tooth.

The diploid complement has 48 chromosomes: 8 pairs of meta-submetacontrics, 15 pairs of subtelo- and acrocentrics (NF = 88), X-chromosome—submetacentric, Y-chromosome—small, acrocentric (Gustavsson. 1971).

Systematic Position

S.I. Ognev (1940) referred the European hare, together with the tolai hare and Cape hare, to the subgenus *Eulagus* Gray (1867). According to the

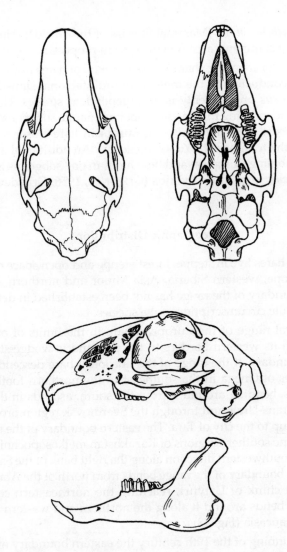

Fig. 47. Skull of the European hare *Lepus europaeus*.

system of A.A. Gureev (1964), the European hare is the only member of the genus *Eulagus*. The main diagnostic character, by which the European hare has been identified under a separate subgenus is the length of the anterior upper incisor, the base of which does not reach the suture connecting the intermaxillary and maxillary bones. The extremely contradictory viewpoint on the status of European hares in the system of the genus *Lepus* was taken by F. Petter (1961), who combined European, tolai and Cape hares in one species *L. capensis* L. 1950 [sic.]. The authors of the monograph

"*Mlekopitayushchie mira*" [Mammal Species of the World*] (Honacki et al., 1982) without comments accepted the system proposed by F. Petter.

An analysis of 18 metric characters of the skull of hares from the northern Palaearctic, conducted by the method of multifactorial statistics, showed that *L. europaeus* undoubtedly is an independent species. However, the premises for identifying this species under a separate subgenus are lacking, since there exist characters bringing European hare closer to blue hare, and also to the tolai hare (E.Yu. Ivanitskaya). An additional argument in favor of combining European and blue hares in one subgenus is the ability of these species to produce hybrids (Grigor'ev, 1956; Gruzdev, 1981, and others).

Geographic Distribution

European hares live in steppe, forest-steppe and open space of the forest zone of Europe, western Siberia, Asia Minor and northern Africa. The southern boundary of the range has not been established in detail in view of the indefinite circumscription of the species.

The natural range of the European hare in the limits of our country extends from its western borders to the east and is wedge-shaped. The northern boundary of the range of the European hare descends from the lower reaches of Onega in the southeast to the northern foothills of the middle Urals, bending around it from the south, ascends in the north to the middle Trans-Urals and through the Sverdlovsk region proceeds east, not reaching up to the city of Tara. The eastern boundary of the main range descends to the southern regions of Kazakhstan melkosopochnik and later follows in a southwestern direction along the right bank of the Sarysu River. The southern boundary of the range bends from north of the Aral Sea along the northern chink of Ustyurt, reaching the northeastern coast of the Caspian Sea, bends around it along the northern and western coasts and enters Transcaucasia (Fig. 48).

212

At the beginning of the 19th century the eastern boundary of the range of the European hare passed roughly along the Ural and Belaya rivers, and the northern—through Yaroslavl' and Peterburg. By the end of the 19th century, European hares were distributed up to the Ural foothills, the Elb River, to the northern coast of the Aral Sea, and to the north—to Vologda and Petrozavodsk. European hares dispersed most rapidly in the southeast. In northern Kazakhstan the tempo of expansion of the range was, on average, 60-80 km/year. The expansion of the range boundary in the northwest was caused by an increase in the areas occupied by fields and

---

*Original title as given in Lit. Cit.—General Editor.

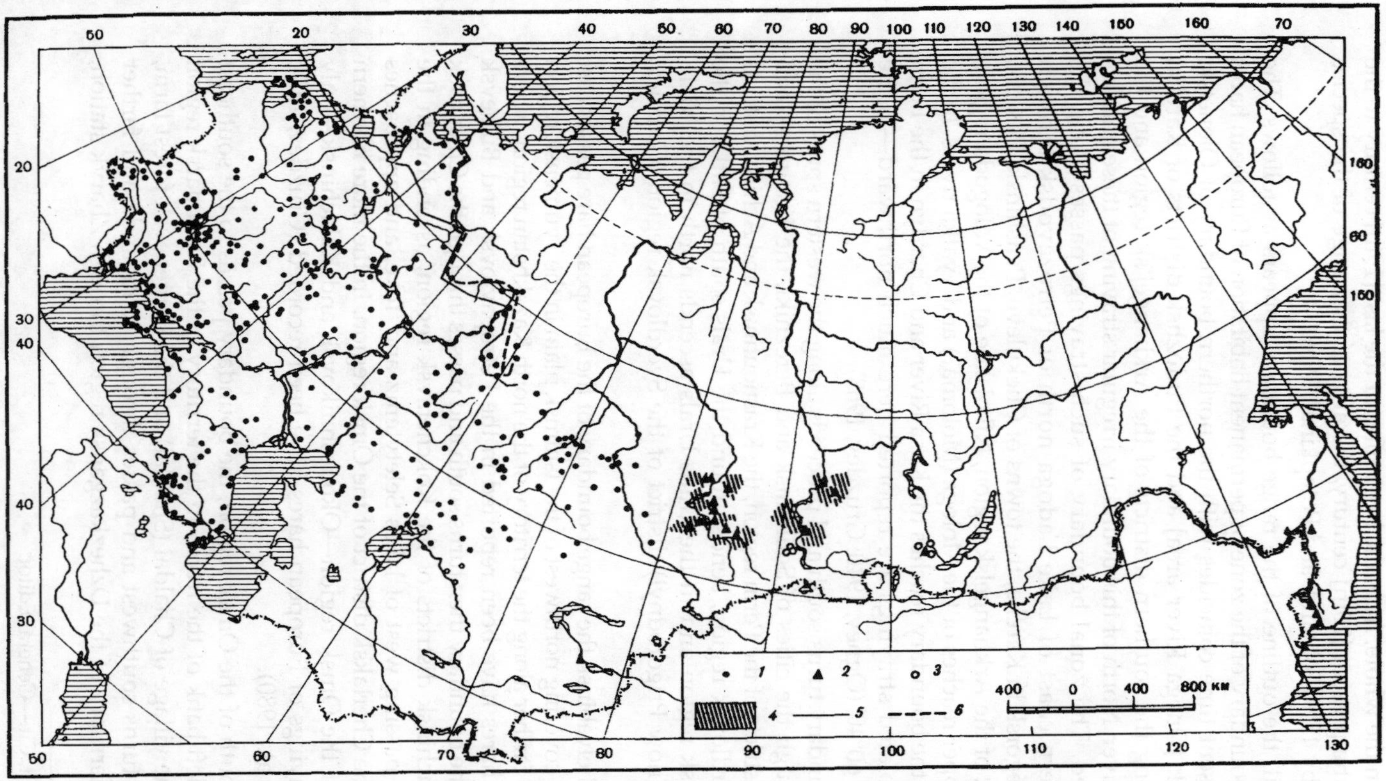

211

Fig. 48. Distribution of the European hare. 1—localities of European hares (mainly from the data of collectors' collections); 2—places of introduction; 3—localities of European hares in newly established populations; 4—region of distribution of artificially established populations (Gruzdev*, 1969); 5—northern boundary of the range of the European hare from the data of 1964 (Gruzdev, 1974); 6—northern boundary of localities of European hares from published data (Gruzdev, 1974).

fellings. On the whole, from the beginning of the last [19th] century to the forties of the present [20th] century, the area of the range of the species increased by 45% (Folitarek and Potapkina*, 1969).

In detail, the outline of the range boundary appears as follows. The northern boundary of the zone of permanent habitation of European hares at the present time coincides with the northern boundary of Olonetsk, Prionezhsk [Onega River area] and the Pudozhsk districts of Karelia, passes along the southern districts of the Arkhangel'sk region and the Chaptsa River. North of this boundary irregular straying of these animals is observed. The zonal boundary of such strayings passes along the northeastern coast of Lake Ladoga north of Petrozavodsk through Medvezh'egorsk in Karelia, the towns of Chekulevo, Penshinsk, Plisetsk, Shenkursk of the Arkhangelsk region, Nikolskoe of the Vologoda region, and the upper reaches of the Mologa (Moloma?) and Vyatka rivers. In the northeast, the boundary reaches the Kosa River and Cherdyn in the Perm region; isolated strayings are reported here much northward—to the latitude of 60°40' (Ognev, 1940; Gruzdev, 1968).

The boundary turns south and proceeds along the western spurs of the Urals through the cities of Solikamsk and Berezniki, the Kungursk and Ordinsk districts of the Perm region, the Krasnoufimsk and Achitsk districts of the Sverdlovsk region, bends around the Urals south of Satka in the Chelyabinsk region and in the Trans-Urals ascends north to the Nizh-netagilsk (now Prigorodnyi) district of the Sverdlovsk region (Gruzdev, 1974).

From Chelyabinsk the range boundary of the European hare passes east to Kurgan, covering northwest of the Ishimsk plain in the Tyumensk region and passes farther along the territory of the north Kazakhstan region, where European hares have been reported in the Petropavlovsk and Bulaevsk districts. The boundary then turns south and passes through the Chkalovsk and Shchuchinsk districts of the Kokchetavsk region, ascending in the northeast, passing west of lakes Seletyteniz and Teke, and approaches Irtysh in the Cherlaksk district of the Omsk region. In the more northern districts of the Omsk region—Okoneshnikovsk and Kalachinsk—only casual strayings of European hares have been recorded (Gruzdev, 1974; Sludskii et al., 1980).

From south of the Omsk region, the boundary passes to the southeast along the left bank of the Irtysh in the territory of the Pavlodarsk region through the village of Chaldai (Shaldai) and the Bergenevsk forest farm, whence it turns southwest and passes through Karaganda and farther along the territory of the Dzhezkazgansk region through Zharyk station,

---

*Not in Lit. Cit.—General Editor.

the village of Karsakpai, the Akshala tract, the eastern part of the Aral Karakums and proceeds to Kzyl-Orda (Kuznetsov, 1948; Gruzdev, 1974; Sludskii et al., 1980).

From Kzyl-Orda the southern boundary proceeds west along the Syr-Darya valley through the Karauzyak and Kazalinsk stations, then along the northern coast of the Aral Sea through Aralsk, dry beds of Batet and western Turangly, along the coast of the Paskevich Gulf and farther in the west along the northern chink of Ustyurt, the valley of Tassai along the Caspian Karakums, to north of Mangyshlak and south of the Gurevsk regions (Ognev, 1940; Sludskii et al., 1980).

213 Beyond the limits of the natural range, extensive measures were undertaken to introduce European hares, as a result of which several new populations were formed in southern Siberia—in the Novosibirsk and Kemerovsk regions, in the Altai territory and in the southern Krasnoyarsk territory (cf. Fig. 48). Individual less numerous populations exist also in the Lake Baikal area, the Trans-Baikal, the Primorsk territory and in southern Khabarovsk territory (Folitarek and Potapkina*, 1969; Gruzdev, 1974; Yudin et al., 1979).

Beyond the borders of our Russia, European hares inhabit steppe, forest-steppe and open spaces of the forest zone of Europe, reaching England, the southern part of Sweden and Finland in the north. In the south, the range of European hares reaches Iran, northern Arabia, Asia Minor and North Africa.

Geographic Variation

Geographic variation in the color and structure of the hair coat, and also dimensions of the body and skull are well manifested in European hares.

For the territory of the USSR, S.I. Ognev (1940), based on the study of variation in the color and structure of fur, and in some cases also body dimensions, identified six** subspecific forms of European hares: *L. e. hybridus* Pallas, 1811 (northern, central and southwestern regions of the European part of the USSR), *L. e. tesquorum**** Ognev, 1924 (forest steppe and steppe zones of the European part of the USSR—from the left bank of the Dnieper to the Volga; in the south—to the northern Caucasus, northern boundary unclear); *L. e. caucasicus* Ognev, 1929 (some areas of the main Caucasus Range and north of it, and also the Black Sea coast of the

*Not in Lit. Cit.—General Editor.
**Only five mentioned—General Editor.
***In Russian original, *tesquorem*—General Editor.

Caucasus), *L. e. cyrensis* Satunin, 1905 (eastern part of Transcaucasia, northern Iran and Asia Minor) and *L. e. caspicus* Ehrenberg, 1830 (Astrakhan and Akhtyubinsk regions, and also west of Kazakhstan).

At the same time, from the descriptions of S.I. Ognev (1940) and examinations of the skins of European hares it follows, the differences in the summer color and structure of fur have a tendency for clinal variation, and geographic variation of the winter color (lighter, absence of change or darkening) has a mosaic nature and is linked with the degree of stability of snow cover. As for the nature of geographic variation of cranial characters, the multifactorial analysis of 18 metric characters of skulls of 346 specimens of European hares showed a strict clinal variation both in the dimensions and forms of the skull (E.Yu. Ivanitskaya).

Thus the reality of existence of all subspecific forms described for the territory of the USSR seems very doubtful; probably, on the territory of the former Soviet Union, the European hare is represented only by the type form, the range of which covers the whole of the European continent.

Biology

Habitat. European hares avoid colonizing places with dense thickets of trees and shrubs, and prefer open areas with dissected relief, sparse vegetation and a small depth of snow cover.

In the zone of mixed forest and forest-steppe, European hares occupy river valleys, forest edges, glades, weedy clearings, forest strips, thinned-out shrubby dry bogs and gullies in fields, glades with herbage and weeds, and also areas with stumps and mounds. However, for European hares the most preferred habitats are fields (Novikov et al., 1970). In the territory of Belorussia, European hares may penetrate deep in the forest, but there they remain in open places—clearings and glades. During the reproductive period European hares are found in crops of vetch, oak, clover, and other cultivated grasses, and less often—in fields with grain and vegetable crops (Serzhanin, 1955; Gruzdev, 1974). European hares living near large inhabited areas often stray in cities; they, more than blue hares, are predisposed to synanthropy. Thus in Leningrad [St. Petersburg] several instances are known about straying of European hares in residential areas and even in industrial areas (Novikov et al., 1970). European hares are common in Moscow territory, especially in park zones, but sometimes they stray deep inside residential areas; for example, European hares were many a time found on wastelands in the vicinity of MGU [Moscow State University] (data of E.Yu. Ivanitskaya).

214 In the Eruslan River sands (Saratov region), European hares are usually found at the border of sand mounds with birch-European aspen groves

and in limans* with rich herbage. However, in the course of the year the preferred habitats of European hares change; this is clearly manifest in young hares. In early spring (to April 15), young most often remain in sun-warmed places: sand mounds, meadows with stacks of unpicked hay, weeds and shrubs. In late spring and summer, young hares are found mainly in fallows, young forest plantations and fields of different crops. By fall young hares concentrate along the edge of massifs with sand mounds (Gruzdev, 1974).

In the Dnieper River area, in the Don basin and Volga-Ural interfluve, European hares are most frequently found on hummocky sands, where there is a favorable combination of a good food base with relief, having many natural shelters convenient also for a fast escape from predators. In open steppes European hares remain along the forest belt and ridges with shrub thickets and weeds. During the reproductive period, the most favorable habitats for them are fields of agricultural crops, areas of young plantations and meadows; moreover the maximum concentration of European hares in these habitats are observed in lowlands, irrigated areas and places infested with weeds (Barabash-Nikiforov, 1947; Gruzdev, 1974).

In the Volga lowlands, in forests and meadows inundated during the time of spring floods, European hares are almost absent in summer. With the onset of thawing, the majority of European hares migrate to steppes and also to irrigated lands. In flood plains of the Volga-Kama territory, European hares are encountered year round. In winter they remain mainly in the central flood plain—along swamps with thickets of osier bed and forest edges, in spring—along riverbanks and in scrub meadows, while in summer, they disperse throughout the flood plain, preferring, however, thickets of osier beds (Table 10.1). European hares frequently visit vegetable gardens and gardens in the vicinity of inhabited places in the second half of winter (V. Popov, 1960).

In regions with intensive land utilization (for example, in the Stavropol' territory) in mid-summer, European hares are found most of all in fields of grain crops and lucerne; in fall they migrate to ravines and clearings, and in early spring they remain in virgin areas of steppe (Table 10.2). Seasonal migrations of European hares are well expressed here, and are linked with the agricultural activity of man. Thus, in April during the period of massive appearance of young, European hares leave winter quarters for more protected virgin areas, and in May, when there is enough undergrowth of winter vegetation and cattle grazing in virgin areas, European hares again came back to fields. In mid-summer with the start of the grain harvest and cattle grazing on greens, European hares migrate to plantations of

*Liman—a drowned river valley—Translator.

Table 10.1. Seasonal variation in biological affinity of European hares in the Volga-Kama territory (in % of number of encounters of hares and traces of their activity) (from V.A. Popov, 1960)

Habitat	Spring IV-V (n = 42)	Summer VI-VIII (n = 38)	Fall IX-XI (n = 50)	Winter XII-III (n = 56)	Total (n = 186)
Vegetable gardens, gardens, threshing floor	2.4	–	8.0	16.1	7.5
Fallow	–	–	14.0	3.6	4.3
Winter crop	9.5	–	26.0	14.3	15.9
Straw and grains	2.4	10.5	4.0	3.6	4.3
Gullies in fields with shrub	16.4	15.8	4.0	9.0	10.7
Pastures and dry paths	4.8	10.5	2.0	7.1	6.0
Flood plain:					
Clear meadow	–	5.2	2.0	–	1.6
Shrub meadows	2.8	10.5	2.0	7.1	6.0
Forest ridges and riverbanks	23.9	7.9	2.0	3.6	9.1
Swamps and osier bed thickets	–	15.8	10.0	10.7	8.6
Forest edges	7.1	7.9	8.0	10.7	8.6
Clearings	2.4	–	2.0	7.1	3.2
Forest and forest glades	2.4	5.2	2.0	3.6	3.2
Other habitats	27.9	8.6	14.0	3.5	10.5

Table 10.2. Seasonal change in the distribution of European hares according to habitats (in % of encounters) in the Stavropol' territory (from Kolosov and Bakeev, 1947)

Habitat	Spring IV (n = 68)	Beginning of summer, V-VI (n = 37)	End of summer, VII-IX (n = 105)	End of winter, I-III (n = 399)
Virgin land	42.6	5.4	18.0	39.3
Scrub	11.6	24.3	55.0	–
Greens	–	32.4	–	16.3
Lucerne	19.2	16.2	–	–
Forest	26.6	21.7	11.0	8.2
Stubble fields	–	–	10.0	0.8
Corn, sunflower, cotton	–	–	6.0	–
Fall-plowed fields	–	–	–	35.4

sunflower and corn, and by the end of August they again appear in fields, where they feed on the grain left over postharvest; at this time European hares often stray into cucurbit fields (Kolosov and Bakeev, 1947).

In the Ukraine, besides grain crops, European hares are often encountered in summer on fallows, hayland meadows and in young forest nurseries. In meadows they prefer to remaln in lowlands with a predominance of herbage. As fallows are progressively plowed and meadows are mowed, European hares migrate to other holdings, most often to fields of different crops (Korneev, 1960). In fields of grain crops, young hares remain most often in weed thickets in lowlands, but the maximum density of young European hares is observed in irrigated vegetable plantations and fields (Gruzdev, 1974).

In Transcaucasia, European hares are found in all landscapes up to an altitude of 3,500-3,700 m above sea level, but the preferred places of their habitation are semideserts and foothill steppes with shrubs and sparse forest, and in the forest belt—forest edges, valleys of rivers and clearings (Kokhiya, 1974; Eigelis, 1980). In the northern Caucasus groups of European hares have been noticed, which differ from one another in the preferred places of habitation—some hares in summer remain in forest edges and forest clearings and in winter stray deep inside the forest; others throughout the year remain in open places and even in winter, when their foraging places are linked to forest edges, they make resting places in fields; a third group of European hares has an affinity for open places with small groves, gardens, and vineyards (Loshkarev, 1970).

In northwestern Kazakhstan, European hares are most numerous in hummocky sands, river valleys and on the banks of lakes with shrubs. In the east (in steppe and forest-steppe zones) European hares live in places with dissected relief, in river flood plains, and also in fields of grain crops. Rubbly-clayey and clayey-saline deserts with poor vegetation and weakly dissected relief are avoided by European hares (Sludskii et al., 1980).

In the Altai and Krasnoyarsk territories, introduced European hares colonize hummocky steppe and forest-steppe habitats in foothills, fields, plowed lands, and neighborhoods of human habitations (Yudin et al., 1979).

In winter, European hares avoid places with deep and loose snow cover, hence they are often found in elevated places, and in depressions, dense thickets of trees and shrubs, and also on level unmowed fields. European hares are often absent in winter or are less numerous. In frosty weather hares lie in small shrubs, and with strong thawings—on the windward side of forest clearings. After heavy snow falls European hares concentrate along roads, where, besides the convenience of movement, they find food. More often than in summer, European hares, at this time of the year, are found in the neighborhood of populated areas. In regions with less snow or snowless winters, the distribution of European hares depends on food reserves, availability of places for resting and number of predators (Gruzdev, 1974).

In Stavropol' territory in the beginning of winter, European hares are most often found in fall-plowed fields, on slopes of ridges, and also at the border of virgin steppes and winter crops. In the second half of winter they remain mainly in cultivated fields, and by the end of winter migrate to groves and forest edges (Kolosov and Bakeev, 1947). A roughly similar nature of winter habitats is observed in European hares in the plains part of the Ukraine (Korneev, 1960).

Population. On the European continent the zone of maximum density of occupation of European hares are the areas bordering Magdeburg, Leipzig, Wroclaw, and Brno; the density of European hares decreases in the direction from Magdeburg to Brno and Budapest. The density of European hares is higher in areas with less fertile forest soils, than in chernozem regions. In agrocenoses of western Europe, maximum numbers of European hares are observed on plowed lands and fields of grain crops. Thus in France, in March-April 1975-1978, in these habitats the population density of European hares reached 470 individuals/1,000 ha (Pepin*, 1986).

In eastern Europe the maximum population density of European hares has been reported in the steppe and south of the forest-steppe zone. Individual areas of high density of European hares are present west of the zone of mixed forests. The distribution of European hares in the range is linked with the height of the snow cover, although differences do exist in population density in areas with similar height of snow cover, which are linked with the peculiarities of relief of the said territory. Areas with a very high population density of European hares are located in regions where the height of snow cover does not exceed 10 cm. Such regions of high density lie west and south of the line passing through Brest—Mogilev-Podol'skii—Poltava—Kotelnikovo—Temir. The second zone, which includes areas with a population density of European hares from high to medium, has the northern and northeastern boundary passing along the line Narva—Vitebsk—Belgorod—Saratov—Kuibyshev—Ural'sk—Aktyubinsk. South of this line the mean decade height of snow cover does not exceed 30 cm. In these two zones inhabit the bulk of European hares of eastern Europe (Gruzdev, 1974).

In the northern part of their range, European hares yield to blue hares in numbers. Thus, in the territory of Leningrad [St. Petersburg] region, according to the data of game husbandry (1871-1960), the number of

---

*Not in Lit. Cit.—General Editor.

European hares in comparison with blue hares was, on average, 1% and fluctuated in different regions from 0.4 to 23.2% (Novikov et al., 1970). Southward, in the territory of Lithuania, at the beginning of the present [20th] century, European hares numerically also yielded their place to blue hares and constituted about 23% of the total population of hares. At present, as a result of the destruction of forests, considerable predominance of European hares has been observed here; their population has increased to 88% (oral communication by N.M. Likyavichene). The maximum density of European hares (20-35 individuals per 1,000 ha) in the territory of Lithuania was reported in the northern, western and southwestern regions (Belova, 1987).

In the center of the European part of Russia more important for European hares is the area of open country and agricultural holdings. Here, the most favorable for colonization of European hares are the Smolensk region, the central part of Tversk and the western part of the Kaluga regions, Orlovsk, Tula, south of Ryazansk, Lipetsk, Tambovsk and the Penza regions, southeast of the Nizhegorod region and the eastern part of Mordovia. In the territory of the Tversk region and in such districts as Bezhetsk, Belskii Goritskii, Zubtsovsk, Kalyazinsk, Pogorel'sk, Sandovsk and Udomelsk, European hares numerically predominate over blue hares. Almost all these districts are within limits of the belt favorable for European hares, for which open areas and dissected territory are characteristic. Moreover, soils here are sandy loam and well drained, which is also no less important for a better state of populations of European hares (Gruzdev, 1974).

In areas with ancient agriculture (such as the Yurev-Polsk and Suzdal'sk districts of the Vladimirovsk region), where at the present time agricultural holdings occupy more than 50% of the territory, the population of European hares throughout the year (in comparison with the neighboring districts) is highest, which is facilitated by dissected relief formed by sprus of the Klinsk-Dmitrovsk Range and not very large areas of fields adjoining forest belts and groves. In the territory of Latvia and Estonia, the areas with the maximum density of European hares are confined to the elevated right bank western Dvina, and in Belorussia—to thinned-out forests on the Belorusskii chain. In Lithuania, even in areas with low population of European hares, their maximum density (100-106 individuals per 1,000 ha) is observed in shrubby and forest habitats (Belova, 1987).

In the Volga-Podol'sk and Dnieper highlands on the right bank of the Dnieper, the population density of European hares is much higher than in the left bank plain, where areas with high density are found only west of Donbass and in the Azov highlands. In the territory of the Voronezh region, the maximum density of European hares (more than 60 individuals per 1,000 ha) has been recorded on the right bank of the Don, where the eastern edge of the Srednerusskii [Middle Russian] highlands juts in, and in the southeastern part of the left bank, where lies the Kalach highland (Barabash-Nikiforov, 1957). Beyond the Volga, with the overall low population of European hares in this part of the range, density increases to the level of medium in the Obshchii Syrt* district.

In territories where the maximum mean decade height of snow cover is more than 30 cm, the population density of European hares is exceptionally low. But small areas with high and medium density are found in the Smolensk-Moscow, middle Russian, Kalachsk and Volga highlands—south of the line from Vitebsk—Kaluga—Uryupinsk. These highland areas dissected by ravines are the most densely populated by European hares.

The number of European hares is also high in lowlands, and they are most often found in river valleys. In Ciscaucasia, for example, the maximum population density of European hares is recorded in the Kuban lowlands, which is dissected by ravines and river streams falling in Kuban and the Azov Sea. In western Stavropol' territory at the beginning of fall, the highest density of European hares is observed on virgin areas (68 individuals per 1,000 ha) and stubble fields (59 per 1,000 ha). However, in the course of winter the number of European hares in different habitats changes significantly, and by the beginning of spring their maximum density is in ravines and in forest (Table 10.3). The population density of European hares in summer differs between different habitats. Thus, according to the observations in the Dyakov forest farm in the Saratov region (Gruzdev, 1974), up until August 15, 1951, the population density of European hares in young forest plantations along fields was 5 times higher than on sandy hummocks overgrown with shrubs: 6 and 1.5 individuals over a 10 km route, respectively. After August 15 the density of European hares on sands became twice higher (2.8 individuals over 10-km route) than in forest plantations near fields (1.3 individuals over 10-km route).

In the Uralsk and Gurevsk regions, the density of European hares is quite high in Narymsk and on the Volga-Uralsk sand massifs. For example in 1961, on the sands and banks of streams in the Kamysh-Samara Lake system and in the vicinity of the town of Novaya Kazanka, and in July 1962—in the vicinity of the village of Urda over a 10 km-route, on average, 6 hares each were found. The lower reaches of the Volga and Ural, the valley slopes of which create conditions of dissection [of relief] are also densely populated with European hares. The density of European hares increases in winter, particularly in valleys. Thus, in the Chapaevsk district in November-December 1961, for every 10 km-route passing through groves

*Syrt—watershed upland—General Editor.

Table 10.3. Change in the population density of European hares in different habitats in the western Stavropol' territory during winter (from Kolosov and Bakeev, 1947)

Habitat	Number of individuals per 1,000 ha (n = 244)		
	January	February	March
Virgin land	42.3	36.3	37.2
Gullies	56.2	7.0	140.0
Stubble fields	7.1	5.5	14.3
Winter crops	123.0	–	–
Fall-plowed field	35.2	71.8	40.0
Vegetable garden	107.1	200.0	–
Forest	103.3	293.7	94.2

of black poplar, white poplar, willows, birch thorn, and buckthorn, on average 18 hares were counted, while along the edge of the flood plain at the border with semidesert—only 8. On the right bank of the Ural, European hares are numerous on weakly fixed sands in the valley of the Uil River. Here in February 1963, there were 7 to 17 hares over a 10-km route (Sludskii et al., 1980).

In the Asiatic part of the range, where, on average for a decade, more than 50 cm snow fell, European hares are less numerous, whereas in the European part the population density of European hares in such areas is quite high. Such a difference can be explained by the more severe climate in the northeast or insufficient dissection of relief in southwestern Siberia, and also by lesser occupancy of this area by people (Gruzdev, 1974).

In montane areas the population density of European hares is usually low, since mountains are usually covered with forest, and in winter the snow is thick there and, in comparison with foothill plains, in mountains the climate is more severe. The population density of European hares is low also in the mountains of central Europe. According to the data of S.S. Kokhiya (1974), in the territory of Georgia in 1967-1968, depending on the place of habitation, the population density of European hares fluctuated from 8 to 9.5 individuals per 1,000 ha, i.e., in the lesser Caucasus European hares are distributed very unevenly. In the Caucasus commercial hunting of hares is almost non-existent; it is somewhat more developed in the Carpathians. In the southern Urals, skins of European hares are generally not processed in several places.

Thus, it can be concluded that the number of European hares in any territory depends on the climate, height of the snow cover, reclammation of the territory by man, and relief of the locality. The population density of European hares in the very same habitat experiences significant changes

219

during the year, which is primarily linked with the abundance and availability of food.

Shelters. European hares are very mobile and, depending on weather conditions, state of the food base, and abundance of predators, are capable of rapidly changing the places of their residence. Feeding predominantly in open habitats, they rest among shrubs and glades. Permanent shelters are absent in European hares. In summer the resting place, particularly in rainy weather, is situated in some shelter—under a shrub, in stubbles, in overgrown boundaries or in a cluster of brushwood. If the substrate [soil] is fairly loose, European hares dig a hole, not using the grass growing nearby, so that lying in such a shelter it is difficult to spot the hare. European hares can use such shelters several times. According to observations in the Saratov region, 58 recorded summer resting places of European hares were under the cover of plants, and 48 of these were either under bushes or on slopes of northern, northeastern and northwestern exposures. Sometimes a concentration of resting places was also observed on shady slopes (Gruzdev, 1974).

The size of the summer resting places depends on the substrate. In the Saratov region the depth of holes, on average, was 29 cm (n = 21), their length varied from 38 to 50 cm (three holes 55, 60, and 67 cm long were an exception—in one case it was two holes merged in one, and two others were holes with niches), and the width of holes varied from 20 to 27 cm. In young European hares the dimensions of holes are less—28-36 cm long and about 17 cm wide (Gruzdev, 1974). In other parts of the range, summer resting places have lesser dimensions. In Hungary the depth, length and width of resting places were 16-20, 30-35, and 14-18 cm; in France, 14-19, 28-33, and 15-18 cm, respectively (Zorner, 1981). In the Stavropol' territory the depth of a hole was, on average, 12 cm (Fig. 49); in the flood plain of the Ural River, the depth of a hole varies from 2 to 10 cm, length—from 37 to 50 cm, and width—from 16 to 19 cm (Sludskii et al., 1980). According to the data of S.I. Ognev (1940) and A.N. Formozov (1959), in the Aral Karakums European hares dig burrows during the reproductive period. In other parts of the range, summer burrows of European hares were not observed.

European hares can build short-duration resting places also in fields of agricultural crops, that is, in places of foraging. In dense thickets of grasses, European hares prefer not to build resting places—in such places they most often lie 10-20 steps from the edge of fields. In sands resting places are usually situated between barhans* on exposed sand or under shrubs, and in strong heat European hares lie atop barhans swept by wind.

---

*Sand dunes—Translators.

218

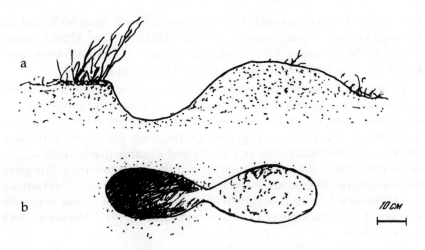

10 CM

Fig. 49. Burrow of a hare in Stavropol' territory.
a—vertical section; *b*—plan (Kolosov and Bakeev, 1947).

In virgin steppes European hares may use as shelters those nests abandoned by marmots, foxes, and badgers (Gruzdev, 1974; Sludskii et al., 1980).

In early spring European hares prefer to lie down on slopes warmed by the sun, while in summer their shelters are situated in shady places. Sometimes under the very same bush, hares make two holes: one, shallow, used for resting in the morning and evening; another deeper, serving as a daytime shelter and always situated on the northern side of the bush. In hot dry weather European hares lie in low places, and in rainy weather— on elevated grounds.

In fall, resting places of European hares may be located on gulley slopes (under overhanging sod and roots of shrubs), and in meadows—under the protection of hay stacks. During crop harvests the protective conditions for European hares deteriorate sharply; young hares disperse to the remaining unmowed parts of the fields, hide in heaps of straw during the day, and upon completion of the harvest and stacking of straw leave for other hiding places like [field] boundaries overgrown with weeds, areas with plantings of corn and in shrubs (Kolosov and Bakeev, 1947; Korneev, 1960).

In winter with the establishment of snow cover, European hares make resting places directly on the snow on the leeward side. In northern Europe, European hares sometimes live for a long time in hay stacks, almost not
220 leaving such shelters. During thawing, they may collect under stumps with roots or under natural overhangs. With deep snow and strong frosts, European hares build burrows, usually situated in places of snow drifts.

In northern Kazakhstan such burrows have a length from 90 to 190 cm (average 140 cm) with an entrance opening width from 15 to 30 cm (average 26 cm). In the vicinity of Aktyubinsk, two burrows of European hares were found, the length of which was 1 and 2 m (Sludskii et al., 1980).

Feeding. In the snowless period of the year, European hares feed mainly on herbaceous vegetation. According to observations on European hares in captive conditions, it has been established that they avidly eat grasses (particularly cultivated), composites (particularly sunflower, Jerusalem artichoke, dandelion, chicory and tansy), and legumes (particularly alfalfa, sweet clover, and clover). Among herbs of the cabbage family, European hares avidly eat bitter cress, shepherd's purse and winter cress, and among the knotweed family [Polygonaceae] the favored ones are prostrate knotweed and buckwheet (Kolosov and Bakeev, 1947; Korneev, 1960; Belova, 1987).

In the territory of Lithuania in the snowless period, European hares feed mainly on fodder grasses and weeds, and also leaves of root and vegetable crops. Summer grain crops are absent intermittently in the ration of European hares—only in the initial stage of growth. On the whole, in this part of the range herbaceous plants in the snowless period constitute 76.3% of their diet, that is, some fraction in the summer diet of European hares is attributed to wood-shrub vegetation (Belov, 1987).

Under natural conditions feeding of European hares from the south-eastern part of the range has been studied in most detail. In the Yeruslan sands (Saratov region) the base of the summer diet of European hares comprised not the plant species growing here (European feather grass, desert wormwood, and euphorbia), but rarer ones—germineous skeletonweed, and baby's breath. These plants are found particularly frequently in the food of European hares in the beginning of summer. In all, in the summer diet of European hares of this area, 30 species of herbaceous plants have been recorded, of which 11 are eaten incidentally. Such species as baby's breath, because of the alkaloids contained in it, is absent in the ration of cattle, but is eaten well by European hares. Practically in all plants only stems are eaten in summer, and only in goat's beard—mainly leaves. A similar set of species has been found in the summer diet of European hares also in the Urdinsk sands (Gruzdev, 1974).

More diverse is the diet of European hares inhabiting the territory of the Ural'sk region (Sludskii et al., 1980). Moreover, analysis of the stomach contents has made it possible to establish that a no less important role in the diet of European hares is played by seeds, the proportion of which in the diet increases toward the end of summer. In the Ural'sk region and in Turgai in summer, the diet of European hares contains wormwood to some extent, while in the Saratov region wormwood is found only in winter diet.

In the Yeruslan sands and in Turgai, one of the more valuable plants in the summer diet of European hares is baby's breath, which in the Ural'sk region is more often found in the winter diet (Solomatin, 1969; Gruzdev, 1974; Sludskii et al., 1980).

In Ustyurt the summer diet of European hares primarily comprises: woody milk vetch*, Caspian onion*, erkek* and large seeded panicum*. In habitats where species composition of herbaceous plants is poor, the diet of European hares may consist of 1-2 species. Thus on Barsakel'mes Island, European hares eat mainly the leaves and shoots of Siberian wheat grass and dzhuzgun [*Calligonum*] (Ismagilov**, 1940; Sludskii et al., 1980).

The foraging places of European hares are determined by the abundance of food, and also safety conditions. According to observations in the Saratov region with a similar abundance of food plants in the virgin and cultivated lands, European hares prefer to feed in open places, which allows them to detect danger in time and hide themselves. In dense thickets European hares avoid feeding even on their favorite plants. A comparison of the feeding behavior of European hares on sandy hummocks covered with bushes and on level forestless areas has shown that hares prefer to feed on hummocks, avoiding places where foxes appear most often (Gruzdev, 1974).

221 In late fall until the formation of snow cover, the dietary base of European hares continues to comprise herbaceous plants and the bark of trees; branches of shrubs are not eaten at this time. In the flood plain of the Ural River in fall, European hares feed mainly on grasses (71% incidence in stomachs), and in this period 38-57% of the diet of European hares consists of dry ramal vegetation (Sludskii et al., 1980). In Turgai in the middle of fall, the stomachs of European hares were filled up to 10-15% of their volume with seeds of weedy plants (Solomatin, 1969).

In winter, depending on the place of habitation, the diet of European hares includes some quantity of ramal food and bark. In the forest zone the proportion of woody vegetation in the winter diet of European hares is higher than in the forest-steppe and steppe zone. For example, in the Leningrad [St. Petersburg] region, European hares in winter feed actively on the seedling growth of eagle-claw maple, English oak, apple and acacia, whereas European hares do not eat willow and raspberry, which are the favorite food of blue hares (Novikov et al., 1970). In European hares living in closed habitats in the territory of Lithuania, their winter diet recorded 25 species of woody-shrub vegetation, while in open habitats—15 species. Here the most preferred foods are shoots of oak, apple, European aspen

*Plant names not confirmed—General Editor.
**Not in Lit. Cit.—General Editor.

and linden, and also daphne willow and raspberry (Belova, 1987). In the Usmansk forest (Vornezh region), European hares in winter most often damage warty bark evonymus, filbert and broom, while European aspen (the principal winter food of European hares in these places) is eaten by them relatively rarely (Barabash-Nikiforov, 1957).

The diet composition of European hares in the beginning of winter and its end is usually different and to a large extent depends on the height of the snow cover (Kolosov and Bakeev, 1947; Sludskii et al., 1980). With shallow snow cover and in the beginning of winter, European hares feed, as a rule, on winter grains at the tops of hummocks and in mowed meadows. However, instances are also known of European hares eating bark and branches of young trees even before snowfall. With high snow and at the end of winter, European hares usually migrate to the forest, where their diet starts to reveal a predominance of sprigs of fodder crops (Table 10.4) or closer to the vicinity of populated areas, where the foraging places are usually concentrated along roads, in gardens, and also near hay stacks and ricks.

Table 10.4. Change of diet composition of European hares in winter in the territory of Lithuania (Belova, 1987) and Belorussia (Gaiduk, 1982)

Type of foods, %	Lithuania		Belurussia	
	December	February-March	December	February
Dry grass and perennial herbaceous ornamental plants	25.0	21.0	–	–
Hay	12.0	3.0	25.0	9.0
Winter crops	28.0	14.0	11.0	15.0
Perennial cultivated grass (clover, etc.)	7.0	19.0	12.0	25.0
Woody-shrubby vegetation	11.0	25.0	4.0	13.5
Needles, and shoots of pine	0.5	3.0	–	–
Other foods	16.5	15.0	48.0	38.5

According to the data of A.M. Kolosov and N.N. Bakeev (1947) in Stavropol' territory, in the beginning of winter the proportion of woody-shrubby vegetation is not high in the diet of European hares—about 10%, and by the end of winter it increases to 70%. A roughly similar character of winter feeding of European hares is observed in open habitat of the Volga-Kama territory (Table 10.5). With shallow snow cover European hares feed not only on leaves, stems and seeds of herbaceous plants, but also can dig out roots, the proportion of which in winter diet may reach 2.7%.

With deep snow hares usually migrate to shrub thickets, but in the absence
222 of a frozen snow crust, they can dig pits or even trenches to 50-60 cm deep
so as to reach herbaceous vegetation.

Table 10.5. Winter diet of European hares in open habitats of
the Volga-Kama territory (from V.A. Popov, 1960)

Type of plant eaten, %	Fields		Flood plains	
	November	February-March	November	February-March
Dry grasses	50.0	40.0	50.0	30.0
Winter crops	35.0	35.0	0	0
Willows	0	5.0	20.0	60.0
Woody vegetation	15.0	20.0	30.0	10.0

In forests and shrub thickets, European hares prefer, as far as possible,
to feed on herbaceous vegetation. Only after the formation of deep snow
cover and a frozen snow crust do they change over to feeding predominantly on woody vegetation. In Stavropol' territory even in January, of 16
species of plants consumed by European hares, only 5 were trees or shrubs.
At this time the main foraging places are concentrated at forest edges, glades
and clearings. By March, when the snow height here reaches 25 cm, woody-
shrubby species begin to predominate in the diet of European hares. On
trees and shrubs, European hares most often gnaw young shoots with
apical buds and thin branches up to 7 mm in diameter. In woody vegetation, as a rule, only the young growth is damaged. Less often bark of
trees and thick branches up to 3-4 cm in diameter and to 70 cm high are
gnawed.

In the forest-steppe zone, for European hares the favorite species of
woody-shrubby vegetation are broom, hawthorn, apricot, eagle-claw maple
and elm. In areas with developed horticulture, European hares cause
damage to plantations of apple, plum, cherry, peach, and also grapes
(Kolosov and Bakeev, 1947; Gruzdev, 1974; Sludskii et al., 1980).

In the Ural region with the formation of snow cover, European hares
dig out green leaves of thistle, wormwood, saltbush, licorice, and also green
and dry grasses from under the snow. In glaze and with deeper snow,
European hares change over to feeding on bark and branches of blackthorn,
poplar and tamarisk, but even at this time nearly 50% of their diet consists
of herbaceous plants. With snow cover less than 10 cm deep, European
hares graze on sunny slopes and, when the thickness of snow cover
surpasses 10 cm, European hares migrate to forest clearings and willow
groves (Sludskii et al., 1980).

According to the observations on European hares in conditions of cage rearing, it has been established that in a day they consume 500-800 g of different food, and with feeding only on greens—up to 1 kg, which constitutes about 15% of their body mass (Zörner, 1981).

European hares avidly drink water, although with an abundance of succulent foods they may avoid and live without water. In the territory of northern Kazakhstan, in summer European hares are found almost exclusively near water springs. In the absence of fresh water, European hares can also drink salty water. Most often they satiate thirst in the morning—up to 9-10 hrs. and in the evening—from 18 to 19 hr (Sludskii et al., 1980).

223 A comparison of the nature of feeding of European hares and blue hares confirms the presence of common features, such as: exclusive herbivory of both species, presence of coprophagy and seasonal changes of foods. However, there do not exist species peculiarities in the diet of these hares. In the first place, it is the sharper seasonal change of foods in blue hares, which in winter practically completely change over to feeding on woody-shrubby vegetation, whereas in European hares a complete change over to ramal food does not occur in winter, and the proportion of woody-shrubby vegetation in the winter diet depends on habitat and weather conditions in the winter period. Moreover in European hares and blue hares there is a clear manifestation of interspecific differences in the preferred foods.

Activity and behavior. European hares are active mainly at dusk and at night. The exception throughout the range is the period of heat, when European hares are active over a longer time. In northern Kazakhstan in spring after the termination of the heat period, European hares feed at 17-18 hr, and in summer somewhat later—about 19 hr; they rest usually at 6-7 hr. The main rhythem of activity of European hares is retained also in the fall; however, during rainy weather they leave for feeding 1-2 hr earlier. In winter during heavy snow falls, European hares do not leave for foraging, while on bright frosty days they begin feeding at 16-17 hr and finish at 8-10 hr (Fadeev, 1966a). According to the data of I.N. Serzhanin (1958), European hares inhabiting the territory of Belorussia begin to rest at midnight, and their morning peak activity starts before sunrise. Night rest has been noticed also in European hares inhabiting the territory of Lithuania—here hares rest from 1 hr in the night to 5 hr in the morning. The most active feeding has been recorded from 19 to 21 hr and 5 to 7 hr (Belova, 1987).

In forest habitats in the northern and northwestern parts of the range, European hares feed also in the day; moreover, their daytime activity is higher in summer. Thus, according to observations in Czechoslovakia, in summer the daytime activity of European hares constitutes about 48% of time (of 10 hours), and in winter and fall—11 to 32%. In winter about 9% of daytime is spent roaming and 12% on such forms of activity as social contacts, comfort behavior, and coprophagy. In summer, 29.4% of daytime activity is spent in foraging, 5.4%—roaming, 9.6%—comfort behavior, and 4.1%—social contacts. In spring and summer European hares remain most closely huddled in shelters from 11 to 15 hr. Thus the winter daytime rest in European hares in all its complexity extends about 8-9 hr, while in spring-summer it is less—5-7 hr (Homolka, 1986). In the territory of Lithuania, the daytime activity of European hares is recorded at 10 and 13 hr. In this part of the range in summer and spring, in 94% of cases European hares can be found from 17 hr to 1 hr in the night and in the morning from 12 to 13 hr. In winter 65.3% of encounters of European hares are in the evening period from 17 hr to 1 hr, and in the morning from 5 to 9-10 hr (Belova, 1987).

Adult European hares never rest deep inside dense vegetation. If the hole is situated under a bush, then its deeper part lies in the center of the bush and the animal lies with its head facing outward and its back in the depression, which provides a fairly good view of the territory. Near resting places there is always a diversion path—open glade, road and pathway continued in the thickets by hares themselves. Such tracts connect resting places with places where hares feed most often. With a sudden advent of danger, diversionary paths are well known to European hares. Rising from resting places in the open, adult European hares usually run at high speed to plantations and, making use of relief of the locality and vegetation, hide rapidly, such that they are visible only for a short time. Frightened in the forest, European hares first run in the open, but then hide again in the forest. Very often while running away from predators or hunters with dogs, European hares use automobile roads and railroad tracts where there exists a possibility to develop maximum speed and, what is no less important, throw-off hunters from pursuit. Sometimes European hares masking their trail swim through rivers, run to the vicinity of cattle yards, or even run through a herd of grazing cattle (Gruzdev, 1974*; Belova, 1987).

224   Depending on the extent of the scare, European hares, rising from rest, describe a circle of greater or lesser diameter and return to their place. While running in a circle the hare disturbs its trace with the help of "duplicates" (moving forward and backward at the same place) and "sweepings" or "jumps" (very long hops to the side of the main track); with a strong scare, European hares run at great speed pressing their ears to the neck. And in places where hares are threatened rarely, they run with ears raised, in

*Not in Lit. Cit.—General Editor.

314

uniform jumps 1-3 m long, often halting in-between. In the snowless period the loops are usually longer, and in winter, particularly in loose snow, loops are shorter. Young European hares (escaping from pursuit, usually do not make large loops and combine short (150-200 m) runs with rest, not moving far from the resting place. Usually young European hares of one brood keep within 80-100 steps of each other (Gruzdev, 1974; Sludskii et al. 1980; Belova, 1987).

As already mentioned, the maximum activity in European hares occurs in the period of heat. At this time hares remain in pairs or groups of 3 to 7 individuals. In such groups most often males are more numerous. During the heat males run in circles, approach and withdraw from females, jump around females and through her chase away other hares, jump on hind feet, and push each other with forefeet. During the period of rest there are naso-nasal contacts, hugging, licking, and mutual biting. In group running the dominant male usually runs in the front; it is he who aggressively threatens rivals. Aggressive attacks of the dominant male end in fights, if subdominant males vie for a female and do not demonstrate a submissive pose. If a female is dominant in a group, she may attack females lower in the hierarchy. In case a pair of dominant female and male of a lower hierarchic rank is formed, the female is more active in courtship play (Belova, 1987).

Before mating a male with lowered tail and raised ears approaches the female; his legs are straight and back is curved. The female initially runs away from the male in even strides. Its ears are pressed toward the back, the tail is raised and somewhat moved sideward. Then the female remains in an excited state and sits down. During mating ears are raised in both the male and female. According to observations on European hares in captive conditions, it has been established that the obligatory elements of sexual behavior are: lording, which lasts for 6 seconds, mounting by the male (about 2 seconds) and actual copulation, which lasts for 5 to 30 seconds (Belova, 1987).

After mating the male produces a sound similar to a dull gurgle. Both partners often remain together for some more time; moreover elements of courtship behavior may be repeated once again. At this time elements of comforting behavior are also characteristic. Mating occurs most often in the evening and in the morning up to midday, and during the second and third heat—sometimes also in the day (Belova, 1987). According to the data of other authors (Caillot and Martinet*, 1983) mating in European hares occurs mainly at night.

Before births, females of European hares spend a large part of their time

---

*Not in Lit. Cit.—General Editor.

in resting places. Unlike blue hares, females of European hares give birth to young on a bed of straw or hay; moreover in cold weather females cover young with hay also from above. Males do not take part in brood care.

Observations in captivity have shown, that the interval between the first and second feeding is usually 3-4 days. At this time newborns sit immobile at the place where they were born, and begin to move about on the 4th-5th day. Females feed young ones sitting and place them between the forepaws or on the side. After completing feeding the female moves away from the young and returns only for the next feeding. During the first two weeks the feeding initiative comes from young, and at the age of two weeks the hungry ones start actively seeking for the female. At the age of two weeks, they may move away up to 80 m from the place of their birth, but return at the onset of dusk (Belova, 1987). The phase of intimate contact between mothers and their young usually lasts 25 days, but sometimes may extend to 40-60 days. During this period, females actively protect their progeny—in the event of danger they cover babies with the hind part of their body. Females of European hares feed young from another mother, only if they do not differ in age from its own babies. The lactation period in European hares lasts, on average, for 45 days (Belova, 1987).

Area of habitation. European hares are territorial animals with relatively well developed settling. Data on the size of individual territories in European hares and peculiarities of their territorial behavior are very few. The area of habitation of European hares depends on the population density (the higher the density, the smaller the size of the territory) and constitutes, on average, 333 ha (Pielowski, 1971). However, this territory is used nonuniformly by hares. The area visited most frequently, according to different data, varies from 16-20 ha (Pielewski, 1972) to 40-66 ha (Homolka, 1985). On this territory are located areas for feeding and resting; here hares spend up to 50% of their time. According to the data of O.P. Belova (1987), the area of such a territory, on average, constitutes 10.2 ± 4.5 ha, and on it are situated 4 to 12 foraging areas. In open habitats at any time of the year, European hares are more mobile, and, on average, run 4 km in a day, and in forest—only 2.8 km.

In forest habitats individual territories of females are, on average, 30 ha, and those of males are 70* ha (Homolka, 1985) or 28.5 ± 5.3 and 41.5 ± 18.6 ha, respectively (Belova, 1987). Smaller areas of individual summer territories of females are also found in Germany: in males, the average is 14.4 ha, in females—5 ha. These differences are attributed, apparently, to greater mobility of males, particularly in the period of heat (Zörner, 1981). However, in open habitats, according to the data of O.P. Belova (1987), the individual territory of females is greater than that in males and, on average

---

*7 = 40 ha—Editor.

female and male territories constitute 43.2 and 21.5 ha, respectively. Large areas of individual territories of females are also reported in the Poznan military area of Poland (Pielowskii, 1972).

Individual territories of both males and females overlap quite considerably, and the degree of their overlapping depends on the time of the year, abundance of food, and shelters. Individual territories overlap particularly strongly in winter, when even "nonfilial" groups may be formed. In such groups there is a quite well developed hierarchic structure, but neutral relationships predominate based on joint exploitation of resources. However, under unfavorable external conditions (winter) or in the mating season, neutral relationships between European hares acquire the nature of reeling. Dominant individuals spend more time in foraging areas and may chase away hares of lower rank. On meeting hares of the same hierarchic level, fights may arise. In groups of European hares, unlike groups of pikas or rabbits, closed primary family cells are absent (Belova, 1987).

European hares mark the boundaries of habitation areas with the secretion of anal glands, and also with the help of excrements and urine. In European hares chin and anal glands are more weakly developed than in rabbits, and inguinal glands, which serve for individual recognition and attracting individuals of the other sex, are better developed in European hares. Although seasonal and sexual differences of glands and their function are more weakly developed in European hares than in rabbits, great activity of chin and inguinal glands is noticed in females of European hares as compared to males (Mykytowycz, 1966; Sokolov and Terekhina*, 1978).

Reproduction. The periods of reproduction of European hares are linked with geographical location of the population and weather conditions. The start of reproduction depends on the level of temperature drops and duration of the photoperiod. Warm weather in winter accelerates gonadal formation and the start of reproduction. Starting from January, with intervals of 7-9 days, the periods of sexual activity repeat alternating with resting phases. In males sexual activity begins earlier than in females. Estrus in females continues for 2-3 days. Ovulation occurs only following successful mating. If the female remains unfertilized, it again becomes ready for covering after 15-17 days (Gruzdev, 1974; Zörner, 1981).

226   Young individuals of the previous year's birth participate in reproduction asynchronously. Hence the start of spring reproduction is usually extended. According to the data of O.P. Belova (1987), females of the previous year's first brood participate in reproduction along with adults, whereas 80% of females of the second generation do not participate in

---

*Not in Lit. Cit., presumably refers to Sokolov and Terekhova, 1978—General Editor.

226

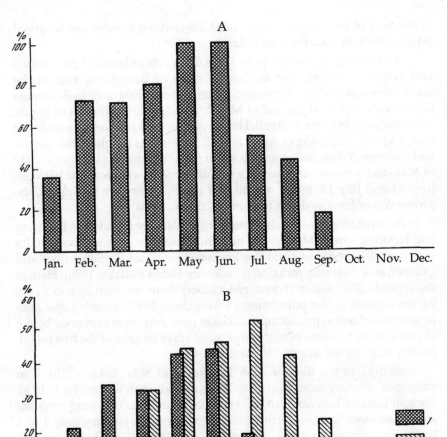

Fig. 50. Dynamics of participation of female European hares in reproduction.
A—Stavropol' territory (from Kolosov and Bakeev, 1974); B—Ukraine:
1—percentage of gestating females; 2—percentage of lactating females
(from Arkhipchuk, 1983).

reproduction, and the brood born in early spring begins to develop reproductive organs by July, and by August their development is halted and regeneration occurs. Mated underyearlings were not observed.

In the west of their range (western and central Europe), the first pregnant females of European hares were recorded in January. In February about 68% of adult females become fertilized, and in April—77%, and in summer, in this part of the range, from 75 to 83.8% covered females are observed (Moller, 1976; Broekhuizen and Masskamp, 1981).

227   In the Volgoda region European hares first come in heat in April (Savino and Lobanov, 1958), and in the territory of Belorussia and in the southeastern part of the chernozem center—in mid-February. In Belorussia first broods appear at the end of March. A large part of broods of spring reproduction (46%) are in April. Hares come in heat here next towards the end of March-beginning of April, and all the females caught in this period had embryos. When they come into heat for a third time is in third decade of May, and a massive appearance of third broods was recorded between June 22 and July 12. In the second half of July-beginning of August, the proportion of females drops to 54.5% (Gaiduk, 1973a).

In the territory of the Ukraine, pregnant females are found from February, and lactating—from March. The maximum number of gestating females was recorded in March, and the lowest—in August. In September and October, as a rule, only foraging females are found (Galaka, 1970). Here in the reproductive season (February-October) there are from 36.6 to 94.6% barren females in the population (Arkhipchuk, 1983). Possibly, the high proportion of non-reproducing females in June-August is explained by the appearance by this time of fairly large numbers of females of the first brood, having attained the size of adult individuals.

According to the data of A.M. Kolosov and N.N. Bakeev (1947), in Stavropol' territory, some European hares reproduce in winter (Table 10.6). Thus, in January-February, out of 58 females investigated, 33 were pregnant and 3 had already given birth; of 29 males caught in January, only 1 had testes in a quiescent state. However, here also the most intensive reproduction occurs in spring: in March the proportion of pregnant females was about 62%, in April—about 80%, and at the end of May all females had participated in reproduction; moreover, more than half of them—already a second time.

228   In Kazakhstan, depending on weather conditions, the start of spring reproduction may be extended or be less harmonious. In the north (flood plain of the Ural River and the Turgai Plateau), European hares begin to reproduce usually in the middle of February, and southward (Barsakel'mes Island) somewhat earlier—in January-beginning of February (Solomatin, 1969; Sludskii et al., 1980). Usually young of first broods in this part of the range appear in March-April, sometimes—in May. Young of the second brood, depending on the geographical location of the population and weather conditions, appear in May-June, and of the third brood, which happens not in all females, appears usually in the middle or end of August (Sludskii et al., 1980).

Over a large part of the range during the reproductive season there are,

Table 10.6. Comparative data on reproduction of European hares in different parts of their range (from Fedeev, 1964, with additions)

Region	Time of appearance of first brood	Number of broods per female for the season	Number of young in brood	Total number of offspring per female in a year	Authors
Central Europe	Mid-March	4-5	4-5	11	Rieck, 1956; Petrusewicz, 1970; Moller, 1976; Pielowski, 1976
Bulgaria	March	4-5	–	10-12	Petrov, 1963
Yugoslavia	March	5	–	10	Valentoncic-Stane*, 1956
Belorussia	February-March	3-4	2-5	8-10	Gaiduk, 1973a
Ukraine	March	3-4	1-6	–	Korneev, 1960
Crimea	End of February	4	1-4	–	Kolosov and Bakeev, 1947
Ciscaucasia	January-February	3-4	1-6	10	Kolosov and Bakeev, 1947
Kyzyl-Agach preserve	January	–	1-3	–	Grekov, 1961
Moscow, Ryazan regions, Bashkiria	April	2	2-8	7-8	Kolosov and Bakeev, 1947
Tataria	April	3	3-6	9-12	V. Popov, 1960
Gur'evsk, Uralsk regions	End March	2-3	1-9	10-12	Fadeev, 1964
Western Siberia	April	1	2-7	2-7	Berger, 1947

*Not in Lin. Cit.—General Editor.

on average, three broods. In the west and southwest of their range, some females may bear up to five broods in a year (cf. Table 10.6). According to the data of L.S. Shevchenko* (1974), in the territory of the Ukraine, almost 100% of females in a year bear 3 broods each, and 30%—4 each. In valleys the number of broods is usually greater than in mountains. In European hares reproducing in captive conditions, there are 2 to 6 broods in a year, here the higher number of broods leads to a decrease in their size (Martinet*, 1976).

In nature the size of a brood depends on weather conditions, time of reproduction and age of female. According to the data of O.P. Belova (1987), the average size of broods of females two years of age (and older) fluctuates from 1.9 to 3.3, whereas in one-year old females the average size of broods is less-from 0.5 to 2.5; moreover, on average, females of the first generation have higher fecundity. As a rule, in females of any age group, second broods happen to be more numerous (Table 10.7). According to the data of analysis of the number of embryos during the first (winter-spring) reproductive period in southern populations of European hares, the number of embryos, on average, was always more in females entering reproduction later (cf. Table 10.8). With fluctuation of the number of embryos per female from 1 to 7 in the Uralsk region in April, 37.5% of females had up to 6 embryos, whereas in March the maximum number of embryos did not exceed 4 (Fadeev, 1964). A roughly similar picture is observed in European hares from Stavropol', although the maximum number of embryos per female was less here (cf. Table 10.8). In the south of their range, European hares most often have 3-4 offspring in one brood (cf. Table 10.8). In the north and west of their range fecundity of European hares, apparently, is roughly in the same. For a reproductive season one female, depending on weather conditions, its age and climatic zone, bears from 7 to 12 young (cf. Table 10.6). On the whole, throughout the range, as also in captive conditions, there is a decrease in the average size with an increase in the number of broods.

According to the data of O.P. Belova (1987), the size of a brood in European hares increases if mating occurs even before parturition. Such a phenomenon (superfetation) is characteristic of European hares, and is often observed in females older than 1.5 years and in the second reproductive period.

229   According to the data of A.M. Kolosov and N.N. Bakeev (1947), resorption of embryos in European hares south of the European part of the range constitutes from 7% (in spring) to 25% (in fall). In the territory of Kazakhstan, the highest embryo mortality in European hares was recorded in February (15%) and the lowest (2.4%) in June (Sludskii et al., 1980). In

*Not in Lit. Cit.—General Editor.

321

Table 10.7. Fecundity of European hares in the forest-steppe part of the Ukraine from the data of 1971-1981 (Arkhipchuk, 1983)

Age groups	First brood			Second brood			Third brood			Total		
	n	Number of off-spring	Per female	n	Number of off-spring	Per female	n	Number of off-spring	Per female	n	Number of off-spring	Per female
Up to 1 year	21	40	1.9	15	40	2.66	5	10	2.0	46	90	4.3
2 years	11	23	2.1	6	18	3.0	2	7	3.5	19	48	4.36
3 years and older	8	19	2.4	6	17	2.83	2	7	3.5	16	43	5.38
Total	40	82	2.1	27	75	2.77	9	24	2.66	81*	181	5.63

*Total should be 40 + 27 + 9 = 76 and not 81 as given in Russian original—General Editor.

Table 10.8. Number of embryos in European hares in the Uralsk Region (Fadeev, 1964) and Stavropol' Territory (Kolosov and Bakeev, 1947) in different seasons

Place	Month	Number of females	Percentage of females with number of embryos							Total embryos	Embryos per female
			1	2	3	4	5	6	7		
Uralsk region	February	16	25.0	43.7	31.3	–	–	–	–	33	2.0
	March	39	17.9	53.8	23.1	5.2	–	–	–	84	2.2
	April	8	–	12.5	–	25.0	12.5	37.5	12.5	40	5.0
	Total	63	17.4	46.0	22.2	6.3	1.7	4.7	17	157	2.5
Stavropol' territory	February-March	?	37	25.0	37.4	–	–	–	–	?	2.0
	May	13	15.0	23.0	31.0	31.0	–	–	–	44	3.4
	June-August	12	–	16.7	25.0	16.6	33.4	8.3	–	47	3.7

European hares inhabiting Belorussia, embryo mortality at the start of the reproductive season was 16%, and in May-June—10.2% (Gaiduk, 1973a).

Growth and development. Observations on European hares in conditions of cage rearing made it possible to establish that pregnancy in them lasts 42 days, although some females may give birth even earlier (36th-41st day) or later (43rd-45th day). Moreover, in the very same female, the duration of pregnancy may vary. Births most often occur in the morning—at 6-10 hr (Arkhipchuk, 1983; Belova, 1987). European hares are born with sight, and with soft dense and curly fur. The weight of newborns varies usually in the range of 80 to 150 g, although there are instances of births of smaller (65 g) and larger (192 g) baby hares (Kolosov et al.*, 1965; Pielowsky, 1971; Zorner, 1978). The mean daily weight gain of baby hares in the first 20 days is 20 g, and by the 20th day their weight increases, on average, four times. In the next 20 days the mean daily weight gain increases to 38 g, and at 1.5 months young attain a weight of 1-1.5 kg. By two months of age the mean daily weight gain decreases to 21 g, and by four months of age— to 19 g. At seven months of age the weight gain of European hares per day does not exceed 6 g. At 8-9 months European hares usually attain the weight of adult individuals (Belova, 1987).

The body weight of baby hares of smaller litters (1-2 baby hares) at birth is more than the baby hares of larger litters, and in the period of milk feeding their weight increases faster, but after the change over to independent feeding, the intensity of growth of these and others evens out and often baby hares of large broods at this time surpass the body weight of their rivals from solitary litters. Baby hares from spring broods are usually larger than those born subsequently and the weight of baby hares born in the summer increases most intensively in the first three months of their life (Arkhipchuk, 1983a; Belova, 1987).

The mean daily body weight gain of European hares reaches a maximum in the first 10-15 days. At one month of age with the change over of young ones to independent feeding and at six months of age, when fall molting begins, there is a fall in the intensity of growth. The maximum growth of limbs is observed in the first 20 days of life of baby hares; by the fifth month the growth of limbs in length usually ceases, but complete development of the pelvic girdle, as also the growth of body, terminates in European hares at one year of age. The most intensive increase in skull dimensions in European hares occurs in the first month of life. By the fourth month of age the intensity of growth of the skull decreases. At seven months of age begins the enlargement of the supraorbital processes, the growth of which, as in other hares, continues throughout life (Arkhipchuk, 1983a).

230

*Not in Lit. Cit.—General Editor.

324

In the first 5 days after birth, in European hares among permanent teeth there are upper and lower incisors, and also upper and lower molars M1 and M2. In newborns the second pair of upper milk incisors and upper M3 lie in alveoles. On the lower jaw, the more developed are the first two milk premolars and one permanent molar (M1), but the second permanent molar tooth is developed considerably weakly. The change of milk teeth begins on the 6th-8th day with the change of the upper and lower premolars. On the 15th day the second pair of milk incisors, third upper premolar, and also the lower first and second premolar teeth are replaced (Arkhipchuk, 1983a).

Starting from 12-15 days of age, young hares gradually change over to feeding on herbaceous vegetation. In this transition period, which lasts 10-15 days, milk teeth are replaced by permanent teeth, and the growth of the facial section of the skull and hind limbs intensifies. At 25-30 days of age, with the transition to independent feeding, body growth slows down and body weight increases. From the 40th day to 5-6 months of age there occurs definitive formation of body proportions of adult animals (Arkhipchuk, 1983a).

After five months of life in European hares, an intensive increase occurs in the weight of the testes and ovaries, which is complete by 8-9 months in individuals born in summer, and by 9-15 months—in those born in February-March. Baby hares of later broods attain sexual maturity at 5-6 months of age, and sometimes—also at 4 months of age. In favorable conditions European hares reproduce up to 4-5 years (Petrusewicz, 1970; Moller, 1976; Broekhuizen and Maaskamp, 1981; Arkhipchuk, 1983a).

Molt. European hares have two seasonal molts—spring and fall. The start and duration of the molt vary throughout their range, and also depend on weather conditions, age of the animal and its build up.

According to the data of I.N. Serzhanin (1955), the spring molt of European hares inhabiting the territory of Belorussia, proceeds quite intensively and continues from the beginning of April to the end of May. In the Voronezh region, European hares molt in the spring apparently somewhat earlier, since here in the middle of April almost entirely freshly clad individuals were found (Barabash-Nikiforov, 1957). In the Astrakhan region European hares begin to molt in mid-March; molting reaches a peak in mid-May, and in June the molting terminates, although individuals with remnants of winter hairs are also found (Gruzdev, 1974). In the west and north of Kazakhstan, the spring molt begins usually in the second half of March-beginning of April, and terminates toward the end of May-beginning of June; south of Kazakhstan the spring molt lasts from the end of February-beginning of March to mid-April (Gruzdev, 1974; Sludskii et al., 1980). According to some data (Larin, 1950; Serzhanin, 1955; Barabash-Nikiforov, 1957), winter fur in European hares changes to summer fur in quite condensed periods; according to others (Gruzdev, 1974; Sludskii et al., 1980)—the spring molt is protracted in character.

The spring molt starts from the face (surface around the nose and eyes), base of the ears and occipital part; later the fore and hind legs, back and sides molt; last to molt is the hip part (Sludskii et al., 1980).

Periods of the fall molt are determined by weather conditions—low temperatures and early snowfall hasten the start of molting. Thus, in the Ural flood plain, snow falling in October 1962 initiated a faster and harmonious course of molting, and by mid-November the majority of European hares were in winter fur. In light winters fall molting ends in the second half of December. To the north of Kazakhstan, in mid-November, European hares usually are in winter fur (Sludskii et al., 1980). To the south of the European part of the range, a complete change of summer fur takes place by the end of November-beginning of December, and in the central and western parts—usually by the middle to end of November (Serzhanin, 1955; Barabash-Nikiforov, 1950).

Winter fur appears first of all on the thighs, then on the rump, back and sides. Last to molt are the snout and individual parts of the back. Thus, the fall molt proceeds in a direction opposite to that of the spring molt (Larin, 1950; Sludskii et al., 1980).

Sex and age structure of population. In the territory of Kazakhstan in European hares, there is a small predominance of female embryos—51.5% (Sludskii et al., 1980), although in different years the ratio of males to females may change somewhat in the embryonic period. Among young of the first brood, caught from May to October, females made up to 53.6% (n = 56) of the population.

The natural sex ratio among adult individuals is difficult to explain, based on the analysis of catches or shooting animals, since during the year mobility of males and females changes differently. In Kazakhstan in spring and summer, the ratio of males to females in sexually mature European hares is usually close to 1:1 (51% males in spring and 54.1% in summer), but in winter among the animals killed females predominate (60%), which possibly is related not only to their high mobility in this period, but also to the high mortality of males (Sludskii et al., 1980). The ratio of males to females in a population of European hares may depend on their population density. An increase in the proportion of females with a decrease in population density also may be explained by higher mortality of females in unfavorable conditions. Thus, some deviations in sex ratio from 1:1 in European hares are linked with weather and food conditions in a year in the given habitat (Galaka, 1970a).

326

The age composition of populations in European hares, as also in lean* ones, changes from year to year, and depends on the number and level of mortality of young which according to some data (Abildgard et al., 1972) constitutes from 2-3% in summer to 10-11% in fall, and according to other data (Broekhuizen, 1979; Kovacs and Ocseny, 1981), it fluctuates from 30% to 81%. The overall mortality of animals of all age groups for the year is about 46%. On average, in populations of European hare, up to 53% consist of the current year's individuals, but in different seasons the proportion of young fluctuates from 32 to 91% (Abildgard et al., 1972; Krik**, 1984).

The increase of population in European hares depends on habitat conditions and numbers—the higher the population density, the lower the level of increase. In the steppe zone of the Ukraine, for one adult hare there are from 1.06 to 3.6 young, and in the forest-steppe zone—from 0.74 to 0.77 (Galaka, 1969). In Denmark, in different years with a change of population density from 320 indivduals to 50 per 1,000 ha, in fall there are from 0.7 to 2.5 young for one adult hare (Abildgard et al., 1972). In Belorussia the coefficient of increase of population varies from 0.9 to 1.4. The second and third broods have the maximum significance for increase of populations of European hares, their contribution being about 60% of all young (Gaiduk, 1973a).

In captive conditions, and according to some data (Abildgard et al., 1972)—and also in nature, their maximum life span is 6 years. Most often in natural conditions, males live not more than 3.5 years, and females— somewhat longer; individual animals may live to 12-13 years (Pielowski, 1971; Arkhipchuk, 1983a). Analysis of histological sections of mandibles showed that the life span of European hares in the northwest of their range (Estonia) is 3-4 years (Krik**, 1984).

Population dynamics. The number of European hares is subject to quite considerable fluctuations. The character of fluctuations of number was studied mainly from the data of skins processed, and on the whole may reflect true dynamics of populations of European hares in our century [20th]. However, in the 70s a decrease in processed skins of European 232 hares is linked with low value and not with decrease in number. The range of fluctuations of numbers in different parts of the range is not uniform, and the peaks of high numbers may not coincide in time. In Europe the full cycle of population dynamics is usually 20-25 years; during this period there occur even fluctuations of lesser amplitude every 2-8 years. The number of hares grows to a culmination point in 2-6 years, and falls to the minimum for 2-4 years (Petrov, 1976).

*Implication not clear, may a misprint for molodnyak—young—General Editor.
**Not in Lit. Cit., presumably a mispring of Kiik—General Editor.

Especially many hares die in years with unfavorable weather conditions, as a result of epizootics and parasitic diseases. The number of European hares decreases after frosty winters with high snow cover and glazing, which, as a rule, are accompanied by intensive killing of European hares by hunters and predators. The most catastrophic for European hares are the years with a dry summer and a cold winter. Summer droughts, typical in the south of the range, lead beside all other things to a decrease in the fertility of females. With spring frosts, high humidity, rains, and floods, there is massive mortality of young hares (Gruzdev, 1974; Krik*, 1983).

Massive mortality of young is observed in regions of intensive agriculture, where with mowing of grasses and harvest of spring crops more hares may die, than from unfavorable weather conditions (Pepin**, 1986). Special mention must be made of the adverse effect of use of poisonous chemicals (direct poisoning and decrease of sexual function) on the number of European hares.

Epizootics are most often observed in European hares in the north and northwest of the range. Here, blue hares, which to a greater extent than European hares, are subject to epizootics. Mortality of European hares from infectious and parasitic diseases increases with higher humidity in the fall-spring period (Sterba, 1982).

Predators affect the number of European hares to a lesser extent than the weather conditions and epizootics. However, animals diseased and emanciated from a shortage of food often fall prey to predators, such that the maximum damage to populations of European hares is inflicted by predators and birds during periods unfavorable for them, particularly in years with low numbers.

Enemies. Among carnivorous mammals there are no species specialized to feeding on European hares, but over the entire range European hares, and particularly their progeny, they are more or less frequently the prey of foxes (common fox, and corsac). The proportion of European hares in the diet of foxes usually does not surpass 15% (Kolosov and Bakeev, 1947; Sludskii et al., 1981). But with lower numbers of small rodents, foxes begin to more actively hunt for hares. Thus in years of low numbers of water voles in the flood plains of the Volga, Kama and Vyatka, the encounters of remnants of European hares in feces of foxes rose to 23-24% (Tikhvinskaya and Gorshkov, 1975). More often European hares fall prey to foxes also in cold snowy winters, when rodents are less available to them, and hares are denied the possibility of fast maneuvering. Moreover, in such winters European hares are usually emaciated because of a shortage of food, which also makes their killing easy (Solomatin, 1969).

---

*Not in Lit. Cit., presumably a mispring of Kiik—General Editor.
**Not in Lit. Cit.—General Editor.

Fewer data are available on the role of European hares in the diet of wolves. Apparently, the extent of damage inflicted on the numbers of European hares by this predator depends on the abundance and diversity of other prey, to which first of all belong ungulates. Moreover, according to the data of P.A. Merts (1953), European hares more often fall prey to wolves in winter, than in spring and summer. In the southwestern part of their range, stone and forest martens may hunt European hares (Bakeev, 1973). European hares are also killed by wandering cats and dogs, the number of which in some places surpasses the number of foxes.

Among predatory birds, European hares are hunted by eagle owls, harriers, golden eagles, white-tailed eagles, sparrow-hawks, goshawks and kites; moreover, the latter mainly attack young hares (Kolosov and Bakeev, 1947; Barabash-Nikiforov, 1957; Sludskii et al., 1980). Young and sometimes adult European hares often fall prey to flocks of hooded crows and magpies.

Competitors. In the territory south of Russia, in the Ukraine and 233 Kazakhstan, the main competitors of European hares are sousliks and domestic cattle. In Kazakhstan herds of saiga may cause considerable damage in foraging places of European hares. In places of sympatry of European and blue hares, between these species competitive interactions are observed for habitat and food resources. European hares, in regions disturbed by man for agriculture, squeeze out blue hares, that is, in such habitats European hares are more competitive. To the south of Kazakhstan, the number of European hares is less in places of their combined habitation with tolai hares, that is, in the southern range of European hares, tolai hare is a more competitive species.

Diseases and parasites. Among ectoparasites on European hares, mites parasitize them most frequently. Their abundance increases toward the south. Infestation of hares living in bairoch* forests and ravines, as a rule, is higher. In all, on European hares 17** species of ticks have been found: *Dermacentor pictus, D. marginatus, Ixodus ricinus, I. persulcatus, I. laguri, Rhipicephalus rossicus, R. pumilio, R. sanguines, R. schulzei, R. turanicus, Haemaphysalis otophila, H. punctata, H. leporis, H. sulcata, Hyalomma plumbeum, H. detritum, Sarcoptes scabiei,* and *Acarus siro* (Kolosov and Bakeev, 1947; Sludskii, 1953; Abdusalimov, 1959; V. Popov, 1960; Dunaev, 1979). The most widely distributed species are *I. ricinus* and *D. pictus*. Up to several tens of mites of different species may parasitize a single individual; and in the southern part of their range in individual years—its several hundreds. In the Volga delta, for instance, the index of abundance of *H.*

---

*Meaning not clear—Translator.
**In Russian original, 18 mentioned—General Editor.

plumbeum varies in different years from 115 to 507.9 (Derevyanchenko and Zheldakova, 1958).

In the Uralsk region fleas have been found on European hares: *Chaetopsylla trichosa, C. globiceps, Amphipsylla kalabukhovi* (Milunov et al., 1964—cited from Sludskii et al., 1980); for eastern Transcaucasia: *Ctenocephalides felis, Xeronopsylla conformis,* and *Rhadinopsylla ukrainica* are characteristic (Isaeva, 1964—cited from E'igilis, 1980), and *Pluex irritans, Ctenocephalides canis,* Cerato*phyllus laeviceps* are common for the Uralsk region and eastern Transcaucasia.

Among helminths, on European hares, 10 species of nematodes have been found: *Protostrongylus terminalis, P. kamenski, Nematodirus aspinosus, Passalurus ambiguus, Graphidium strigosum, Trichostrongylus retortaeformis, T. probolurus, T. colubriformis, Trichocephalus leporis* (found only in European hares) and *Micipsella numidica.* Two species of cestodes—*Mosgovoyia pectinata* and *Taenia pisiformis,* and two species of trematodes: *Dicrocoelium lanceatum* and *Fascicola hepatica* have also been reported (Kolosov and Bakeev, 1947; Gvozdev, 1964).

Infectivity of European hares with nematodes and trematodes, in comparison with blue hares, is usually less, although with such nematode species specific to hares as *N. aspinosus,* in some places European hares may be infected to 100% (Gvozdev, 1964). According to the data of A.M. Kolosov and N.N. Bakeev (1947), in the Ukraine more than 45% of European hares are infected with *T. leporis;* moreover in individual animals the degree of infestation reaches 2,000 helminths. Nematodes of the genus *Protostrongylus* are found in European hares in the northern part of their range, while in Bashkiria, and in the steppe part of Ciscaucasia, and in the Ukraine, they are practically absent. Infectivity of European hares with cestodes in the southwestern part of their range changes considerably from year to year and has two seasonal peaks—spring and fall. With a high degree of infectivity with cestodes, the intensity of infection increases, and cases of deaths due to this parasite are possible (Kolosov and Bakeev, 1947).

Coccidosis is the widely distributed and frequent parasitic disease of European hares. In the Ukraine, Caucasus, north of Kazakhstan and in Bashkiria, in the intestine and liver of European hares, coccids *Elmeria leporis* and *E. stiedae* have been recorded (Kolosov and Bakeev, 1947; Sludskii et al., 1980). To the west of Kazakhstan, about 12% of European hares were found infected with coccids, *E. hungarica, E. robertsoni* and *E. europaea* (Sludskii et al., 1980). The intensity of coccid infection is higher in spring and fall. Among protozoan diseases also widespread is toxoplasmosis, the intensity of infection of which in European hares is higher in winter (Dunaeva, 1979).

European hares are carriers of pathogens of such diseases as mite rickettsiosis, tularemia, dicrocoeliosis and, apparently, mite encephalitis and Crimean hemorhagic fever (Dunaeva, 1979).

Practical Significance

Skins of European hares are processed for producing felt and its various items. The meat of European hare is valued, which in gastronomic qualities is not inferior to that of domestic rabbit. The European hare is also an important object of game hunting.

Like other species of hares, European hares play a no less important role in the diet of valuable fur-bearing animals, the numbers of which (particularly foxes) depend on the number of hares.

European hares may cause damage to some agricultural crops. In summer they most often damage cucurbitaceous crops, and in winter— garden crops. In winter, plantations of some broad-leaved species of trees suffer due to European hares.

European hares have a definite epidemiological significance, since they are a source of infection of tularemia, brucellosis and other infectious diseases of man. They can also be carriers of toxoplasmosis (Kolosov and Bakeev, 1947; Sludskii et al., 1980).

Tolai Hare
Lepus (Lepus) tolai Pallas, 1778

1778. *Lepus tolai* Pallas. Nov. Spec. Quadr. Glir. Ord.: 17. Buryat ASSR, Selenga River (Ognev, 1940: 162).

1841. *Lepus tibetanus* Waterhouse. Proc. Zool. Soc. London: 7. India, Jammu and Kashmir, Baltistan, Lower Tibet, upper reaches of the Indus River.

1861. *Lepus aralensis* Severtzov. Akklimatizatsiya, 2, 2: 49. "Aral Sea area" (from tautonymy). Nom. nudum (without description).

1873. *Lepus lehmani* Severtzov. Izv. O-va Lyubit. Estestv., Antropol., Etnogr., 8, 2: 83. Kazakhstan, Kzyl-Ordinsk region, lower reaches of the Syr Darya River, Kazalinsk.

1875. *Lepus pamirensis* Gunther. Ann. Nat. Hist., 16: 229. Tadzhikistan, Gorno-Badakhshan A.[utonomous] R.[egion] ("Pamir"), "Lake Sary-Kul".

1882. *Lepus butlerovi* Bogdanov. Ocherki prirody Khivinskogo oazina [Notes on nature of Khivinsk oasis]: 68. Uzbekistan, Amu Darya River. Nom. nudum.

1882.   *Lepus kensleri* Bogdanov, ibid.: 69. Uzbekistan, Amu Darya River. Nom. nudum.

1907.   *Lepus zaisanicus* Satunin. Ezhegodn. Zool. Muzeya Akad. Nauk, 11 (1906): 161. Kazakhstan, eastern Kazakhstan region, Zaisan and Kenderlyk.

1912.   *Lepus quercerus* Hollister. Proc. Biol. Soc. Washington, 25: 182. Altai territory, Gorno-Altai A.[utonomous] R.[egion], Chui steppe.

1922.   *Lepus tolai buchariensis* Ognev. Ezhegodn. Zool. Muzeya Akad. Nauk, 23, 3: 475-476. Uzbekistan, Surkhandarinsk region, east of Termez, Khatin-Robot; SE Bukhara, Buzachi.

1922.   *Lepus bucharensis* Ognev. Biol. Izv. 1: 102.

1928.   *Lepus tolai desertorum* Ognev et Heptern. Zool. Anz., 75: 262. Turkmenia, Askhabad region, Gyaursk district, Annau.

1934.   *Lepus europaeus turcomanus* Heptner. Fol. Zool. Hydrobiol., 6: 21. Turkmenia, Krasnovodsk region, 60 krn north of Dzhebel, Ak-Kuyu.

Diagnosis

Dimensions medium.

Hair coat soft. Color of upper side of the body brownish-gray, ochreous-gray or gray with dark ripple. Winter fur in the eastern part of the range ochreous, gray or light bluish-gray with well-developed ripple or striated pattern. In the western part of the range, winter fur is similar in color to summer fur. Tail is black or black-brown above, white beneath.

Skull of light construction, with bulged cranium, with relatively narrow base of rostrum. Ratio of width of frontal bones behind the supraorbital processes to width of the base of the rostrum is 47-67% (average—57%). Zygomatic width in the anterior part 32.3-38.9 mm (average—35.6 mm). Tooth row short—alveolar length of the upper tooth row 14.6-17.0 mm (average—15.8 mm). Supraorbital processes not expanded posteriorly. Base of the upper anterior incisor reaches the region of the suture connecting the intermaxillary and maxillary bones. Length of the diastema of the lower jaw in relation to length of the lower tooth row more than 80%. Articular processes of the mandible strongly bent backward.

235

Description

Body length from 387 to 580 mm; tail length—75-116 mm; length of hind feet—106-135 mm; ear length—83-119 mm.

Color of the upper part of the head is usually somewhat darker than the back; sides of the head are lighter. The throat and the lower part of the head are white. The occiput and the upper side of the neck are ochreous-rustry or ochreous-brown. The front part of the chest and the section of the neck adjoining it are ochreous-brown or reddish-brown-brown, often with a whitish bloom.

The end of the snout has an ochreous or yellowish tinge. The anterior surface of the ears is always darker than the head—brownish-gray or ochreous-brown in color. The back of the ears is lighter—whitish or whitish-gray. The tip of the ear with a dark, often black-colored spot (Fig. 51).

The lower side of the body and the inner surface of the legs are white. Distal sections of the legs are ochreous-brown. The upper surface of the tail is black or black-brown; lower—white. The tip of the tail has a well-developed tuft of white stiff hairs.

Winter color of the fur, particularly in the region of the pelvis, in tolai hares from the northern and montane regions, is lighter than in summer and has well-developed speckles.

The skull is not large. Intermaxillary bones are relatively short and project slightly from the anterior margin of the nasal bones. Lateral sides of the maxillary and squamosal bones have a well-developed reticular

Fig. 51. Tolai hare *Lepus tolai*. Chu-Talass Myumkums
(Photo by A.A. Karpov).

structure. The frontal bones are depressed in the interorbital region, but the supraorbital processes are raised upward. The occiput is broad and bulged.

237   The tympanic bullae are large and rounded in form. The anterior upper incisors are steeply lowered down; their bases reach the suture connecting the maxillary and intermaxillary bones.

The mandible is realtively massive, with a short diastema and weakly bent backward articular process (Fig. 52).

236

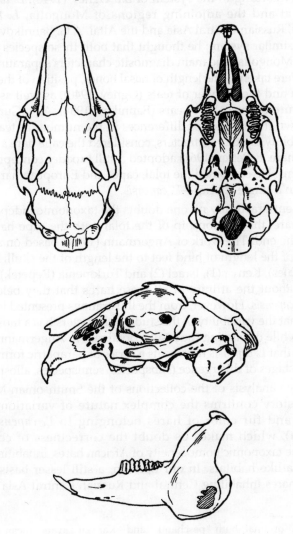

Fig. 52. Skull of the tolai hare *Lepus tolai*.

The diploid complement has 48 chromosomes: 8 pairs of meta-submetacentrics, 15 pairs of subtelocentrics and acrocentrics (NF-78). X-chromosome—submetacentric, Y-chromosome—small, acrocentric (Vorontsov and Ivanitskaya, 1969).

Systematic Position

So far there is no unanimous opinion on the taxonomic status of the tolai hare. According to the system of S.I. Ognev (1940), *L. tolai* inhabits Trans-Baikal and the adjoining regions of Mongolia, *L. tibetanus*—Kazakhstan, Russian Central Asia and the Altai. A.G. Bannikov (1954) also supports a similar system; he thought that both these species inhabit the territory of Mongolia. The main diagnositc characters separating tolai and cape hare were taken as the length of nasal bones, position of the first lower molar tooth and the number of teats (Ognev, 1940), as well as the length and the nature of the color of ears (Eannikov, 1954). A.A. Gureev (1964), having shown the absence of differences in the number of teats, and the wider variability of other characters, considered the cape hare a subspecies of the tolai hare. F. Petter (1961) adopted a still broader concept in relation to these forms. He combined the tolai, cape* and European hares with the cape* hare in a single species—*L. capensis.*

If at present practically no one doubts the taxonomic independence of European hare, the relationship of the tolai with the cape hare remains unclear to the end. The work of Angermann (1983), based on an analysis of the ratio of the length of hind feet to the length of the skull, hares from Iran (5 samples), Kenya (1), Israel (2) and Turkmenia (Repetek), reached a conclusion about the affinity of all these hares that they belong to one species—*L. capensis*. However, from the illustrations presented in the work, it follows, that the variability of the studied characters has a mosaic nature. Moreover, while discussing the results obtained, R. Angermann makes an assumption that *L. capensis* represents a series of parapatric forms standing at different stages of divergence (subspecies, semispecies, allospecies).

A tentative analysis of the collections of the Smithsonian Museum of Natural History confirms the complex nature of variation of metric characters and fur color of hares belonging to *L. capensis* (E. Yu. Ivanitskaya), which makes us doubt the correctness of conclusions regarding the taxonomic homogeneity of African hares, inhabiting savanna and savanna-like habitats. In our opinion, a still lesser basis exists for combining hares inhabiting Central and Russian Central Asia with cape

---

*In Russian original, both "peschanik" and "Kapskii zayats" mean cape hare—General Editor.

hares. Thus, based on the aforementioned, it is necessary to recognize for tolai hare the status of an independent species. To explain the southern boundary of distribution of the tolai hare, a more detailed analysis of a large number of samples of hares inhabiting north Africa, Asia Minor and southwest Asia is necessary.

Geographic Distribution

Tolai hares inhabit desert, semidesert and montane regions from the eastern coast of the Caspian Sea to southeastern Trans-Baikal. The northern part of the range of the tolai hare can be divided into four areas: Kazak-Russian Central Asia, Altai-Tuva, Buryat and eastern Trans-Baikal (Fig. 53).

The northern boundary of the Kazakh-Russian Central Asian area of the range of the tolai hare passes roughly at 48° N. lat. along the southern regions of Kazakhstan—from the northeastern coast of the Caspian Sea, through the lower reaches of the Ural and Emba rivers, the Ustyurt Plateau, the Bolshie [Great] Barsuki sands, the Aral Sea area of the Karakums, Aryskum, the lower reaches of the Sarysu River, north of Betpak-Dala, the northern Lake Balkhash area and reaches west to the spurs of the southern Altai. The western boundary of the range of the tolai hare passes along the eastern coast of the Caspian Sea, in the south entering Iran and Afghanistan.

239 The Altai-Tuva area of the range of tolai hares from the west is bordered by the Chui steppe, from the north—by the Kuraisk, Chikhachev, Tsagan-Shibe'tu, western and eastern Tannu-Ola ranges, in the east it reaches the Sengilen Range, and in the south enters Mongolia.

The Buryat part of the range of the tolai hare mainly lies west of the Khamar-Daban Range. In the west it is limited by the Malyi [Lesser] Khamar-Daban Range; in east the boundary passes along the Selenga River, and in the south the range enters Mongolia. The place of habitations of tolai hares in the eastern Sayan—in the upper reaches of the Irkut River and in the vicinity of the village of Mondy—are somewhat isolated in relation to this part of the range (Ognev, 1940).

The eastern Trans-Baikal part of the range, which includes the southern steppes of the Chita region, extends east of the Altan River in the Kyrinsk district to the banks of the Argun River. The southern boundary of this part of the range passes into the territory of China. From the north the distribution of tolai hares is restricted by the Mogoitui and Ononsk ranges, upper reaches of the Borzya River, and the northern spurs of the Nerchinsk and Klinchkinsk ranges (cf. Fig. 53).

According to the data available at present, the northern distributional

238

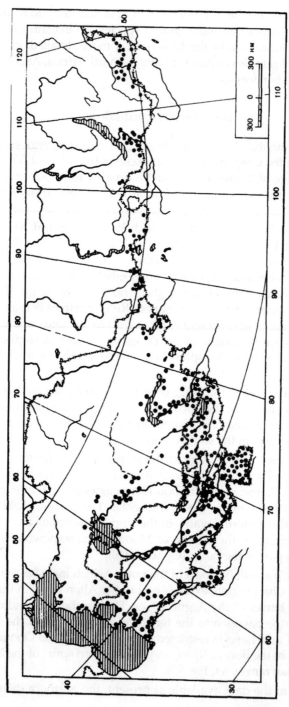

Fig. 53. Localities of the tolai hare in Kazakhstan, Russian Central Asia and Russia.

boundary of tolai hares in the northwest of their range passes along the northern coast of the Caspian Sea (region of the Volga-Ural sands), runs east to the lower reaches of the Emba River and then southeast to the central and southern regions of the Ustyurt Plateau. From the Ustyurt Plateau the boundary turns northeast, where tolai hares are found in the Bolshie [Great] Barsuki sands and the northern Area Sea area (Sludskii et al., 1980; collections of ZM MGU). The boundary then turns south and follows in a southeastern direction along the right bank of the Syr Darya through Karaozek (Kara-Uzyak) station, the valley of the Surumbai River, Lake Ayak-Kul' (Ayakkol') and the vicinity of Kyzyl-Orda (Ognev, 1949; collections of ZM MGU). The boundary proceeds to the Aryskum sands and along the coast of the lakes Telekol', Kul'kol', Sorkol', and the lower fork of the Sarysu River descending south to the Sarysu Muyumkums (Sludskii et al., 1980; collections of ZM MGU and Nizhegorod Institute). From the Sarysu Muyunkums the northern boundary of the range passes east along the northern part of Betpak-Dala to the northern Lake Balkhash area, where tolai hares are found along the valleys of the Mointy and Tokrau rivers, and farther east—to the village of Ayaguz. Ascending northeast, the boundary passes along the valley of the Kokpekta River and the town of Samarskoe to the town of Mirolyubovka on the left bank of the Irtysh (Ognev, 1940, Sludskii et al., 1980; collections of KazGU, Alma-Ata). The town of Mirolyubovka is the extreme northeastern locality of tolai hares in the limits of the Kazakh-Russian Central Asian part of the range.

From here the distributional boundary of tolai hares passes southeast along the Zaisan basin in the foothills of the Kurchumsk Range, along the banks of the Kara-Itrysk and Kenderlyk rivers to the Saur foothills, which is the extreme southeastern distributional limit of tolai hares in the Kazak-Russian Central Asian part of the range (Ognev, 1940; Sludskii et al., 1980). In the direction east to west and south of Kazakhstan, tolai hares are reported in the Alakol' basin, in the flood plain of the Tokty River (area of the Dzhungarian gates), in the valley of the Usek River (20 km from Dzharkent—now Panifilov), in the vicinity of Dzharkent in the Uigursk district of the Alma-Ata region, in the Karatau mountains, along the valleys of the Malyi [Lesser] and Bolshoi [Great] Kokpak rivers and along the spurs of the Terskei Alatau Range (Ognev, 1940; Gvozdev, 1949; Sludskii et al., 1980; collections of ZM MGU and Russian Central Asian Antiplague Institute, Alma-Ata).

East and south of Kirgizia tolai hares live in the central Tien Shan and the Sary-Dzharsk syrts* in the flood plain of the Atbashinka River; southward—in the Altai valley (Aizin, 1979; collections of ZM MGU and Nizhegorod University).

---

*Syrt—watershed upland—General Editor.

In Gornyi [Montane] Badakhshan tolai hares are found along the western spurs of the Sarykol'sk Range, along the northern slopes of the Vakhansk Range, the southern slopes of Yuzhno [southern] Alichursk and Shakhdarinsk ranges, and along spurs of the Ishkashimsk, Yazgulemsk, Vanchsk and Darvaz ranges. Thus east of Gornyi [Montane] Badakhshan, tolai hares are common in the Markansu tract (southern slopes of the Trans-alai Range), in the basin of Karakul', Shorkul' and Rangkul' lakes, and in the vicinity of the towns of Shadput, Takhtamysh and Shaimak. Tolai hares are particularly numerous in the extreme southeast of the region—in the 240 valley of the Karakol River. South of Gornyi [Montane] Badakhshan tolai hares have been recorded in the vicinity of the village of Kyzyl-Rabat, in the Dzhartygumbez tract in the northeast of the southern Alichursk Range, in the area of Zorkul' and Karantul' lakes, along the valley of the Karadzhidga River, in Kumda Pass, the Kokbai tracts and in the valley of the Shakhdara River. In the western Pamir tolai hares are encountered near the mouth of the Bodomadara River, in the lower reaches of the Shakhdar River, in the valleys of the Gunt and Vanch rivers, in the upper reaches of the Obimazor River (Darvaz Range) and on the Peter I Range in the vicinity of the village of Lairun (Ognev, 1940; Davydov, 1974; Odinashoev, 1987; collections of ZM MGU, IZIP Akad. Nauk Tadzh SSR).

South of the western part of Tadzhikistan in the Kulyabsk and Kurgan-Tyubinsk regions, tolai hares are recorded along the Khozratishok Range, in the valley of the Kyzylsu River, along the valley of Vaksh River, and in the vicinity of Kurgan-Tyube (Davydov, 1974; collections of Kiev and Nizhnegorod universities). In the Surkhandarin region tolai hares inhabit the territory of the "Tigrovaya balka" preserve, in the vicinity of Khatyn-Rabad (east of Termez) and in the Aral-Paigombar preserve (Ognev, 1940; Volozheninov, 1972; Davydov, 1972; collections of KGU and IZIP Akad. Nauk Tadzh.).

Southeast of Turkmenia tolai hares were caught in the Charshanginsk district in the Darai-Dara ravines (collections of KGU). Data are not available on habitations of tolai hares southwest of the Karakums. Westward, in the area of Badkhyz, the southernmost localities of tolai hares in Turkmenia are in the vicinity of the villages of Takhta-Bazar and Kushka (collections of ZM MGU and Institute of Zoology, Ashkhabad). In Kopetdag tolai hares are encountered near the village of Sulyukli and in the vicinity of Germob. On the southwestern spurs of Kopetdag, tolai hares were recorded along the valley of the Chandyr River in the vicinity of the town of Sharlouk of the Kizil-Bazarsk district and the town of Madau Kazandzhissk district (Ognev, 1940; collections of ZM MGU, ZIN, Institute of Zoology, Akad. Nauk TurkmSSR and KGU). Farther southwest tolai hares are found 35 km east of Gasan-Kul' and in the valley of the Artek River, the lower reaches of which are the extreme southwestern place of habitation of tolai hares in the Kazakh-Russian Central Asian part of the range (Ognev, 1940).

In the west the distributional boundary of tolai hares extends along the eastern coast of the Caspian Sea. Southwest of Turkmenia tolai hares were found in the vicinity of the villages of Gasan-Kul' and Chikshilyar. Northward tolai hares enter deep in the desert and are found 60 km north of the village of Meshed (Meshkhed), 20 km west of Bala-Ishema in the Greater Balkhans, and 60 km north of the village of Dzhebel' (Ognev, 1940; collections of ZM MGU and Institute of Zoology, Turkmenia). Northwest of Turkmenia tolai hares live on the coast of the Caspian Sea in the vicinity of Krasnovodsk and along the coast of the Kara-Bogaz-Gol Gulf (Ognev, 1940; collections of ZM MGU). On the Mangyshlak Peninsula tolai hares inhabit the Karatau, Senek, and Toumak tracts; and they were caught 60 km from the village of Eralievo (collections of "Mikrob" Institure, Saratov). Northward tolai hares were recorded on the Buzachi Peninsula (Ognev, 1940).

The Altai-Tuva part of the range of tolai hares includes southeast of the Gornyi [Montane] Altai and south of Tuva (cf. Fig. 53). The western boundary of this part passes along the valley of Chagan-Burgazy, mouth of the Elangash River, the Chagan-Uzun River, the right bank of the Chui River and the western slopes of the Kuraisk Range. From here, the boundary turns south and descends to the village of Kosh-Agach, then turns east and proceeds along the valley of the Tobozhok River to the Chikhachev Range (Ognev, 1940; Firstov, 1957; Yudin et al., 1979). South of the Gornyi [Montane] Altai, tolai hares were recorded along the slopes of the Sailyugem Range: in the Bolshie [Greater] Saragoby, in the vicinity of the village of Tashanta, and along the valley of the Ulan-Dyrg River in the flood plain of the Chagan-Burgazy River (Firstov, 1957; collections of BINZ).

In extreme western of Tuva, tolai hares inhabit mountain steppes of the Chikhachev Range. Eastward (in the Bai-Taiginsk district), the boundary passes along the Mungun-Taiga mountain, the vicinity of the village of Mugur-Aksy, the valley of the Karga River, and along slopes of the Tsagan-Shibe'tu Range (Yudin et al., 1979; collections of ZM MGU and Institute of Zoology, Kiev). Farther east, tolai hares were caught along the southern slopes of the western and eastern Tannu-Ola, particularly in the flood plain of the Torgal'yg-Khem River, in the vicinity of the village of Ak-Chira. In the Tes-Khemsk district, tolai hares are found in the vicinity of the village of Khol-Ozhu, and southward—near the village of Shara-Sur. On the northeastern slopes of the eastern Tannu-Ola Range, tolai hares were caught in the vicinity of the village of Balgazyn of the Tandinsk district. Southeast of the Altai-Tuva part of the range, tolai hares are reported in the Erzinsk district near Lake Tere-Khol' and on the southwestern slopes of the Sangilen upland (Yudin et al., 1979; collections of BINZ and Institute of Zoology, Kiev).

In the Buryat part of the range, tolai hares are confined to isolated parts in the upper reaches of the Irkut River and in the area of the Tunkinsk goltsy [bald mountains] on the southern slopes of the eastern Sayan (Fig. 53). The western distributional boundary of tolai hares in the southeastern Buryat part of the range passes along the Dzhidinsk district. From the town of Khuldag, situated on the right bank of the Dzhida River and being the extreme southwestern locality of tolai hares within these limits, the boundary turns northeast, reaches the town of Ulzar and becomes the northern limit. Proceeding in a northeastern direction along the southern slopes of the Malyi [Lesser] Khamar Daban, the boundary passes along the left bank of the Dzhida River through the village of Iro and in the vicinity of the villages of Udunga and Tamcha, in the area of Lake Gusinoe, along the Ubukun River to Ulan-Ude, where lies the extreme northeastern locality of tolai hares (Fetisov, 1935; Ognev, 1940; collections of ZM MGU, BIN and "Mikrob" Institute, Saratov).

The eastern boundary of this part of the range of tolai hares passes along the Selenga River. On the northern bank of the Selenga, tolai hares were recorded in the vicinity of the town of Ganzurino, farther south— along the Khilok River and in the town of Ust'-Kiran in the Kyakhtinsk district. Ust'-Kiran, apparently, is the extreme southeastern locality of tolai hares in the Buryat part of the range (Fetisov, 1935; Ognev, 1940).

In southeastern Trans-Baikal west of the Altan River, apparently tolai hares are not found (Nekipelov, 1961). The western distributional boundary of tolai hares within the limits of the eastern Trans-Baikal part of the range passes along the Altan River (cf. Fig. 53). The northern boundary of the said part of the range from the Altan River along the steppes of the Akshinsk district proceeds east and northeast. Along the Aginsk steppe north of the Aga River (not reaching the Buryatskaya station and the town of Bereya), the boundary passes southeast and proceeds through the town of Kapaksar and Ust-Unda, then passes west of the town of Dolgokycha and Antiya to the town of Olonda. From here the boundary passes east through the towns of Kurunzulai, Kutugai, Man'kovo, the settlement of Aleksandrovsk factory, descends south to the town of Puri, later ascends northeast, and proceeds through the towns of Shara, Sharakon, Don and Kalga to the steppe parts of the Nerchinsko-Zavodsk district, where tolai hares are very rare. Farther east the range of tolai hares enters China. South of the eastern Trans-Baikal part of the range, tolai hares are found in the steppes of Dauria, near Kulusutai, in the vicinity of the town of Staryi [old] and Novi [new] Chindant, and also in the steppes south of the Onon River (Ognev, 1940; Nekipelov, 1961).

Geographic Variation

A.A. Gureev (1964), while combining tolai and cape hares in one species, recognized the existence of two sub-species, *L. tolai tolai* and *L. t. tibetanus*, in the territory of the USSR. In the system of S.I. Ognev (1940), these forms are considered independent species with one subspecies of tolai hare and four* subspecies of cape hare: *L. tibetanus lehmani* Severzov, 1872 (southern Kazakhstan—from Usturt and the northern Aral Sea area to Alma-Ata and the southern part of the Semipalatinsk regions; Kirgizia, Uzbekistan—south of Fergana and Tashkent—excluding the Chui steppe); *L. t. buchariensis* Ognev, 1922 (valley areas of Tadzhikistan and northeast of Afghanistan); *L. t. pamirensis* Blanford, 1875 (Pamir); and *L. t. desertorum* Ognev et Heptner, 1928 (deserts of Russian central Asia—from the southeastern coast of the Caspian Sea to the Amu-Darya). The infraspecific system used by S.I. Ognev was developed mainly on the basis of color variation and structure of the fur.

The morphometric analysis of 18 skull characters of 485 specimens of tolai hares (including from the territory of Mongolia) conducted by me, confirmed a considerable range of variation of metric characters. However, the character of variation, as in other species of hares, is clinal in nature, which makes it possible to doubt the validity of the existence of subspecies recognized by S.I. Ognev (1940). Moreover, above analysis confirms the absence of gaps in the clines between the populations of hares referred by A.A. Gureev (1964) to subspecies *L**. t. tolai* and *L. t. tibetanus*. An examination of skins of tolai hares showed that the variation of color does not bear a clearly expressed clinal character, but with sufficiently high individual variability of this character, a definite link is seen between the color and type of habitat. Hares living in similar habitats in different parts of the range, as a rule, have identical character of color. Since the morphological analysis of skulls of hares conducted by me confirms the affinity of these two forms to one species, tolai hares and the subspecies forms of cape hares are considered here within the circumscription of one species.

242

Biology

Habitat. Tolai hares live in plain deserts and semideserts, and in the eastern part of the range—also in dry steppes. In mountains they ascend to an altitude of 4,900 m above sea level, where they colonize mountain steppes, xerophytic sparse forests and high-mountain deserts. The character of the distribution of habitats of tolai hares is determined primarily by

---

*Editor's note: 5 subspecies listed.
**In Russian original, Z.—General Editor.

the presence of shelters (bushes, large stones, marmot burrows).

In Mangyshlak and Ustyurt tolai hares inhabit slopes of hills with boulders and outcrops of rocks, sandy areas with bushes of tamarisk and depressions with thickets of saxaul, and in Mangyshlak—coastal dunes with well-developed vegetation (Sludskii, 1953).

In the Kyzylkums, north of the Karakums and Muyunkums, tolai hares are most often encountered on hummocky sands in thickets and thinned-out saxaul groves, and also in drags of sandy massifs at the border with irrigated lands. They willingly colonize edges of oases and populated areas. In summer tolai hares are usually concentrated near water bodies and springs (Sludskii, 1953; Krivosheev, 1959; Reimov, 1985). Southeast of the Karakums the favorite places of habitations of tolai hares are black saxaul forests alternating with open areas with wormwood-ephimer and wormwood-blue grass cover (Nurgel'dyev, 1969). According to the data of Yu.F. Sapozhenkov (1964), east of the Karakums in spring, tolai hares are concentrated on fixed sands with saxaul thickets, which are most convenient during the period of feeding young. At other times of the year, tolai hares are found roughly uniformly often both on fixed and shifting sands with shrub vegetation. Tolai hares avoid open rubbly-clayey deserts, large massifs with shifting sandy barhans, and dense riparian, bulrush and saxaul thickets (Perevalov, 1953; Reimov, 1985).

In Betpak-Dala tolai hares are most numerous in intermontane depressions with thickets of chee grass and semishrubs; in the central part of Betpak-Dala, where rubbly deserts almost devoid of shrub-vegetation and chee grass predominate, tolai hares are found only in thickets of *Spireanthus* (Ismagilov, 1961).

In valleys of such rivers as the Syr-Darya, Chu, Ili, Karatal, Lepsy, Aksu and Chernyi [Black] Irtysh, tolai hares inhabit saxaul groves, thickets of salt trees and tamarisk, edges of riparian forests, meadows with reed grass, licorice and reed, and also cultivated and abandoned fields. According to the observations on tolai hares in the lower reaches of the Chu River, in spring the animals are most numerous in thickets of salt trees, and in fall—in thickets of tamarisk (Sludskii et al., 1980). In the lower reaches of the Ili, in all seasons, tolai hares almost uniformly colonize sand-hummocks (here the population density of hares in spring is higher than in other habitats, because of the earlier appearance of ephimerals) and salt tree thickets; in winter, some hares migrate to riparian forests and reed thickets; more often than at other times, hares are found in winter in solonchaks (Table 11.1). In the Alakul' depression tolai hares are found in wormwood-grass semidesert among chee grass thickets and sparse shrubs of salt tree, and in the valley of the Tentek River—on gravel-beds overgrown with osier beds and skeletonweed (Sludskii, 1953).

Table 11.1. Encounters of tolai hares (in %, n = 320) in different habitats in the lower fork of the Ili, from census data of 1949-1959 (from Perevalov, 1953)

Time of the year	Habitat and census area				
	sandy hummocks (180 ha)	salt tree thickets (20 ha)	riparian forests (50 ha)	seed thickets (400 ha)	solonchaks (1,200 ha)
Spring	68.7	25.0	–	6.3	–
Summer	54.0	38.4	–	–	7.6
Fall	36.3	54.5	–	–	9.2
Winter	25.0	43.7	6.3	12.5	12.5

243   In valley regions of southwestern Tadzhikistan, the most typical places of habitation of tolai hares are riparian forests, unused lands and fallows. In fall tolai hares are often found in semidesert areas, and in summer—in cotton fields (Stroganov and Stroganova, 1944).

In the mountains of Tadzhikistan and Badakhshan, tolai hares colonize steppefield slopes of southern and northern exposures with thinned-out juniper forests. In the upper limits of the forest belt and on levelled areas of river terraces and slopes, tolai hares prefer places with large boulders, and also abandoned burrows of long-tailed marmots. In mountain meadows tolai hares remain in willow thickets and sea buckthorns (Davydov, 1974). In the eastern Pamir the places of habitation of tolai hares are the foothills of ranges and ancient moraines with rock fragments or large stony detritus with quite poor vegetation, among which predominate winter fat and wormwoods. The vertical distributional boundary of tolai hares in the Pamir is at an altitude of 4,900 m above sea level (Abdusalyamov, 1965).

In the Tien Shan tolai hares inhabit valleys of mountain streams, where they remain in riparian forests and along gravel beds with thickets of shrubs and chee grass. Tolai hares are most numerous along the edges of spruce and juniper forests; they are common also on gentle slopes covered with thinned-out coniferous forests. Here, the vertical boundary of distribution of tolai hares lies at an altitude of 3,000 m above sea level (zone of the subalpine belt and upland steppes). In the Kirgizian Alatau, tolai hares reach an altitude of 2,000 m above sea level and colonize thickets of common apple, uryuk*, barberry, and rose. In the Talass Alatau tolai hares remain in steppe-like meadows and glades among juniper stands and do not ascend above 1,500 m above sea level. In winter they are numerous in plantations and shrubs along railroads (Perevalov, 1953; Sludskii, 1953; Sludskii et al., 1980).

---

*Local name—General Editor.

344

In the Gornyi [Montane] Altai, tolai hares are found in the alpine belt, along valleys of rivers and in lake basins, and in the southeast—in the tundra-steppe zone of ranges bordering the Chui basin (Koneva, 1983). In Tuva they most often colonize river flood plains, mountain gorges and edges of larch forest massif on stony slopes, and in the high mountain zone—in meadow and mountain steppes (Zonov, 1974).

In the southern and western Trans-Baikal, typical habitats of tolai hares are steppe areas with thickets of pea shrub or chee grass, river valleys with willow, sea buckthorn, pea shrub and other shrub thickets, and also slopes of hills with stone detritus or hummocky steppes with outcrops of bedrocks. Moreover, in summer tolai hares may migrate to humid places with thickets of iris or nettle. In open steppes and larch groves on northern slopes of hills, tolai hares are rarely found (Fetisov, 1935).

In Mongolia tolai hares inhabit steppes, forest-steppes, and at places—also mountain meadows at an altitude of 3,300 m above sea level. In Khe'ntei and Khangal tolai hares are found along river valleys with shrub thickets, in light larch groves on mountain slopes (to the altitude of dwarf arctic birch and alpine zone, where blue hares inhabit), and in stony ravines. In the desert zone tolai hares remain mainly around oases; they are also encountered on hummocky sands with tamarisk and saxaul (Bannikov, 1954).

Thus, it can be concluded that under conditions of plain and foothill landscapes, the optimal habitats of tolai hares are river valleys and lake basins with areas of dense woody-shrubby (tugai [riparian]) vegetation. Tolai hares do not enter deep in thick riparian forests, but very readily colonize their edges. In southern deserts the population of tolai hares is not linked to the type of substrate, but it is positively correlated with the adundance of shrub vegetation. In northern deserts the optimal habitats for tolai hares are the neighborhoods of springs and chee grass thickets. In high-mountain deserts the affinity of tolai hares for ancient moraines and undulating slopes is characteristic.

Usually tolai hares do not perform distant migrations, but in the course of a year they may change their habitat, which is linked with the seasonal change of food and their search for shelters during the period of reproduction. Tolai hares may perform short-durational migrations during the period of massive flights of blood-sucking insects. The distance of migrations depends on the time of the year, habitat and the number of hares. For example, there are no clearly manifested seasonal migrations in hares inhabiting thickets of salt tree (cf. Table 11.1). In the eastern Karakums the most conspicuous migrations occur in spring, when hares migrate to fixed sands with saxaul thickets (here the places for rearing progeny are most protected), and in summer, when from fixed sands a majority of hares migrate in search of food to shifting sands with thinned-out shrubs and growing herbaceous vegetation (Sapozhenkov, 1964).

In the high mountains of the Pamiro-Alai and Tien Shan in snow-rich winters, tolai hares descend to 1,000-1,500 m, where the depth of snow cover facilitates unobstructed food acquisition. In the forest belt of mountains, migrations of tolai hares in the vertical direction are less manifested—here in winter tolai hares migrate to slopes of southern exposures and to less snowy areas (Davydov, 1974). In the Alai valley, in summer tolai hares quite uniformly colonize its western part, and in winter form groups in shrub thickets along the Surkhob River. In the Chui valley in summer, tolai hares are found very rarely, but by winter they descend here from the mountains and remain along streams and irrigation ditches. In the Sary-Dzhas summer habitats of tolai hares are steppe areas, and in winter they relocate to the edges of spruce forests (Kuznetsov, 1948a; Yanushevich et al., 1972).

Thus, in the absence of distant migrations, seasonal migrations are characteristic of tolai hares, which are more pronounced in montane habitats.

Population. In individual areas the tolai hare is a common and periodically numerous species. Given the considerable and quite sharp fluctuations in numbers, there is no sense to conduct an estimate of population density and relative abundance for specific areas. It may, however, be noted, that with the rise in numbers in different parts of the range, its identical levels are recorded—on the order of 10-30 individuals per a 10 km route of census, and a population density in the range of 10-30 individuals per 100 ha (Perevalov, 1953; Yanushevich et al., 1972; Ochirov and Bashanov, 1975; Sludskii et al., 1980; Reimov, 1985).

In Turkmenia tolai hares are numerous in the northwest of the republic and in the valley of the Amu Darya, and also in individual areas of the western Kopetdag and in dense black saxaul thickets of the southeastern Karakums, while in the Trans-Unguz and central Karakums, along the old irrigation ditches of the ancient delta of the Amu Darya, in Karabil and Badkhyz, tolai hares are frequent, but do not attain a high level of numbers (Kostin, 1956; Sapozhenkov, 1964; Nurgel'dyev, 1969).

In Uzbekistan, over a large part of the Karakums, tolai hares are frequent but, as a rule, less numerous. High numbers of tolai hares were recorded in individual years in some areas adjoining the delta of the Amu Darya and in the area of Nuratau (Pavlenko and Gubaidulina, 1970; Reimov, 245 1985). In the south of the republic, where tolai hares are less numerous, but occur throughout, their maximum density is observed on hummocky fixed and semifixed sands with saxaul-shrub and herbaceous associations. In riparian habitats the population density of hares is less (maximum 19 individuals per 100 ha). The least densely inhabited are the sandy-clayey areas with level relief and compact soil—here, even in years of high numbers, population density remains less than 1 individual per 100 ha (Volozhennikov, 1983).

In Kazakhstan tolai hares are particularly numerous in valleys of the Syr Darya and its old riverbed (Kuvan Darya, Zhana Darya), Chu, Talassa and Ili; high numbers (though not that frequent) were also recorded in several areas of central Mangyshlak, the Aral Sea area, Karakums, sands of the southern Lake Balkhash area, and valleys of the Karatal and Trans-Ili basin (Perevalov, 1953; Sludskii et al., 1980).

In Kirgizia the highest population density of tolai hares is observed in the western part of the Issyk Kul' basin, and also in the syrts of Naryn and Sarydzhas (Zimina, 1964; Yanushevich et al., 1972). Earlier, in Tadzhikistan tolai hares were numerous in riparian forests of the valleys of the Amu Darya and Kafirnigan, but in recent years, as a result of agricultural development of this region and intensive encroachment, the number of tolai hares here is gradually decreasing. Their fairly high number is conserved in juniper forests of mountain ranges of the central part of Tadzhikistan, and in the high-montane Badakhshan and eastern Pamir. Thus, according to the census data of 1959-1968, in the juniper forests of central Tadzhikistan, the population density of tolai hares was 3 to 10 individuals/100 ha, and in the eastern Pamir (Rangkul' basin) at the end of the reproductive period, up to 23 hares were encountered over a 10-km route (Davydov, 1974).

In Tuva and the adjoining regions of northwestern Mongolia, tolai hares are numerous in semidesert parts of the Ubasnur, Ure'gnur and Achitnur basins. The maximum population density (10-29 individuals per 100 ha) is recorded here along the valleys of rivers, but is declining in the Lake Ubsa-Nur region and in steppes with pea shrub (Obukhov, 1973; Ochirov and Bashanov, 1975; Boyarkin, 1984). In southwestern Trans-Baikal tolai hares are numerous in the valley of Selenga, and the highest population density is characteristically found in the Gusinoozersk basin (Fetisov, 1935).

Shelters. According to the data of A.A. Sludskii (1953) and V.P. Kostin (1956), tolai hares usually do not dig burrows, but dig oval pits (35-40 cm in diam., 15-20 cm in width and 5-17 cm in depth), in which they settle for daytime rest. In the lower reaches of the Syr Darya, tolai hares make resting places in holes up to 60 cm deep, which they dig under bushes on northern slopes of hummocks. In sands resting places are most often located under shrubs of saxaul and camel's thorn and in riparian forests—in thickets of salt tree, tamarisk, winter fat and reed. By the end of summer and in fall, in some places tolai hares prefer to make resting places in meadows and fields (Baitanaev, 1974). During rains or moist snow, hares gather in dense thickets of bushes. During rest tolai hares lie with their head facing the wind and with a change of its direction, change their pose. In winter they respond particularly sensitively to wind (Sludskii et al., 1980).

In the Karakums resting places of tolai hares are located most often in thickets of saxaul, kandym [*Calligonum*] and cherkez [*Salsola richteri*]; and in barhan sands—in bushes of joint fir and selin [*Selinum*]. Usually tolai hares do not rest in hollows, but in interbarhan depressions and on shrubby hummocks. Newborn hares initially remain in dense shrub thickets or in burrows of cape hares and sousliks (Sapozhenkov, 1964).

In the areas of Nuratau, tolai hares use old burrows of foxes and tortises as shelters or dig shallow burrows with openings of 14 x 10 cm (Pavlenko and Gubaidulina, 1970).

In montane regions of Tadzhikistan and in the Tien Shan, shelters of tolai hares are most often located under stones and rock fragments, and in mountain riparian forests and gravelly shoals—under bushes of willows and sea buckthorn (Yanushevich et al., 1972; Davydov, 1974). The most diverse, and often also the most complexly built, holes are found in tolai hares living in the Pamir. Here, besides the usual depressions, tolai hares may make holes in a heap of stones or between two large stones, such that an over-hang is formed. Shelters are formed under large boulders, in the form of a through corridor, and in the forest belt of the Pamir, holes are dug under stones to a depth of 45 cm (Fig. 54). In the Pamir and Tien Shan, tolai hares often use old burrows of red long-tailed marmots as shelters. In such burrows tolai hares not only rest but also hide from predators, and possibly also feed their young (Abdusalyamov, 1956; Davydov, 1974).

In the Gornyi [Montane] Altai, tolai hares make resting places in the

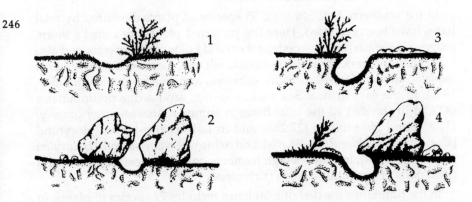

Fig. 54. Types of shelters of tolai hares in the Pamirs. 1—resting place of tolai hares under a bush; 2—resting place between rocks; 3—hole made under a bush; 4—hole under a rock (Abdusalyamov, 1945).

form of small pits under bushes of pea shrub or dig shallow burrows in cliffs of terraces along banks of dried mountain streams (Kolosov, 1939).

In flood plain areas of the Tuva part of the range, the daytime resting places of tolai hares are usually located in willow groves at the border with fields, and on mountain slopes under large stones or in old burrows of marmots. In mountain steppes in the southwest of Tuva, the main shelters of tolai hares are marmot burrows, in which hares raise their progeny (Obukhov, 1973; Zonov, 1974). In tolai hares living in the Trans-Baikal and in northern Mongolia, resting places are most often located in bushes of pea shrubs and in open steppes and slopes of hills and, as in Tuva, hollows under large stones or marmot burrows; sometimes expanded burrows of sousliks may serve as shelters (Fetisov, 1955; Bannikov, 1954).

Feeding. Tolai hares use leaves, shoots, bark, wood, seeds, inflorescences, rhizomes, and bulbs of plants as their food. In the arid zone of the Kazakhstan part of their range, in spring the diet of major significance consists of the growing parts of ephemerals and ephemeroids. At this time the preferred foods are sedges, cheatgrass brome, *Schismus arabicus* and *Eremopyrum orientale*. In summer hares change over to feeding on *Ammodendron connolyi*, sagebrush [wormwoods], "biyurgun"*, camel's thorn and desert grasses. In late fall and winter, the basic food consists of branchlets of tamarisk, salt tree, old-world winter fat and sharp-fruited evonymus. In all, in the arid zone of Kazakhstan, 118 species of plants have been recorded in the diet of tolai hares, of which there are 66 species of grasses, 38 of shrubs and semishrubs, and 14 of woody plants. However, the number of the most frequently devoured species is much less—about 20 (Perevalov, 1958; Sludskii et al., 1980).

In the southern Aral Sea area, 35 species of plants devoured by tolai hares have been recorded. Here the preferred plants are camel's thorn, sedges, sagebrush [worm-woods], tubers and bulbs of ephemerals, and also branchlets and bark of saxaul, tamarisk, salt tree, locoweed, *Ammodendron connolyi*, cherkez [*Salsola richteri*], saltworts and swamp poplar (Reimov, 1985). In the eastern Aral Sea area in spring, sand sedge predominates (87.6%) in the diet of the tolai hare; in summer *Ammodendron connolyi* (31.8%) and *Horaninovia* (27.2%); and in fall—coarse-leaved glorybind [*Convolvuus subhirsutus*] (36.8%) and white Persian saxaul [*Haloxylon persicum*] (19.2%). In years of high numbers, tolai hares eat vegetative and floral parts of agricultural crops (Krivosheev, 1959).

In the Karakums the diet of tolai hares includes 72 species of plants. In the beginning of spring, hares eat mainly chestgrass brome and sedge

---

*Local name of plant, not traceable in our source materials—General Editor.

[*Caryx physodes*] (up to 95% of stomach contents). In the end of April-beginning of May, tolai hares begin to leave saxaul thickets to feed in barhan sands, where their main food is locoweed, heliotrope, *Salium* and kumarchik [*Agriophyllum*]. In July-September, besides vegetative parts, their diet contains tubers, bulbs and rhizomes of such plants as Lehman's eminium [*Eminium lehmani* or *Helicophyllum lehmani*], desert candle, long-stemmed iris, luk gusinyi* [*Gagea*], Litwinow's giant fennel, *Selium* and *Ammodendron connolyi*. In fall and winter the main food of tolai hares (to 80% mass of stomach) consists of leaves, twigs and bark of *Ammodendron connolyi* (Sapozhenkov, 1984). In years with humid falls with massive vegetation of sedge [*Caryx physodes*] and grasses—these plants may constitute up to 50% of the tolai hares (Ishadov, 1974).

In the Karshinsk steppe up to the second decade of April, young shoots of semishrubs, sagebrush, *Ammodendron connolyi*, kovrak* and some other grasses predominate in the diet of tolai hares. From the end of April to the third decade of May, the main food of tolai hares are the seeds of bluegrass, which constitute up to 90-100% of the stomach contents. In fall seeds of various plants constitute up to 80% of the stomach contents (Salikhbaev et al., 1967).

In southern Uzbekistan (Aral-Paigambar preserve), the diet of tolai hares recorded 48 species of plants (Volozheninov, 1983). In fall and winter hares feed here mainly on fruits of lokh* [*Elaeagnus*], dry herbaceous plants, twigs of shrubs, and sometimes—bark of trees and shrubs. From February to October herbaceous plants predominate in the diet of tolai hares, and the vegetative period of any species determines the composition of diet in the given period.

In the valley regions of southwestern Tadzhikistan during spring, summer and fall, tolai hares mainly feed on herbaceous vegetation, and only in winter twigs, and in spring bark and rhizomes appear in their diet. The following plants are eaten most frequently: horsetail, mammoth wild rye, brooms, bulbous bluegrass, couch grass, sedges, salt brush, "shiritsa*," sweet clover, lucerne, licorice, camel's thorn, sea lavenders, nightshade and Asiatic poplar. In all, 63 species of plants have been recorded here, which are eaten by tolai hares, among which 10 species are cultivated crops and 29 species—weeds (Stroganov and Stroganova, 1944).

More diverse is the feeding of tolai hares inhabiting the montane forests of Tadzhikistan, where the preferred plants are: fescue, bluegrass, feather grass, barley, knot-weed, summer cypress, buttercup, sagebrush, locoweed, giant fennel, onion, lucerne, and among shrubs—leaves, shoots and fruits of honeysuckle, barberry, willows and sea buckthorn. In winter, particularly

*Common name could not be found—General Editor.

with high snow cover, tolai hares feed mainly on branches and bark of trees and shrubs (Davydov, 1974).

In high mountains in spring and summer, grasses, Cyperaceae and legumes constitute the main food of tolai hares. Most frequently found in their stomach are kobresia, winter fat, summer cypress, sweet vetch, crazyweed and sagebrush. In fall, hares prefer legumes, among which crazyweed and locoweed play the maximum role. At this time they often enter fields where they eat rye and barley. In winter twigs of winter fat and sagebrush predominate in the diet of tolai hares. In heavy-snow winters hares descend in the valleys and often approach inhabited areas, where they feed on near hay stacks (Abdusalyamov, 1965; Davydov, 1974).

In the mountain-steppes of Tuva, herbaceous plants and their rhizomes predominate in the diet of tolai hares. Even in flood plains and ravines rich in shrubby vegetation, hares prefer to dig out grass and roots from under the snow. In search of areas convenient for foraging with non-frozen and loose ground, hares may move from resting places to a distance of 2 km. Paths are laid out to the permanent places of foraging (Zonov, 1974).

In natural conditions one feeding tolai hare consumes about 120 g of plant mass (Sapozhenkov, 1964). In desert conditions tolai hares may go without water, but at watering holes they drink water avidly.

Activity and behavior. The character of activity of tolai hares depends on the time of the year. In summer, tolai hares are active mainly at dusk and night, and in spring (during heat) also in the brighter part of the day. According to the observations of Yu.F. Sapozhenkov (1964), tolai hares in the Karakums do not lead a strictly nocturnal mode of life—in summer they are active from 20-21 hr to 6 hr; in May their activity starts from 17 hr; in dull cloudy weather, hares come out for feeding also in the daytime. In any weather young hares are more active in daytime. The activity of tolai hares falls in windy dusty weather.

In the lower reaches of the Chu River, the maximum activity of tolai hares is observed in February-April (start of reproduction and migration from winter habitats) and in September-October (higher activity of young). In winter, tolai hares are most active from 17 to 24 hr and from 5 to 8 hr. In spring and in first half of summer, the maximum activity is observed in the morning hours; when mating occurs, nocturnal activity increases in summer and fall. Females are least active in the last stage of pregnancy; this time they almost avoid resting places (Sludskii et al., 1980).

While chasing, tolai hares usually run in a straight line, not making a loop, and having run to the shelter, they disappear. During running tolai hares jump a length of 2.5-3 m. When in an undisturbed state before going for their daytime rest, tolai hares disturb their traces leading to the shelter; here, like other species of hare, they perform sideways jumps, and make double or triple loops (Sludskii et al., 1980).

Area of habitation. Data on the area of individual territory of tolai hares are very scanty. It is known that in fall in adult tolai hares living in the territory of Kazakhstan, the area of individual territory, on average, constitutes 2.1 ha (n = 11), and in young (n = 4)—1.7 ha. In the confines of individual territory, there are several resting places and foraging areas (Sludskii et al., 1980).

Reproduction. In the desert part of the range, tolai hares first come into heat usually at the end of January-beginning of February, and in warm winters north of the Karakums—in the beginning of January. In mountains and in the north of their range, the reproductive cycle begins at the end of February-beginning of March. The periods of first heat are usually long, since females of different age groups do not simultaneously take part in reproduction. Females, which did not reach the age of one year by the beginning of heat, take part in reproduction only in the spring or summer. For example, in the plains of Kazakhstan in January-February, not more than 75% of females are fertilized. In winter males are reproductively more active; over 90% of them have intensive spermatogenesis. In the lower reaches of the Ili, massive pregnancies are observed in March, when the proportion of pregnant hares is 69-78%. In April-May the wave of repeat matings begins, which happens soon after births. The proportion of pregnant females in this period increases appreciably; in the lower reaches of the Chu and Amu Darya, it is 95-97%. In June-August, some old females bear a third brood and young of the previous year's birth (depending on the generation), give birth for the first or second time. In June, in the lower reaches of the Ili, the proportion of pregnant females is about 44%, and by August this index falls to 6% (Perevalov, 1956; Baitanaev, 1974; Sludskii et al., 1980; Reimov, 1985).

Termination of the reproductive period depends on the habitat and weather conditions. In desert habitats, where feeding conditions deteriorate by the beginning of summer, reproductive activity of both males and females falls appreciably by June, and by August practically ceases (Fig. 55). In other habitats reproduction continues most often to July-August. Thus, the reproductive period in tolai hares over a large part of their range extends for about 5 months (Sapozhenko, 1964). Under favorable weather conditions (for example, humid summers in the desert) individual females may bear progeny in fall—in October or even in November. Thus, pregnant females were caught in November, 1976, in the Muyunkums, and in October 1979 in the northern Kyzylkums (Sludskii et al., 1980); in the Khorezmsk region in October a female that had just given birth was caught (Ostapenko, 1963). South of Uzbekistan two waves of reproduction were reported—one from January to May; another, in October (Volozheninov, 1983).

249

247

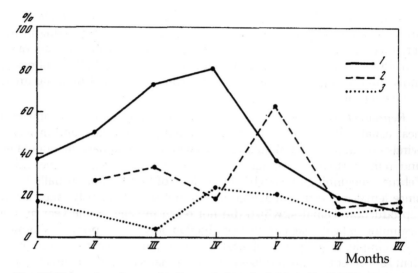

Fig. 55. Dynamics of reproductive activity of tolai hare in the Karakums.
1—percentage of pregnant females; 2—percentage of females lactating and
recently given birth; 3—percentage of females with resorbed embryos
(from Sapozhenkov, 1964).

According to the data of A.A. Perevalov (1956) and A.A. Sludskii et al.
(1980), individual old hares during the reproductive season may bear up
to four broods. An indirect confirmation of the possibility of four broods is
given by the above-cited information of finding pregnant females in the
fall period. Usually in tolai hares there are two or three broods in a year,
and in the southern Aral Sea area, only individual females give birth three
times (Reimov, 1985).

Fecundity of tolai hares, studied on the basis of the analysis of the
number of embryos, is almost similar over the entire range—usually they
have 1 to 6 embryos. The maximum number of embryos (9) was recorded
in the lower reaches of the Ili and on the Karshinsk steppe, where the high
and average fecundity is 5.6 and 6.2 embryos per female (Salikhbaev et al.,
1967; Sludskii et al., 1980). High fecundity is also typical of tolai hares
from Tuva (Table 11.2)*, but here there is no third brood, and only 60% of
females give birth twice (Boyarkin, 1984). Two broods are typical also of
tolai hares from the Trans-Baikal area, where 5 of 29 pregnant females had
7 embryos each, and the rest—5.6 each (Fetisov, 1935). Thus, here, as in
Tuva, with limited periods of reproduction, general fecundity of tolai hares
is apparently at the same level, as in other parts of the range.

In any type of habitat, the size of first brood, in comparison with the

---

*In Russian original, given as Table 12.2—General Editor.

second, is usually small. The number of baby hares in the third brood (if present) is usually more than in the first, but less than in the second (Table 11.2). The largest broods occur in April-May, when feeding conditions are the most favorable almost over the entire range. At this time the lowest level of embryo resorption is recorded. Embryo mortality is relatively high in winter and in early spring. For example, in the lower reaches of the Ili, in March 1951, in seven pregnant females more than 30% of embryos were resorbed (Perevalov, 1956). According to the data of G.S. Davydov (1974), in early spring, the proportion of females with resorbed embryos was five times higher than during the first and third reproduction waves. Often resorption of embryos occurs in young and emaciated females, the size of broods of which is most often smaller than in old and well-fed hares. It is shown that in the territory of Tadzhikistan, females with a body weight of 1,610 g are found in a pregnant state three times more than females with a weight of 1,200-1,600 g, and their fecundity is roughly 1.5 times higher. The same pattern was noticed in other parts of the range (Davydov, 1974; Sludskii et al., 1980).

Data on the number of embryos, obtained for 5 years in the lower reaches or the Chu, confirm that the fecundity of tolai hares may change from year to year (Table 11.3*).

Growth and development. In tolai hares pregnancy varies from 45 to 50 days (Sludskii et al., 1980).

In tolai hares, baby hares are born with sight, and their body is covered with quite dark fur with a very distinct dark stripe on the back. The weight of newborns is 85-100 g; length, 110 to 140 mm (Perevalov, 1956; Sapozhenko, 1964). For some time after birth, young apparently lose some weight, which is confirmed by a comparison of the weight of embryos and newborns in the lower reaches of the Chu River (Sludskii et al., 1980).

Baby hares feed on milk for 10-15 days. They change over to plant food completely at the age of 20-25 days.

Population dynamics. In different parts of the range and in different habitats the amplitude of fluctuations in numbers of tolai hares has its own peculiarities. In river valleys, as a rule, there are no sharp fluctuations in numbers. In the north of their range and in high mountains, a sharp decrease in numbers of tolai hares occurs in years with cold and heavy snow in winter, when feeding conditions deteriorate. Mass mortality of hares occurs during glazing. With low temperatures a large part of young hares of the first brood die, which also adversely affects the number of tolai hares. In the desert zone numbers decrease during the period of spring-summer droughts. The years when dry summers alternate with cold

*In Russian original, this table is missing—General Editor.

Table 11.2. Fecundity of tolai hares in different parts of their range

Time of observations	Place of observations	Number of females investigated	Number of embryos (range)	Average number of embryos	Authors
January	Karakums	17	1-3	1.6	Sapozhenkov, 1964
February	Karakums	18	1-4	2.6	Sapozhenkov, 1964
March	Karakums	21	2-6	3.5	Sapozhenkov, 1964
February-March	Lower reaches of the Ili River	72	1-6	1.8	Perevalov, 1956
		48	1-6	2.4	Sludskii et al., 1980
	Lower reaches of the Chu River	10	1-4	2.3	Sludskii et al., 1980
	Valley of Vaksh River	6	1-6	2.2	Davydov, 1974
	Mountains of central Tadzhikistan	5	1-6	2.2	Davydov, 1974
April	Karakums	26	2-6	4.0	Sapozhenkov, 1964
	Pamir, Badakhshan	15	1-5	2.8	Davydov, 1974
	Southern Aral Sea area	12	3-6	4.4	Reimov, 1985
May	Karakums	19	2-5	3.7	Sapozhenkov, 1964
	Southern Aral Sea area	37	1-6	3.7	Reimov, 1985
	Pamir, Badakhshan	8	2-6	4.0	Davydov, 1974
	Tuva	–	2-7	5.3	Boyarkin, 1984
April-May	Lower reaches of the Ili River	43	1-7	4.2	Perevalov, 1956
		42	1-9	4.6	Sludskii et al., 1980
	Lower reaches of the Chu River	122	2-8	5.6	Sludskii et al., 1980
	Kyzylkums (north)	28	3-6	4.3	Sludskii et al., 1980
	Kyzylkums (northwest)	24	3-6	4.1	Reimov, 1984
	Valley of Amu Darya River	46	2-6	4.3	Reimov, 1984

Contd.

Table 11.2 contd.

Time of observations	Place of observations	Number of females investigated	Number of embryos (range)	Average number of embryos	Authors
	Sarykamysh	8	3-5	3.7	Reimov, 1984
	Trans-Unguz Karakums	12	2-5	4.0	Reimov, 1984
	Valley of the Vaksh.	11	2-6	4.5	Davydov, 1974
	Mountain of central Tadzhikistan	11	2-5	4.8	Davydov, 1974
	Southern Ustyurt	7	3-6	4.4	Sludskii et al., 1980
	Chink of Ustyurt	26	2-7	4.2	Reimov, 1984
June	Karakums	25	1-4	1.2	Sapozhenkov, 1964
	Southern Aral Sea area	19	3-5	3.5	Reimov, 1984
	Tuva	–	3-8	6.8	Boyarkin, 1984
July	Southern Aral Sea area	16	2-6	3.2	Reimov, 1984
August	Southern Aral Sea area	18	1-5	3.0	Reimov, 1984
July-August	Lower reaches of the Ili River	45	1-8	4.2	Perevalov, 1936
		24	1-6	3.5	Sludskii et al., 1980
	Lower reaches of the Chu River	74	2-4	3.9	Sludskii et al., 1980
	Valley of the Vaksh River	13	1-5	3.8	Davydov, 1974
	Mountains of Central Tadzhikistan	15	2-6	4.2	Davydov, 1974
	Badakhshan, Pamir	17	1-6	3.5	Davydov, 1974

356

winters are particularly unfavorable for tolai hares. Observations on numbers of tolai hares in 1959-1973 in the lower reaches of the Chu River, the Muyunkums and the Kyzylkums showed that change in weather conditions in these regions led to a 2-3-fold fluctuation in number of hares. Moreover, it is shown that temperature and humidity affect the distribution of hares in habitats (Baitanaev, 1974; Bekenov*, 1974).

Epizootics, massive trade, predators and the economic activities of man influence numbers of hares to a somewhat lesser degree, than the weather conditions. Tolai hares, in comparison with other species of hare, are least adapted to living in regions reclaimed by man. For example, in the dry 252 steppes of the foothill plains of the Terskei Alatau, numbers of tolai hares appreciably declined as a result of agricultural development in these places (Zimina, 1964).

Competitors. In the mountain and high-mountain regions of Tadzhikistan, competitors for food for tolai hares, the especially in winter, are fewer. In summer long-tailed marmots, pikas, and voles compete with tolai hares for the same foods (Davydov, 1974). In the plains in spring and winter, cape hares and sousliks are competitors of tolai hares for food, and among wild ungulates—saiga, sheep and other domestic cattle are competitors of tolai hares everywhere.

In places, where the range of tolai hares coincides with the ranges of blue hares and European hares, competitive relationships possibly exist between the different species of hares. In any case, in the north and northwest of the range of tolai hare, where the southern distributional boundary of European hare passes, numbers of both species are appreciably lower, than in other regions (Fadeev, 1966).

Enemies. Wolves, jackals, foxes, karakal lynxes, wild cats and perevyazka** hunt tolai hares. Domestic cats and dogs also catch tolai hares. The main enemies of tolai hares practically throughout their range are foxes, in the diet of which tolai hares constitute usually 1 to 9% (Sapozhenkov, 1964; Palvaniyazov, 1974). A large number of tolai hares are destroyed by wolves, in the diet of which, in years of high numbers, tolai hares constitute over 37% (Palvaniyazov, 1974).

Many predatory birds feed on tolai hares: eagle owls, golden eagles, imperial eagles, steppe eagles, white-tailed eagles, spotted eagles, long-legged buzzards, Egyptian vultures, sparrow-hawks, little owls, kestrels, moor harriers, eagle owls***, eared owls and desert owls. Predatory birds most intensively exterminate tolai hares during the period of feeding

---

*Not in Lit. Cit.—General Editor.
**Marbled polecat—General Editor.
***Repeated in the Russian original—General Editor.

fledglings and in winter. Sometimes monitors attack tolai hares (Sapozhenkov, 1964; Ishunin and Pavlenko, 1966).

Diseases and parasites. Twenty-two* species of ixodid mites have been recorded: *Ixodes redikorzevi, I. crenulatus, I. kazakstani, I. persulcatus, Haemophysalis*** *numidiana, H. punctata, H. caucasica, H. concinna, H. warburtoni, Dermacentor marginatus, D. daghestanicus, D. pavlovskyi, Rhipicephalus turanicus, R. pumilio, R. leporis, R. schulzei, R. bursa, R. sanguineus, Hyaloma asiaticum, H. plumbeum, H. anatolicum.* Of these, the most widely distributed are: *H. asiaticum, H. turanicus, D. daghestanicus, D. marginatus* and *R. pumilio*, the last 3 species being as a rule, the most numerous. Among gamasid mites, *Hypoaspis marinus* and *H. aculeifer* have been found on tolai hares, and among the trombuculid mites—*Neotrombicula angulata, N. obscura, Halenicula* sp. and *Trombicula tantula* (Gvozdev, 1949; Dubinin and Bregetova, 1952; Sapozhenkov, 1964; Davydov, 1974; Sludskii et al., 1980).

The time of massive appearance of mites on tolai hares depends on weather conditions and the type of habitat. In the lower reaches of the Ili—it is last days of April-beginning of May. The maximum intensity of infestation with mites occurs in June. By fall, the number of mites on tolai hares decreases, and from November to March hares are free of them.

Fleas are quite numerous on tolai hares; not only *Hoplopsyllus gracialis* is specific to hares, but such species as *Xenopsylla conformis, X. gerbilli, X. hirtipes, Echidnophaga oschanini, Coptopsylla lamellifer, C. beiramelensis,* and *C. olgae* are characteristic of rodents. Five species of lice have been found on tolai hares: *Haemodipsus lyriocephalus, H. conformalis, H. gracilis, Frontopsylla glabra* and *Hoplopleura acanthopus;* moreover the first of these species is most widely distributed and is common to tolai, European and blue hares (Gvozdev, 1949; Dubinin and Bregetova, 1952; Sapozhenkov, 1964; Davydov, 1974; Sludskii et al., 1980).

Tolai, as also other species of hares, are quite intensively infected with endoparasites. About 25 species have been recorded in tolai hares. *Dicrocoelium lanceatum* is a widely distributed (with the exception of desert habitats) and frequently found trematode, while *Fascisola hepatica* is found only in the Kochkorsk valley in Kirgizia. Periodically high is the infection of tolai hares with cestodes: *Mosgovoyia pectinata, Taenia pisiformis, Multiceps serialis, Gvoasdevilepis fragmentata, Andrya cuniculi, A. rhopalocephala* and *Echinococcus granulosus.* Among nematodes, the following have been recorded in tolai hares: *Dermatoxys veligera, Passalurus ambiguus, Protostrongylus termiinalis, Nematodirus aspinosus, Trichostrongylus*

252

---

*In Russian original, only 21 species listed—General Editor.
**In Russian original, *Haemophisalis*—General Editor.

retortaeformis, T. colubriformis, Micipsela numidica (Gvozdev, 1948, 1948a; Gafurov et al., 1971; Yanushevich et al., 1972).

It has been shown that tolai hares living on sands, as a rule, are infected with helminths less intensively than hares from river valleys. For example, in the Muyunkums the level of helminth infection is in the range of 8%, whereas in the valley of the Chu River this index reaches 70% (Baitanaev, 1974).

Tolai hares are a source of the tularemia pathogen. Infectivity of hares with the tularemia microbe in the lower reaches of the Chu River fluctuates from 0.9 to 14.0% (Sludskii et al., 1980). Most often diseased tolai hares are found in spring—in March-May; less often, in early fall.

In areas of natural foci of plague, tolai hares may be the casual carriers of the plague microbe, and in the southern Lake Balkhash area the pathogen of brucellosis has been isolated from tolai hares (Sludskii et al., 1980).

Practical Significance

In places with high numbers (southern regions of Kazakhstan and Russian central Asia), tolai hares are the object of commercial and amateur hunting. Skins of tolai hares were particularly actively processed toward the end of the 20s-beginning of the 30s. At present, skins of tolai hares are almost not processed on a commercial scale, and hare hunting is done almost exclusively for meat. The tolai hare plays a significant role in the food of valuable commercial species. Since tolai hares are quite strongly infested with mites, which are carriers of several diseases of animals and man, they play a definite role in the spread of some infections. Cases of human tularemia infection have been recorded during processing of skins and use of meat of tolai hares as food. Tolai hares do not inflict appreciable damage to agricultural crops and pastures.

Manchurian Hare
Lepus (Allolagus) mandshuricus Radde, 1861

1861. *Lepus mandshuricus* Radde. Melange Biol., Bull. Acad. St. Petersb., 35: 684. Khabarovsk territory, Bureinsk Range.

1922. *Lepus mandshuricus* sbph. *melanonotus* Ognev. Ezhegodn. Zool. Muzeya Akad. Nauk, 23, 3: 489. Primorsk [Pacific coastal] territory, Sikhote-Alin Range, "15 versts* from Sol'sk" (proposed as "subphase").

---

*1 verst = 1.067 km—General Editor.

Diagnosis

Head in relation to body quite large. Length of ear, measured from occiput, equals length of skull or scarcely surpasses it. Length of tail (with terminal hairs) is about 60-70% of the length of the hind feet (excluding claws).

Tail, in winter and summer, gray with brownish-black dorsal part.

Fur quite soft, guard hairs erect, thicker than in tolai hares. Underfur wavy-curly. Winter fur somewhat lighter than in summer, more gray. Winter fur in melanic individuals remains black-brown.

Skull with a relatively narrow and weakly bulged cranium. Ratio of width of the frontal bones behind the supra-orbital processes to width of the base of the rostrum from 51 to 57% (average—53%). Length of the bony palate equal to or slightly more than the width of choana. Alveolar length of the upper tooth row 16.6-18.5 mm (average—17.5 mm). Zygomatic arches massive and wide apart—zygomatic width in the anterior part 37.1-44.6 mm (average—40 mm). Supraorbital processes narrow and short—not reaching the posterior margin of the frontal bones. Mandible short and massive. Length of the lower diastema in relation to the length of the lower row of teeth constitutes 80%. Articular process short, steeply raised upward (Fig. 57).

Description

Color highly variable but almost the same throughout the year. Back and upper part of head usually ochreous-brown or ochreous-gray with a dark ripple formed by black tips of guard hairs. Sides of the head are somewhat lighter in color, with large whitish spots in the anterior and lower parts. Dark band below eyes (Fig. 56). Sides of body, limbs and breast are of a pale yellow color, and belly is dirty white. In the southern Primorsk [Pacific coastal] territory, about 0.5% are partial melanists, in which the head, back, sides and paws are black; the throat is pale yellow and the belly is white. In the valley of the Bira and Bidzhan rivers, and in south of the Amur region partial melanists are found still more frequently (Yudakov and Nikolaev, 1974). In the territory of the Khingan preserve, very light, almost white, individuals are found (Chernolikh, 1973).

Body length—430-490 mm; length of hind feet—110-130 mm; ear length—75-90 mm; body weight—1,700-1,900 g.

The diploid complement in the type form has 48 chromosomes: 8 pairs of meta-submetacentrics, 15 pairs of subtelo- and acrocentrics (NFa = 78); X-chromosome—sub-metacentric, Y-chromosome—small acrocentric (E.Yu. Ivanitskaya).

Fig. 56. Manchurian hare *Lepus mandshuricus*
(Photo by I.G. Nikolaev).

Systematic Position

S.I. Ognev (1940) included the Manchurian hare in a separate genus *Allolagus*. To this genus he referred hares living in Japan. A.A. Gureev (1964) combined the Manchurian hare with *brachyurus* in one species and included it in the genus *Caprolagus*, the members of which (12 species) inhabit Asia (from Afghanistan to Japan), Africa and Central America (Mexico). A.A. Gureev does not mention diagnostic characters while combining all these species, whereas Ognev (1940) lists seven characters, on the basis of which members of the genus *Allolagus* differ from *brachyurus**. Angermann (1966), based on the analysis of morphometric characters and the structure of the fur of hares, referred by Gureev to the genus *Caprolagus*, and their comparison with members of the genus *Lepus*, came to the conclusion that Manchurian hare is a separate species. Moreover, she concluded that the Manchurian hare and *brachyurus* belong to the genus *Lepus*. Only one species—*C. hispidus*—is included in the genus *Caprolagus*.

---

*Capitalized in original.

254

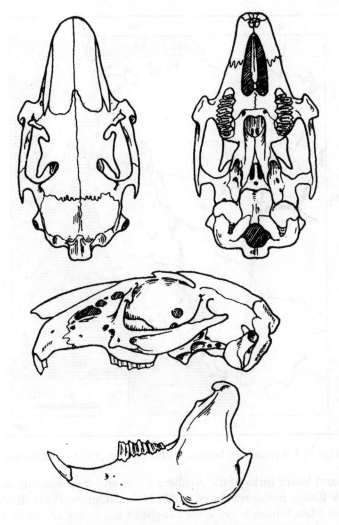

Fig. 57. Skull of the Manchurian hare *Lepus mandshuricus*.

256

Geographic Distribution

In the territory of Russia, Manchurian hares live in the Primorsk [Pacific coastal] territory and in the basins of the middle and partly lower Amur River (Fig. 58).

In the northwest of their range, Manchurian hares are found in the vicinity of Blagoveshchensk (Yudakov and Nikolaev, 1974). From here the distributional boundary of the Manchurian hare proceeds east along the Amur River cutting across the Zeya-Bureya plain; it passes along the

362

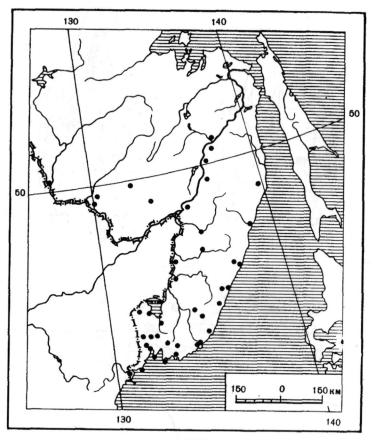

Fig. 58. Localities of Manchurian hares in the territory of Russia.

middle and lower forks of the Arkhara River and then ascends along the Bureinsk Range in the northeast to 50° N. lat. (Ognev, 1940). Toward the southeast Manchurian hares were caught in the valley of the In River in the basin of the Amur (collections of Institute of Biological Problems of the North, Magadan). Later, the northern boundary passes along the right bank of the Amur to the town of Srednee [Middle] Tambovskoe, which is the extreme northern locality of the Manchurian hare (Ognev, 1940).

Along the western slopes of Sikhote-Alin, the distributional boundary of the Manchurian hare turns south and following the left bank of the Amur passes through the town of Verkhnee [Upper] Tambovskoe, the mouth of the Gur River, cuts across the lower fork of the Anyui River, in the middle— the Khor and Bikin Rivers, and through the Sikhote-Alin' Range enters its eastern slopes in the region of the upper reaches of the Serebryanka River (Ognev, 1940; Yudakov and Nikolaev, 1974). From here the northern boundary turns northeast and follows along the eastern slopes of Sikhote-Alin' through the upper reaches of the Bolshaya [Greater] Kama and Maksimovka rivers to the mouth of the Samarga River (Yudakov and Nikolaev, 1974). The extreme northeastern locality of the Manchurian hare is the basin of the Koppi River (Dul'keit, 1956).

257   The eastern distributional boundary of the Manchurian hare in the south passes along the coast of the Sea of Japan. Manchurian hares were reported south of the Samarna River in the vicinity of the village of Ternei in the basin of the Rudnaya River, in the valley of the Arzamazovka River, at several points in the Sudzukhinsk preserve, in the area of the Tikhaya Zavod Inlet (Vostok Gulf), on Putyatin Island, near Vladivostok, in the mouth of the Suifun River (Gulf of Amur), and on the Gamov Peninsula (Gulf of Peter the Great) right up to Lake Khasan, being the southernmost locality of Manchurian hares (Ognev, 1940; Dul'keit, 1956; Yudakov and Nikolaev, 1974; collections of ZM MGU and BIN). In the south and west, the range of Manchurian hares passes outside the limits of Russia.

Along the state border in the area of the Primorsk [Pacific coastal] and Khabarovsk territories, in a south to north direction, Manchurian hares were reported in the "Kedrovaya Pad" preserve on the western bank and Lake Khanka—in the vicinity of the town of Troitsk and in the valley of the Komissarovka River, in the area of the village of Imanskii, in the basin of Ussuri (south of the mouth of the Bikin River) and in the vicinity of Khabarovsk (Ognev, 1940; Yudakov and Nikolaev, 1974; collections of ZM MGU).

The localities of Manchurian hares within the range are not distributed uniformly (Fig. 58). In the Primorsk [Pacific coastal] territory this species is found throughout, except in the high-montane Sikhote-Alin', and in the Amur region and Khabarovsk territory (except the valley of the Arkhara River and the territory of Evreisk A [utonomous]. R [egion]). Manchurian hares are found rarely.

Beyond the borders of our country [Russia], Manchurian hares live in northern and northeastern China, the Korean Peninsula and in Japan (Kyu-Shu, Shikoku, Hondo, Dogo, Oki and Sado islands).

Geographic Variation

Apparently, Manchurian hare is a monotypic species.

Biology

Habitat. The distribution of habitats of Manchurian hares is confined to broad-leaved forests with a Manchurian type of vegetation, where

364

Mongolian oak, Siberian filbert, actinidia, Amur grape, Chinese magnolia vine and shrub lespedeza predominate. The most characteristic plant in habitats of the Manchurian hare is the shrub lespedeza—beyond the limits of its range in the Far East, Manchurian hares are not found. Most often Manchurian hares may be found on the southern slopes of mountains and spurs along valleys of rivers and streams with dense thickets of trees and shrubs. Favorable for them are flood plains of rivers with willow thickets and piles of driftwood, as well as low water-divide areas with rock outcrops and thickets of Daurian rhododendron, Amur deutzia, thin-leaved mock orange [*Philadelphus tenuifolius*] and different lianas. Manchurian hares willingly inhabit forest fellings and burned-out forests. In the Khankai plain and in the valley of the Razdol'naya River, Manchurian hares live in ravines and atop spurs overgrown with bushes and low trees (Yudakov and Nikolaev, 1974).

Manchurian hares practically do not enter the zone of fir-spruce forests of high mountains, which is explained by the absence here of well-developed underbrush and poor herbaceous vegetation. Moreover, thick snow cover in such forests obstructs food procurement and free movement of Manchurian hares (Yudakov and Nikolev, 1974).

In Manchuria, Manchurian hares inhabit mainly mixed forests with a predominance of broad-leaved species and thick underbrush, river valleys with willow thickets and foothills with stony slopes. They are encountered in the vicinity of human habitations, in meadows and bogs, and thickets of grasses and sedges. They often enter open places in summer. In the taiga zone of the Khingan Range, Manchurian hares are not encountered (Chernolikh, 1973).

Distant migrations have not been reported in Manchurian hares. Usually they migrate short distances in search of food, and males during heat may 258 move a distance of 3 km in search of females. During fall floods hares migrate from river valleys to adjoining hill slopes. In winter (December-January) in search of places better protected against the wind, hares also abandon broad river valleys. In all such cases the distance covered by Manchurian hares does not exceed a few kilometers (Dul'keit, 1956).

Population. In the Primorsk [Pacific coastal] territory, where Manchurian hares occur almost everywhere, their number is constantly high. In the vicinity of Ussuriisk, for example, in the early 50s the density of Manchurian hares was about 27 individuals per 100 ha, and in the Gamov Peninsula—to 55 per 100 ha (Dul'keit, 1956). The southern and southeastern slopes of Sikhote-Alin' (from the basins of the Batal'yanza and Artemovka rivers in the south to the basin of the Rudnaya River in north), the southwestern part of the Primorsk [Pacific coastal] territory (from Lake Khanka to Lake Khasan) and the basin of the Ussuri River are the areas of commercial density of Manchurian hares. In the Amur region and in Khabarovsk territory, with the exception of the valley of the middle and lower forks of the Arkhara River and the territory of the Evreisk A [utonomous]. R [egion]., the number of Manchurian hares is insignificant (Yadakov and Nikolaev, 1974).

Feeding. The summer diet of Manchurian hares inhabiting the Primorsk [Pacific coastal] territory includes 111 species of herbaceous plants (including 2 species of fern and 1 species of horsetail) and 29 species of woody, shrubby and liana vegetation: among them there are 18 species of composite, 15 species of legume, 13 of the rose family, 9 of the buttercup family, 8 sedges, 5 each of cereals and labiates, and 3 species of umbelifer. Members of the Fabaceae are found most frequently (57% of cases) in their diet. In any habitat, hares willingly eat stoloniferous and red-petioled wormwoods, and also 4 species of vetch, uniparous/unijugate, Amur, Japanese and pleasant. In damp valleys and ravines, the main food of hares is Tatarinov's rattlesnake root, and in open valleys and burned-out forests—Thunberg's meadow rue. In summer in habitats with well-developed herbaceous vegetation, the diet composition of Manchurian hares almost does not include woody and shrubby species. However, hares living in thickets of lespedeza bushes and large-winged evonymus [*Evonymus macropterus*] feed on these plants both in winter and summer. Moreover, in summer Manchurian hares willingly eat shoots of Amur maackia [*Maackia amurensis*] and lianas, and less often—thin leaved mock orange and Amur deutzia (Yudakov and Nikolaev, 1974).

In such species of herbaceous plants, angelica, rattlesnake root, and tassel flower, Manchurian hares eat mainly the stems and petioles of leaves; more so Manchurian hares nibble at rattlesnake root and tassel flower most often at the base, and eat only the upper part of the stem. In meadow rue, only the tips of stems are eaten, but in this case hares do not break stems but bend them. In sedges, they eat the leaves and flowering stems. In grapes, actinidia and magnolia vine saplings, petioles of leaves and stems are eaten. At the end of summer, fruits of Siberian apple, Amur cork tree, hawthorn and kolomikta vine appear in the diet of Manchurian hares. Hares persistently nibble at cast off horns of axis deer and roe deer. Sometimes hares come to the sea coast, where they dig out the washings of black kelp and common eelgrass (Dul'keit, 1936; Yudakov and Nikolaev, 1974).

The winter diet of Manchurian hares recorded 19 species of herbaceous plants and 26 species of woody-shrubby species and lianas. On the whole, in the Primorsk [Pacific coastal] territory, the most frequent (79% of cases) are the shrub lespedeza (41%) as also willows, Manchurian and Siberian filbert, large-winged and few-flowered [pauciflorus] evonymus. In river valleys the main foods are the shoots of willows and Amur maackia. In the case of maackia, apple, lespedeza and willows, besides shoots, hares nibble at the bark and sometimes the wood. On steep hill slopes, where there is less snow in winter, herbaceous plants constitute a major part in the diet of Manchurian hares: sagebrush, vetches, cereals, Pacific gypsophila, clematis, meadow rue, burnet, horsetail and others (Yudakov and Nikolaev, 1974).

Shelters. The nature of distribution and number of Manchurian hares depend not only on the quality of the food base, but also on the presence of shelters suitable for them. Unlike other species of hare, Manchurian hares, in the event of danger, reluctantly leave their resting places. There are three types of resting places in Manchurian hares: open, closed and shelters (Dul'keit, 1956). Open resting places are usually made in shrub thickets. These resting places are temporary, being used in good weather or for short periods. Most often, hares use closed resting places, which are made in dense shrubs under deadwood and breaks, and also permanent shelters in the hollow of fallen trees, in old badger nests, in caves of overhangs or in niches under stones. Places with such reliable hideouts are most densely populated by Manchurian hares. Open winter resting places are often located in stacks of dry grass (Dul'keit, 1956; Vassiljev et al., 1956; Yudakov and Nikolaev, 1974).

Activity and behavior. Manchurian hares are most active after the onset of darkness and before dawn. They also feed at night. Hares do not venture in open places, but move in thickets of grasses and shrubs, where they lay paths and holes. Usually the nocturnal outing of hares does not extend beyond 0.5 km, and the maximum distance which the hare covers, moving from one foraging area to another, does not exceed 3 km. Males in search of females in heat may cover a similar distance (Dul'keit, 1956).

During resting, Manchurian hares leave shelters only in the event of direct danger. When in a shelter, hares allow humans to come within a distance of 1 m (more often 3 to 7 m). Hares less willingly leave well-protected burrows even during snowfall. During continuous rains, snow storms and glazing, hares do not leave shelters for several days. With snow not more than 40 cm high, Manchurian hares can feed for a long time under snow, laying long tunnels (Dul'keit, 1956).

Before leaving for resting, Manchurian hares usually obliterate their tracks. Their looping maneuvers are more similar to those of blue hares than to European hares. In the event of danger (with the availability of a reliable hideout nearby), a hare runs straight to the shelter and does not make a loop. In winter while running away from a chaser, hares usually keep to their beaten paths, although they move quite nicely on loose snow (Dul'keit, 1956). During running the length of jumps of Manchurian hares usually varies from 64 to 87 cm, but sometimes may reach even 104 cm

(V.V. Gruzdev), and the distance, to which hares jump during marking, surpasses 1.5 m. A scared hare initially runs 50-100 m in a straight line, and then begins to make circles of 150-200 m in diameter (Vassil'ev et al., 1965).

Scared Manchurian hares often produce well audible sounds similar to sneezing (Dul'keit, 1956).

Area of habitation. The size of individual territories have not been studied in Manchurian hares. It is only known that the diameter of the area on which a hare has permanent shelters and foraging places does not exceed a few hundred meters (Yudakov and Nikolaev, 1974).

Reproduction. Data on reproduction of Manchurian hares are very meager. It is known that in the southern Far East, the beginning of heat in them is in February-begnning of March. The first pregnant females are recorded in the first decade of March. Actively reproducing females were found up to August 24 (Dul'keit, 1956; Yudakov and Nikolaev, 1974). Thus, the reproductive period in Manchurian hares apparently lasts six months. During this time some females may bear 2-3 broods. Each brood may have up to 6 young, most often 3 to 5.

Growth and development. Not studied.

Molt. Manchurian hares molt twice a year. The spring molt begins in the end of February. The duration and course of the spring molt have not been studied. Summer fur has roughly the same color as winter fur, but is shorter and stiffer. The fall molt terminates roughly in the second half of November, but is somewhat prolonged in females born at the end of summer (Vassilev et al., 1965).

Sex and age structure of population. Not studied.

Population dynamics. The population dynamics of Manchurian hares has practically not been studied. According to the data of G.D. Dul'keit* (1965), the population of Manchurian hares is more or less stable. According to other data (Vassil'ev et al., 1965; Yudakov and Nikolaev, 1974) there are 3-5-year cycles of fluctuation in populations of this species. The main reasons for mortality of Manchurian hares are, apparently, high floods, fires, heavy snow, cold winter and flashes of helminth infections.

It has been reported that felling of virgin forests favorably influences the process of dispersal and rise in numbers of Manchurian hares, since forest fellings (and fires) produce rich herbaceous vegetation and broad-leaved species of trees (Bromlei, 1964).

Enemies. The main enemies of Manchurian hares are foxes, wolves, Far Eastern forest cats, and kharza**, leopard and lynx, Siberian weasels also

260

---

*Not in Lit. Cit.—General Editor.
**Common name unknown—General Editor.

hunt Manchurian hares, particularly in years of their high numbers. Instances are known of attacks of wild boars and bears.

Among predatory birds, golden eagle, white-tailed eagle, gray and Ural owls, sparrow-hawks, buzzards and eagle owls prey on Manchurian hares (Yudakov and Nikolaev, 1974).

Competitors. Competitive interactions of Manchurian hares with other species of mammals have not been studied. In places of combined habitation with blue hares, competition for food resources is possible, but the differences and the nature of preferred habitats leads to minimum overlapping of trophic niches in these species.

Diseases and parasites. In the summer larvae and nymphs of mites *Haemophysalis japonica, H. neumanni, H. concinna, Dermacentor silvarum, Ixodes persulcatus* and *I. angustus* have been found on Manchurian hares. Up to 1,127 *H. japonica, H. concinna, I. persulcatum* (predominantly nymphs) have been found on one hare. The index of abundance of these mites reaches 56. In winter, for *H. japonica,* Manchurian hares are the main hosts. At this time up to 290 mites were collected on the backside of the ear conchae of one hare (Belyaev and Soldatov, 1969).

Seven species of helminths have been recorded in Manchurian hares and the infectivity may reach 95%. The most pathogenic are the lung nematodes *Protostrongylus terminalis* and stomach nematodes *Obeliscoides leporis.* These parasites can be the cause of mortality among hares. Moreover, in Manchurian hares, *Longistriata leporis* and *L. kurenzovi,* specific for Primorsk [Pacific coastal] territory, and also the widely distributed cestodes *Mosgovoyia pectinata* and the trematodes *Dicrocoelium lanceatum* have been detected in Manchurian hares. In the territory of Primorsk [Pacific coastal] territory, all species of helminths, except *Trichostrongylus retortaeformis* found only in blue hares, are common for Manchurian and blue hares (Sadovskaya, 1953).

Practical Significance

Despite a fairly high population density of Manchurian hares in some districts of the Primorsk [Pacific coastal] territory, there is no commercial exploitation of this species, since in Manchurian hares the skin is thin, rips off readily, and traders prefer to catch other species of hare. In the Primorsk [Pacific coastal] territory, at present, 1-2 thousand skins of blue and Manchurian hares are processed (separate statistics on processing of these species is not reported). On the Korean Peninsula, the area of which is much less than that of the Primorsk territory, up to 8 thousand Manchurian hares are processed in a year (Yudakov and Nikolaev, 1974).

The species composition of vegetation at places with a higher population density of Manchurian hares, because of intensive selective eating, may experience significant changes. This is particularly noticeable in limited local areas, for example, in insular forest areas, peaks of slopes, in ravines surrounded by plowed lands. In such places, shoots of Amur maackia, winged evonymus, lespedeza bush as also stems of grasses, legumes and composites may be eaten away almost completely (Yudakov and Nikolaev, 1974).

LITERATURE CITED

261-   Abdusalimov, N.S. 1959. Fauna, geograficheskoe rasprostranenie i nekotorye
272   voprosy ekologii kleshchei sem. Ixodidae v Azerbaidzhane [Fauna,
       geographic distribution and some problems in the ecology of ticks of the
       family Ixodidae]. *Izv. Akad. Nauk AzSSR*, No. 1, pp. 41-52.

Abdusalyamov, I.A. 1962. Nekotorye materialy po ekologii i morfologii
       bolsheukhoi peshchukhi *Ochotona macrotis* na Pamire [Some material] on
       the ecology and morphology of long-eared pika *Ochotona macrotis* in the
       Pamir]. *Izv. Akad. Nauk TadzhSSR, Otd. Biol. Nauk*, No. 2(9), pp. 86-91.

Abdusalyamov, I.A. 1965. Materialy po biologii pamirskogo zaitsa [Materials on
       the biology of Pamir hare]. ibid., No. 2(19), pp. 58-73.

Abe, H. 1971. Small mammals of central Nepal. *J. Fac. Agr. Hokkaido Univ.*, Vol.
       56, pt. 54, pp. 368-423.

Abildgard, F., J. Andersen and O. Barndorf-Nielsen. 1972. The hare population
       (*Lepus europaeus* Pallas) of Illumo Island, Denmark: A report of analysis of
       the data from 1957-1970. *Dan. Rev. Game Biol.*, Vol. 6, No. 5.

Afanas'ev, A.V. 1960. Zoogeografiya Kazakhstana (na osnove rasprostraneniya
       mlekopitayushchikh) [Zoogeography of Kazakhstan (based on the
       distribution of mammals)]. Izd-vo AN KazSSR, Alma-Ata, 258 pp.

Afanas'ev, A.V. and A.M. Belyaev. 1953. Kratkii obzor gryzunov Pavlodarskoi
       oblasti [A brief review of the rodents of Pavlodar region]. *Tr. In-ta Zoologii
       Akad. Nauk Kaz. SSR*, Vol. 2, pp. 31-40.

Afanas'ev, A.V. and P.S. Varagushin, 1939. Ocherk mlekopitayushchikh
       Kazakhskogo nagor'ya [Outline of mammals of the Kazakh upland]. *Izv.
       Kaz. Fil. Akad. Nauk SSSR, Ser. Zool.*, Vyp. 1, No. 1, pp. 5-30.

Afanas'ev, A.V. and A.A. Sludskii. 1947. Materialy po mlekopitayushchim i ptitsam
       Tsentral'nogo Kazakhstana [Materials on the mammals and birds of central
       Kazakhstan]. ibid., Vyp. 6, pp. 48-64.

Afanas'ev, Yu.G. and Yu.A. Grachev, 1970. Raspredelenie i chislennost' zaitsa-
       belyaka i zaitsa-rusaka v raionakh sosnovykh lesov Kazakhstana
       [Distribution and population of blue hare and European hare in areas of
       pine forests of Kazakhstan]. In: *Materialy Nauch.-Prioizv. Soveshch. po
       Okhotnich'emu Promyslu i Zverovodstvu v Kazakhstane*, Alma-Ata, pp. 15-25.

Afanas'ev, Yu.V. 1962. O sobole (*Martes zibelina averini* Bashanov, 1943) v
       Kazakhstane [On the sable (*Martes zibelina averini* Bashanov, 1943) in
       Kazakhstan]. *Tr. In-ta Zoologii Akad. Nauk Kaz. SSR*, Vol. 17, pp. 144-166.

Agafonov, V.A. 1975. Zashchitnye i kormovye svoistva razlichnykh ksnykh
       ugodii dlya zaitsa-belyaka [Sheltering and foraging characteristics of different

lands for blue hare]. In: *Okhotnich'e Khozyaistvo v Intensivnom Kompleksnom Lesnom Khozyaistve*, Girionis: Kaunas, pp. 147-148.

Agafonov, V.A. 1982. Sutochnaya aktivnost' zaitsa-belyaka i faktory ee opredelyayushchie [Daily activity of blue hare and factors influencing it]. In: *Promyslovaya Teriologiya*, Nauka, Moscow, pp. 231-238.

Aizin, B.M. 1979. Gryzuny i zaitseobraznye Kirgizii: Ekologiya, rol' v podderzhanii prirodnykh ochagov nokotorykh zabolevanii [Rodents and Lagomorphs of Kirgizia: Ecology and role in supporting natural foci of some diseases]. Ilim, Frunze, pp. 183-199.

Aksenova, T.G. and P.K. Smirnov, 1986. Osobennosti stroeniya genitalii Lagomorpha [Peculiarities of the structure of genitalia of Lagomorpha]. In: *IV S'ezd. Vsesoyuz. Teriol. O-va*, Moscow, pp. 5-6.

Aliev, L.V., B.A. Galaka, A.P. Fedorenko and L.S. Shevchenko, 1972. O vliyanii yadokhimikatov na razmnozhenie zaitseov-rusakov [On the effect of toxic chemicals on the reproduction of European hares]. *Vestn. Zoologii*, No. 2, pp. 58-61.

Alina, A.V. and N.F. Reimers, 1975. Nazemnye mlekopitayushchie (Mammalia) Ayanskogo poberezh'ya Okhotskogo morya [Land mammals (Mammalia) of the Ayan coast of the Sea of Okhotsk]. *Tr. In-ta Biol. In-t SO AN SSSR*, Vyp. 23, pp. 127-140.

Amanguliev, A. and M. Sapargel'dyev, 1970. K izuchenii fauny kleshchei krasnotelok ryzhevatoi pishchukhi [On the study of spinning mites of Afghan pika]. In: *XVIII Nauch Konf. Prof.-Prepod. Sostava TGU im. Gor'kogo, Posvyashch. 100-letiyu so Dnya Rozhdeniya V. I. Lenina: Tez. Dokl.*, Askhabad, p. 238.

Andrushko, A.M. 1952. Gryzuny vostochnoi chasti Karagandinskoi Oblasti (Kazakhstan) i vliyanie ikh na rastitel'nost' pastbishch [Rodents of the eastern part of Karaganda region (Kazakhstan) and their influence on the vegetation of pastures]. *Uch. Zap. LGU*, No. 141, Vyp. 28, pp. 45-109.

Angermann, R. 1966. Beitrage zur Kenntnis de Gattung *Lepus*. 2. Der taxonomische Status von *Lepus brachyurus* Temminck und *Lepus mandshuricus* Radde. *Mitt. Zool. Mus. Berlin*, Bd. 42, No. 2, pp. 321-335.

Angermann, R. 1967. Beitrage zur Kenntnis de Gattung *Lepus* (Lagomorpha, Leporidae). 3. Zur Variabilitat palaearkitscher Schneehasen. *Ibid.*, Bd. 43, No.1, pp. 161-178.

Angermann, R. 1983. The taxonomy of Old World *Lepus*. *Acta zool. fenn.*, No. 174, pp. 17-21.

Argyropulo, A.I. 1932. Materialy po faune gryzunov Srednei Azii. 2. Pishchukhi (*Ochotona* Link) Srednei Azii i Kazakhstana [Materials on the rodent fauna of Soviet Central Asia. 2. Pikas (*Ochotona* Link) of Soviet Central Asia and Kazakhstan]. *Tr. Zool. In-ta Akad. Nauk SSSR*, Vol. 1, pp. 31-57.

Argyropulo, A.I. 1935. Zametki o zveryakh severo-vostochnoi Mongolii po sboram Mongol'skoi ekspeditsii Akad. Nauk SSSR 1928g. [Notes on the animals of northeastern Mongolia from the collections of the Mongolian expedition of the Academy of Sciences of the USSR in 1928]. *Tr. Azerb. Mikrobiol. In-ta*, Vol. 5, Vyp. 1, pp. 245-264.

Argyropulo, A.I. 1948. Obzor retsentnykh vidov sem. Legomyidae Lilijeb. 1886 (Lagomorpha, Mammalia) [Review of recent species of the family

Lagomyidae Lilijeb. 1886 (Lagomorpha, Mammalia)]. *Tr. Zool. In-ta Akad. Nauk SSSR*, Vol. 7, pp. 124-128.

Arkhipchuk, V.A. 1983a. Metodicheskie Ukazaniya po Razvedeniyu Zaitsa-rusaka v Usloviyakh Pitomnika [Methodological Notes on Breeding European Hare in Nursery Conditions]. Moscow, 65 pp.

Arkhipchuk, V.A.* Razmnozhenie i postnatal'noe razvitie zaitsa-rusaka [Reproduction and postnatal development of European hare]. Diss. Kand. Biol. Nauk, Kiev.

Arkhipchuk, V. and V. Gruzdev, 1986. Dikie kroliki na Ukraine [Wild rabbits in the Ukraine]. *Okhota i Okhotnich'e Khoz-vo*, No. 12, pp. 8-10.

Aspisov, D.A. 1936. Zayats-belyak [Blue hare]. In: *Raboty Volzhsko-Kamskoi Okhotopromyslovoi Biostantsii, Kazan*, Vyp. 4, pp. 3-180.

Averin, Yu.V. 1948. Nazemnye pozvonochnye vostochnoi Kamchatki [Terrestrial vertebrates of eastern Kamchatka]. *Tr. Kronotskgo Gos. Zapovednika*, Vyp. 1.

Babaev, Ya. and M. Sapargel'dyev. 1970. Gel'mintofauna ryzhevatoi pishchukhi i nekotorye ee osobennosti v razlichnykh raionakh Kopetdaga [Helminth fauna of Afghan pika and some of its peculiarities in different areas of Kopetdag]. *Izv. Akad. Nauk TSSR, Ser. Biol. Nauk*, No. 1, pp. 58-65.

Baitanaev, O.A. 1974. Zayats-tolai (*Lepus tolai* Pall.) v tugainom ochage tulyaremii doliny r. Chu Dzhambul'skoi oblasti [Tolai hare (*Lepus tolai* Pall.) in the tugai tularemia focus of the Chu valley of Dzhambulsk region]. *Izv. Akad. Nauk KazSSR, Sere Biol.*, No. 5, pp. 49-54.

Bakeev, N.N. 1973. Kamennaya kunitsa: Kavkaz [Stone marten: Caucasus]. In: *Sobol', Kunitsy Kharza*, Nauka, Moscow, pp. 213-219.

Bakeev, Yu.N. and N.N. Bakeev, 1973. Lesnaya kunitsa: Ural i Zapadnaya Sibir' [Forest marten: the Urals and western Siberia]. ibid., pp. 172-186.

Bakhaeva, A.V. 1960. Fauna kleshchei Ixoidea Zapadnoi Turkmenii [Ixoidea ticks fauna of Western Turkmenia]. In: *Voprosy Prirodnoi Ochagovosti i Epizotologii Chuma v Turkmenii*, Ashkhabad, pp. 349-356.

Banfield, A.W.F. 1974. *The mammals of Canada*. Univ. Toronto press: Toronto.

Bannikov, A.G. 1954. Mlekopitayushchie Mongol'skoi Respubiliki [Mammals of the Mongolian Republic]. Moscow, 670 pp. (*Tr. Mongol. Komis.*, Vyp. 53).

Barabash-Nikiforov, I.I. 1957. Zveri Yugo-vostochnoi Chasti Chernozemnogo Tsentra [Animals of Southeastern Part of Chernozem Center]. Kn. Izd-vo, Voronezh.

Baranov, P.V. 1984. Nekotorye materialy k ekologii pishchukhi Severnogo Khenteya [Some materials on the ecology of pika of northern Khentei]. In: *Melkie Mlekopitayushchie Zapovednykh Territorii*, Moscow, pp. 79-83.

Bazhanov, V.S. 1955. K voprosu istoricheskogo izmeneniya ekologii nekotorykh pishchukh [On the question of evolutionary changes in the ecology of some pikas]. *Tr. In-ta Zoologii Akad. Nauk KazSSR*, Vyp. 4, pp. 83-96.

Bedak, A.L. 1940. Zayats-belyak v Zapadnoi Sibiri [Blue hare in western Siberia]. *Tr. Biol. In-ta Tom. Un-ta*, Vol. 6, pp. 62-79.

Belkin, V.V. 1984. Opredelenie vozrasta zaitsa-belyaka i analiz vozrastnoi struktury ego populyatsii [Determination of the age of blue hare and analysis

---

*In Russian original, year of citation missing, presumably 1983b—Translator.

of age structure of its population.] In: *Registriruyushchie Struktury i Opredelenie Vozrasta Mlekopitayushchikh: Tez. Dokl. Vsesoyuzn. Konf.*, Moscow, p. 3.

Bell, D.J. 1980. Social olfaction in lagomorphs. *Olfactions Mammals Prov. Symp.*, London, Nov. 24-25, 1978. London. New York, pp. 141-164.

Belova, O.P. 1987. Ekologiya i povedenie zaitsa-rusaka v nevole i pri reintroduktsii [Ecology and behavior of European hare in captivity and on reintroduction]. Diss. ... Kand. Biol. Nauk, Kaunes, 265 pp.

Belyaev, A.M. 1933. Spisok Gryzunov Kazakhstana [Checklist of Rodents of Kazakhstan]. Alma-Ata, 72 pp.

Belyaev, V.G. 1968. Severnaya pishchuka (*Ochotona alpina* Pallas) v Magadanskoi oblasti [Northern pika (*Ochotona alpina* Pallas) in Magadan region]. *Izv. Irkutsk. N.-I. Protivochum. In-ta*, Vol. 27, pp. 60-68.

Belyaev, G.M. and G.M. Soldatov. 1969. Ob ekologii zaitsev i ikh parazitofauna v Primorskom krae [On the ecology of hares and their parasitic fauna in Primorsk territory]. *Dokl. Irkut. Protivochum. In-ta*, Vyp. 8, pp. 280-282.

Beme, L.B. 1952. K poznaniyu gryzunov Kazakhstoi, skladchatoi strany [On the rodents of Kazakh folding area]. *Byull. MOIP, Otd. Biol.*, Vol. 57, Vyp. 2, pp. 45-53.

Berger, N.M. 194. Opyt introduktsii zaitsa-rusaka v Zapadnoi Sibiri [The experiment of introducing European hare in western Siberia]. *Uchen. Zap. Novosib. Gos. Ped. In-ta*, Vyp. 5, pp. 3-18.

Berman, D.I., I.F. Kuzmin and L.G. Tikhomirova. 1966. Royushchaya deyatel'nost' zhivotnykh na ravninnykh past-byshchakh yugo-vostochnoi Tuvy [Digging activity of animals in plain pastures of southeastern Tuva]. *Vopr. Geografii*, Vol. 69, pp. 60-75.

Bernshtein, A.D. 1963. Materialy po ekologii krasnoi pishchukhi (*Ochotona rutila* Sev.): Obraz zhizni i pitanie [Materials on the ecology of red pika (*Ochotona rutila* Sev.): Mode of life and feeding]. *Byull. MOIP, Otd. Biol.*, Vol. 68, Vyp. 4, pp. 24-36.

Bernshtein, A.D. 1964. Razmnozhenie krasnoi pishchukhi (*Ochotona rutila* Sev.) v Zailiiskom Alatau [Reproduction of red pika (*Ochotona rutila* Sev.) in Trans-Ili Alatau]. ibid., Vol. 69, Vyp. 3, pp. 40-48.

Bernshtein, A.D. 1970. Ekologiya bolsheukhoi pishchuki (*Ochotona roylei macrotis*) [Ecology of long-eared pika (*Ochotona roylei macrotis*)]. *Fauna i Ekologiya Gryzunov*, Vyp. 9, pp. 62-109.

Bernshtein, A.D. and G.A. Klevezal'. 1965. Opredelenie vozrasta krasnoi i bolsheukhoi pishchukhi. [Determination of age of red and long-eared pikas]. *Zool. Zhurn.*, Vol. 44, No. 5, pp. 787-789.

Bibikov, D.I. and I.I. Stogov. 1963. Materialy po mlekopitayushchim Chingiztau [Materials on mammals of Chingiztau]. *Byull. MOIP, Otd. Biol.* 68, Vyp. 4, pp. 14-23.

Boag, B. 1986. Observations on the localized distribution of wild rabbit (*Oryctolagus cuniculus*) with non-agouti coat colouring. *J. Zool. A.*, Vol. 210, No. 4, pp. 640-642.

Boifani, L. and C.P. Lieckfeld. 1989. Kaninchen: Fruchtbar und Fruchtsam. *Natur*, No. 4, pp. 54-61.

Bondarenko, A.A., V.P. Klimov and Yu.M. Astashin, 1971. Novye dannye po epizootologii chumy v Gornom Altae [New data on epizootic of plague in Gornyi Altai]. *Dokl. Irkut. Protivochum. In-ta*, Vyp. 9, pp. 22-24.

Bondar', E.P. 1956. Materialy po mlekopitayushchim Betpak-Daly i yugo-zapadnoi chasti Kazakhskogo nagor'ya [Materials on the mammals of Betpak-Dala and southwestern part of Kazakh upland]. *Tr. Sredneaz. N. I. Protivochum. In-ta*, Vyp. 3, pp. 107-121.

Bondar', E.P. and I.V. Zhernovov, 1960. Ekologo-faunisticheskii ocherk gryzunov Zapadnoi Turkmenii [Ecological and faunistic outline of rodents of western Turkmenia]. In: *Voprosy Prirodnoi Ochagovosti i Epizootologii Chumy v Turkmenii*, Ashkabad, pp. 211-391.

Boyarkin, I.V. 1984. Sravnitel'naya Ekologiya Zaitseobraznykh i Gryzunov Zapadnoi Chasti Tuvinskoi ASSR [Comparative Ecology of Lagomorphs and Rodents in Western Part of Tuva ASSR]. Izd-vo Irkut. In-ta, Irkutsk, 170 pp.

Boyd, I.L. 1985. Investment in growth by pregnant wild rabbits in relation to litter size and sex of the offspring. *J. Anim. Ecol.*, Vol. 54, p. 147.

Broekhuizen, S. 1979. Survival in adult European hares. *Acta Theriol.*, Vol. 24, No. 32/36, pp. 465-474.

Broekhuizen, S. and F. Maaskamp. 1981. Annual production of young in European hares (*Lepus europaeus*) in the Netherlands. *J. Zool.*, Vol. 193, No. 4, pp. 499-516.

Brom, I.O. 1952. Pitanie korsaka v Zabaikal'e [Feeding of korsak in Trans-Baikal]. *Izv. Irkut. N. I. Protivochum. In-ta*, Vol. 10, pp. 19-21.

Brom, I.P. 1954. Materialy po biologii zabaikel'skogo khor'ka [Materials on the biology of the Trans-Baikal pole cat). ibid., Vol. 12, pp. 27-29.

Bromlei, G.F. 1964. O printsipakh okhrany nazemnykh mleko-pitayushchikh na yuge Dal'nego Vostoka [On the principles of conservation of land mammals in the southern Far East). In: *Okhrana Prirody na Dal'nom Vostoke*, Vladivostok, No. 3, pp. 95-102.

Buslaeva, N.N. and A.K. Fedosenko. 1964. Blokhi, paraziti-ruyushchie na melkikh mlekopitayushchikh v vysoko-gor'yakh Zailiiskogo Alatau [Fleas parasitizing small mammals in Trans-Ili Altai high-mountains). *Tr. In-ta Zoologii Akad. Nauk. KazSSR*, Vol. 22, pp. 177-183.

Chernikin, E. 1965. Zayat-belyak na Kamchatke [The European hare in Kamchatka]. *Okhota i Okhotnich'e Khoz-va*, No. 10, pp. 23-24.

Chernolikh, L.N. 1973. Mlekopitayushchie Khinganskogo zapovednika [Mammals of the Khingan Preserve]. In: *Voprosy Geografii Dal'nego Vostoka. Ser. Zoogeogr.*, Vol. 2, pp. 125-136.

Chernyavskii, F.B. 1984. Mlekopitayushchie Krainego Severo-vostoka Sibiri [Mammals of the Extreme Northeast of Siberia). Nauka, Moscow, 385 pp.

Chernyavskii, F.B., N.E. Dokuchaev and G.E. Korolenko. 1978. Mlekopitayushchie srednego techeniya reki Omolon [Mammals of the middle reaches of the Omolon River]. In: *Fauna i Zoogeografiya Mlekopitayushchikh Severo-Vostoka Sibirii*, Vladivostok, pp. 26-65.

Chugunov, Yu.D. 1961. Mongol'skaya-pishchukha v Gobi-Altae [Mongolian pika in Gobi Altai]. *Byull. MOIP, Otd. Biol.*, Vol. 60, Vyp. 3, p. 173.

Corbet, G.B. 1978. *The Mammals of the Palaearctic Region. A Taxonomic Review*. Brit. Mus., London, Ithaca, 314 pp.

Cowan, D.P. 1987. Aspects of the social organization of the European wild rabbit (*Oryctolagus cuniculus*). *Ethology*, Vol. 75, No. 3, pp. 197-210.

Danzan, G. 1978. K izucheniyu fauny Yel'mintov (Cestoda, Acanthocephala) zaitseobraznykh i gryzunov MNR [On the study of helminth fauna (Cestoda, Acanthocephala) of lagomorphs and rodents of M[ongolian] P.[eoples] R[epublic]. In: *Epidemiologiya i Profilaktika Osobo Opasnykh Infektsii v MNR i SSSR*, Ulan-Bator, 169 pp.

Darskaya, N.F. 1957. Blokhi daurskoi pishchukhi (*Ochotona daurica* Pall.) [Flexs of Daurian pika (*Ochotona daurica* Pall.)]. In: *Fauna i Ekologiya Gryzunov*, Izdvo MGU, Moscow, pp. 163-170.

Davydov, G.S. 1972. Razmnozhenie predstavitelei otryada zaitseobraznykh (Lagomorpha, Mammalia) [Reproduction of members of the order of lagomorphs (Lagomorpha, Mammalia]. In: *Voprosy Zoologii Tadzhikistana*, Dushanbe, pp. 180-191.

Davydov, G.S. 1974. Fauna Tadzhikskoi SSR T-20, Ch. 1, Mlekopitayushchie (Zaitseobraznye, susliki, surki) [Fauna of Tadzhik SSR, Vol. 20, Pt. 1. Mammals (lagomorphs, susliks and marmots)]. Donish, Dushanbe, pp. 69-257.

Dawson, M.R. 1967. Lagomorph history and the stratigraphic record. In: *Essays Paleontol. and Stratigr. Raymond Moore Comm. Vol.* pp. 287-316.

Demin, E.P. 1960. Gryzuny khrebtov Sailyugem i yuzhnoi chasti Chikhacheva [Rodents of the Sailyugem Ranges and southern part of Chikhachevo]. *Izv. Irkut. N. I. Protivochum. In-ta*, Vol. 23, pp. 206-213.

Demin, E.P. 1962. K biologii mongol'skoi pishchukhi [On the biology of Mongolian pika]. ibid., Vol. 24, pp. 303-307.

Demina, G.I., E.P. Demin and Z.I. Shchekunova, 1961. Epizootiya chumy na pishchukhakh v Zapadnoi Mongolii [Epizootic of plague in pikas in western Mongolia]. *Dokl. Irkut. N. I. Protivochum. In-ta*, Vyp. 2, pp. 31-39.

Derevshchikov, A.G. 1971. O rasprostranenii mlekopitayushchikh i ptits v verkhov'yakh basseina reki Buguzun [On the distribution of mammals and birds in the upper part of the Buguzun River basin]. In *Priroda i Prirodnye Resursy Gornogo Altaya*, Gorno-Altaisk, pp. 308-310.

Derevshchikov, A.G. 1975. Osobennosti raspredeleniya i nekotorye prichiny depressii chislennosti pishchukh v yugovostochnom Altae [Peculiarities of distribution and some causes for depression of population of pikas in southern Altai]. In: *Mezhdunarodnye i Natsional'nye Aspekty Epidnadzora pri Chume*, Irkutsk, Pt. 2.

Derevyanchenko, K.I. and K.A. Zheldakova, 1958. K roli iksodovykh kleshchei v prirodnom ochage tulyaremii v del'te Volgi [On the role of ixodid ticks in the natural focus of tularemia in the Volga delta]. *Tr. Astrakh. Protivochum Stantsii*, Vyp. 2, pp. 301-311.

Derviz, D.G. 1982. Organizatsiya poselenii pishchukh [Organization of pika habitation]. In: *III S'ezd Vsesoyuz. Teriol. O-va*, Moscow, Vol. 1, pp. 183-184.

Derviz, D.G., N.V. Derviz and A.V. Chabovskii, 1983. Vliyanie uslovii obitaniya na povedenie mongol'skoi i daurskoi pishchukh [Effect of habitat conditions

376

on the behavior of Mongolian and Daurian pikas]. In: *Povedenie Zhivotnykh v Soobshchestve*, Nauka, Moscow, Vol. 2, pp. 236-239.

Derviz, D.G. and N.S. Proskurina, 1983. Vzaimootnoshenie mongol'skoi i daurskoi pishchukh v sovmestnom poselenii [Interrelationship of Mongolian and Daurian pikas in combined habitation]. ibid., Vol. 2, pp. 159-161.

Dmitriev, P.P. 1985. Svyaz'mezhdu kustarnikami mongol'skoi stepp i koloniyami mlekopitayushchikh [The relation between shrubs of Mongolian steppe and colonies of mammals). *Zhurn. Obshchei Biologii*, Vol. 46, pp. 661-669.

Dmitriev, P.P. 1991. Rastitel'nost' nor daurskoi pishchukhi i ee znachenie v razvitii stepnykh ekosistem [Vegetation of burrows of Daurian pika and its significance in the development of steppe ecosystems]. In: *Ekologiya Pishchukh Fauny SSSR*, Nauka, Moscow, pp. 5-13.

Dmitriev, P.P. and R.P. Guricheva, 1978. Melkie mlekopita-yuschie v pastbishchakh biotsenozakh Vostochnogo Khan'jaya [Smaller mammals in pasture habitats of eastern Khangai]. In: *Geografiya i Dinamika Rastitel'nogo i Zhivotnogo Mira MNR*, Nauka, Moscow, pp. 124-131.

Dubinin, V.B. and M.N. Dubinina, 1951. Parazitofauna mlekopitayushchikh daurskoi stepi [Parasitic fauna of mammals of Daurian steppe]. In: *Fauna i Ekologiya Gryzunov*, Izd-vo MOIP, Moscow, pp. 98-156 (Materialy po Gryzunam, Vyp. 4).

Dubinin, I.B. and N.G. Bregetova, 1952. Paraziticheskie krovososushchie kleshchi pozvonochnykh zhivotnykh Turkmenia [Parasitic blood-sucking ticks of vertebrate animals of Turkmenia]. *Tr. ZIN Akad. Nauk SSSR*, Vol. 10, pp. 45-60.

Dubrovskii, Yu.A. 1959. Zimnie zapasy stepnoi pishchukhi v Aktyubinskikh stepyakh [Winter reserves of steppe pika in Aktyubinsk steppes]. *Tr. In-ta Zoologii Akad. Nauk KazSSR*, Vol. 10, pp. 254-257.

Dubrovskii, Yu.A. 1963. Rasprostranenie stepnoi pishchukhi i nekotorye cherty ee ekologii na zapade Kazakhstana [Distribution of steppe pika and some features of its ecology in the west of Kazakhstan]. *Byull. MOIP. Otd. Biol.*, Vol. 68, Vyp. 4, pp. 44-49.

Dul'keit, G.D. 1956. Man'chzhurskii zayats v Ussuriiskom krae [Manchurian hare in the Ussuri Territory]. *Zool. Zhurn.*, Vol. 35, No. 6, pp. 916-921.

Dul'keit, G.D. 1964. Okhotnich'ya Fauna, Voprosy i Metody Otsenki Proizvoditel'nosti Okhotnich'ikh Ugodii Altaisko-Sayanskoi Gornoi Strany [Game Fauna Problems and Methods of Evaluating Productivity of Game Areas of Altai-Sayan Montane Countryside]. Krasnoyarsk, 352 pp. (*Tr. Gos. Zapovednika "Stolby"*, Vyp. 4).

Dunaeva, T.N. 1979. Sistematicheskii obzor mlekopitayushchikh kak nositelei boleznei: Semeistvo Leporidae Gray, 1821—zayach'i [Systematic review of mammals as disease carriers: Family Leporidae Gray, 1821—hares]. In: *Meditsinskaya Teriologiya*, Nauka, Moscow, pp. 50-68.

Dvoryadkin, A.V. and G.D. Dzyuba, 1988. Sostayanie populyatsii zaitsa-belyaka Yuzhnogo Pribaikal'ya [The state of blue hare population in southern Cis-Baikal]. In: *Problemy Ekologii Pribaikal'ya. Tez. Dokl. III. Vsesoyuzn. Nauch. Konf.*, Irkutsk, Pt. 4, 90 pp.

Dzhoiloev, A. 1959. K ekologii bolsheukhoi pishchukhi [On the ecology of long-eared pika]. *Tr. Sredneaz. N. I. Protivochum. In-ta*, Vyp. 6.

Egorov, O.V. 1965. Dikie Kopytnye Yakutii [Wild Ungulates of Yakutia]. Nauka, Moscow, 259 pp.

Egorov, O.V. and Yu.V. Labutin, 1964. Materialy po pitaniyu krupnykh khishchnykh mlekopitayushchikh Verkhoyan'ya [Materials on feeding of large predatory mammals of the upper Yana River area]. In: *Pozvonochnye Zhivotnye Yakutia, Yakutsk*, pp. 51-59.

Eigelis, Yu.K. 1980. Gryzuny Vostochnogo Zakavkaz'ya i Problema Mestnykh Ochagov Chumy [Rodents of Eastern Transcaucasia and the Problem of Local Plague Foci]. Saratov, 262 pp.

Ellerman, J.R. and T.C.S. Morrison-Scott, 1951. *Checklist of Palaearctic and Indian Mammals, 1758 to 1946*. London, 810 pp.

Erbaeva, M.A. 1988. Pishchukhi Kainozoya [Cenozoic Pikas]. Nauka, Moscow, 222 pp.

Eshelkin, I.I. 1978. Gnezdovye vzaimootnosheniya ploskocherepnoi polevki (*Alticola strezovi*) i mongol'skoi pishchukhi (*Ochotona pricei*) v gorno-altaiskom chumnom ochage [Nesting interactions of flat-skulled vole (*Alticola strelzovi*) and Mongolian pika (*Ochotona pricei*) in Gornyi Altai plague focus]. *Zool. Zhurn.*, Vol. 57, No. 9, pp. 1403-1408.

Eshelkin, I.I., B.D. Lazarev and S.M. Prutov, 1968. Materialy po razmnozheniyu gryzunov-chumonositelei v Gornom Altae [Materials on multiplication of rodents—carriers of plague in Gornyi Altai]. *Izv. Irkut. N. I. Protivochun. In-ta*, Vol. 27, pp. 7-12.

Eshelkin, I.I. and S.M. Prutov*, 1971. Prichiny zmeneniya chislennosti mongol'skoi pishchukhi v gorno-altaiskom ochage chumy [Reasons for change in population of Mongolian pika in Gornyi Altai plague focus]. *Dokl. Irkut. N. I. Protivochum In-ta*, Vyp. 9, pp. 196-197.

Fadeev, V.A. 1964. Razmnozhenie zaitsa-rusaka v zapadnom Kazakhstane [Reproduction of European hare in western Kazakhstan]. *Tr. In-ta Zoologii Akad. Nauk KazSSR*, Vol. 23, pp. 150-168.

Fadeev, V.V. 1966a. Osobennosti rasprostraneniya zaitsarusaka i peschanika v kazakhstane [Peculiarities of distribution of European and tolai hare in Kazakhstan]. ibid., Vol. 26.

Fadeev, V.A. 1966b. Pitanie i sutochnaya aktivnost' Zaitsa-rusaka v Zapadnom Kazakhstane [Feeding and daily activity of European hare in western Kazakhstan]. ibid., Vol. 26.

Fedorov, K.P. and A.F. Potankina. 1975. Gel'minty pishchukh (Ochotonidae) yuga Zapadnoi Sibiri [Helminths of pika (Ochotonidae) in south of western Siberia]. In: *Sistematika, Fauna, Zoogeografiya Mlekopitayu-shchikh i Ikh Parazitov*, Nauka, Novosibirsk, pp. 203-211.

Fedosenko, A.K. 1974. O nekotorykh morfologicheskikh osobennostyakh pishchukh (*Ochotona*) [On some morphological peculiarities of pikas (*Ochotona*)]. *Zool. Zhurn.*, Vol. 53, No. 3, pp. 485-486.

*In Russian original, Purtov—General Editor.

378

Feng Zuojian ana Zheng Changlin, 1985. Studies on the pikas (genus *Ochotona*) of China—Taxonomic notes and distribution. *Acta Theriol. Sin.*, Vol. 5, No. 4, pp. 269-289.

Fetisov, A.S. 1935. Biologicheskie nablyudeniya za Zabaikal'skim zaitsem-tolaem *Lepus tolai* Pall. 1835* [Biological observations on Trans-Baikal tolai hare *Lepus tolai* Pall. 1835]. *Izv. Vost. Sib. S.-Kh. In-ta*, Vyp. 1, pp. 138-148.

Fetisov, A.S. 1936. Materialy po sistematike i geografii rasprostraneniya mlekopitayushchikh Zapadnogo Zabaikal'ya [Materials on the systematics and geography of distribution of mammals of western Trans-Baikal area]. *Izv. Irkut. N. I. Protivochum. In-ta*, Vol. 3, pp. 86-119.

Fetisov, A.S. 1940. Opredelitel' gryzunov Pribaikal'ya i Zabaikal'ya [Keys to rodents of Cis-Baikal and Trans-Baikal Area]. Obl. Izd-vo Irkutsk, 92 pp.

Fetisov, A.S. 1942. Novye issledovaniya po faune gryzunov Zapadnogo Zabaikal'ya [New studies on the rodent fauna of western Trans-Baikal area]. *Izv. Biol.-Geogr. NII pri Vost. Sib. Un-te*, Vol. 9, Vyp. 3/4, pp. 121-144.

Firstov, N.I. 1957. O rasprostranenii gryzunov na yuzhnoi granitse Altaya [On the distribution of rodents at the southern boundary of Altai]. *Izv. Irkut. N. I. Protivochum. In-ta*, Vol. 16, pp. 102-109.

Flerov, K.K. 1927. Pischukha Severnogo Urala [Pikas of the Northern Urals]. In: *Ezhegordnik zoologicheskogo Muzeya Akad. Nauk SSSR*, Moscow.

Flerov, K.K. and I.M. Gromov, 1934. Mlekopitayushchie Doliny Sumbar i Chandyra [Mammals of the Sunbar valley and Chandyr]. *Tr. Kara-Kalinskoi i Kizyl Atrekskoi Parazitol. Ekspeditsii*, Vyp.**, pp. 291-237 [sic.].

Flux, J.E. 1970. Life history of the mountain hare (*Lepus timidus scoticus*) in northeast Scotland. *J. Zool.*, No. 1.

Folitarek, S.S. 1940. Geograficheskoe rasprostranenie zaitsa-rusaka *Lepus europaeus* Pall. v SSSR [Geographical distribution of European hare *Lepus europaeus* Pall. in the USSR]. *Tr. In-ta Evoluyuts. Morfologii i Ekologii Zhivotnykh, Akad. Nauk SSSR*, Vol. 3, Vyp. 1, pp. 14-21.

Fomushkin, V.M., I.V. Zhernovov and G.D. Svidenko. 1967. O biologii ryzhevatoi pishchukhi (*Ochotona rufescens* Gray) [On the biology of Afghan pika (*Ochotona rufescens* Gray)]. In: *Ekologiya Mlekopitayushchikh i Ptists*, Nauka, Moscow, pp. 28-31.

Formozov, A.N. 1929. Mlekopitayushchie Severnoi Mongolii po Sboram Ekspeditsii 1926 g.[Mammals of Northern Mongolia from Collection of the 1926 Expedition]. Izd.-vo Akad. Nauk SSSR, Leningrad, 144 pp.

Formozov, N.A. 1981. Adaptivnost' povedeniya pishchukh k zhizni v kamennistykh biotopakh [Adaptivity of behavior of pikas to life in stony habitats]. In: *Ekologiya, Struktura Populyatsii i Vnutividovye Kommunikativnye Protsessy u Mlekopitayushkhikh* Nauka, Moscow, pp. 245-263.

Formozov, N.A. 1986. Zony simpatrii altaiskoi i severnoi pishchukh [Saympatric zones of Altai and northern pikas. In: *IV S'ezd. Vsesoyuz. Teriol. Ob-va*, Moscow, Vol. 1, pp. 146-147.

---

*In Russian original, 1935—Translator.
**Omitted in Russian original—Translator.

Formozov, N.A. and A.A. Nikol'skii. 1979. Polozhenie ural'skoi pishchukhi v gruppe *"alpina"* (Bioakusticheskii analiz) [The status of Ural pika in the *"alpina"*—group (Bioacoustic analysis)]. In: *Mlekopitayushchie Ural'skikh Gor: Inform. Materialy*, UN Ts. Akad. Nauk SSSR, Sverdlovsk, pp. 80-82.

Gafurov, A.K., G.S. Davydov and S.A. Mukhamadiev, 1971. K izucheniyu gel'mintofauny gryzunov i zaitseobraznykh Tadzhikistana [On the study of helminth fauna of rodents and lagomorphs of Tadzhikistan]. *Izv. Akad Nauk TadzhSSR, Otd. Biol. Nauk*, No. 4(45), pp. 97-99.

Gaiduk, V.E. 1970. Ekologo-morfologicheskie osobennosti Zaitsa-belyaka Belorussii [Ecological and morphological peculiarities of blue hare of Belorussia]. In *Populyatsionnaya Struktura Vida u Mlekopitayushchikh*, Moscow, 106 pp.

Gaiduk, V.E. 1973a. Zaitsy v Belorussii, [Hares in Belorussia]. *Okhota i Okhotnich'e Khoz-vo*, No. 10, pp. 12-14.

Gaiduk, V.E. 1973b. Razmnozhenie Zaitsa-rusaka (*Lapus europaeus* Pall.) v Belorussi [Reproduction of European hare (*Lepus europaeus* Pall.) in Belorussia). *Vestn. Zoologii*, No. 3.

Gaiduk, V.E. 1975. Vliyanie nekotorykh abioticheskikh faktorov nalin'ku i pigmentatsiyu volosyanogo pokrova zaitsa-belyaka (*Lepus timidus*) [Effect of some abiotic factors on the molting and pigmentation of the hair coat of blue hare (*Lepus timidus*)]. *Zool. Zhurn.*, Vol. 65, No, 3, pp. 425-431.

Gaiduk, V.E. 1982. Nekotorye aspekty ekologii simpatricheskikh blizkorodstvennykh vidov zaitseobraznykh [Some aspects of closely related sympatric species of Lagomorphs]. *Ekologiya*, No. 5, pp. 55-60.

Gaiskii, N.A. and N.D. Altareva, 1944. Daurskaya pishchukha, (*Ochotona daurica* Pall.) kak nositel' chumnoi infektsii na territorii Zabaikal'sko-Mongol'skogo enzooticheskogo ochaga [Daurian pika (*Ochotona daurica* Pall.) as a carrier of plague infection in the territory of trans-Baikal-Mongolian enzootic focus]. *Izv. Irkut. N. I. Protivochum. In-ta*, Vol. 5, pp. 135-147.

Galaka, B.A. 1969. O polovom i vozrastnom sostave populyatsii zaitsa-rusaka v stepnoi i lesostepnoi zonakh USSR [On the sex and age composition of populations of European hare in steppe and forest-steppe zones of Ukr. SSR]. In: *Izuchenie Resursov Nazemnykh Pozvonochnykh Fauny Ukrainy*, Kiev, pp. 32-35.

Galaka, B.A. 1970a. Nekotorye dannye po biologii razmno-zheniya zaitsa-rusaka *Lepus europaeus transsylvanicus* Matsch. [Some data on the reproduction biology of European hare *Lepus europaeus transsylvanicus* Matsch.]. In: *Fauna Moldavii i Ee Okhrana*, Kishinev, pp. 162-163.

Galaka, B.A. 1970b. Ob optimal'noi strukture populyatsii zaitsa-rusaka [On the optimal population structure of European hare]. In: *Populyatsionnaya Struktura Vida u Mlekoptitayushchikh*, Moscow, pp. 106-107.

Galkina, L.I. A.E. Potapkina and B.S. Yudin. 1977. Ekologo-faunisticheskii ocherk melkikh mlekopitayush-chikh (Micromammalia) yugo-vostochnoi Tuvy [Ecological and faunistic outline of small mammals (Micro-mammalia) of southeastern Tuva]. In: *Fauna i Sistematika Pozvonochnykh Sibiria*, Novosibirsk.

Galuzo, I.G. and V.F. Novinskaya. 1961. Tripanozomy zhivotnykh Kazakhstana. 2. Tripanozomy gryzunov [Trypanosoma of animals of Kazakhstan. 2.

380

Trypanosomaes of rodents]. In: *Prirodnaya Ochagovost' Boleznei i Voprosy Parazitologii*, Alma-Ata, Vyp. 3, pp. 152-172.

Garbuzov, V.K. and I.N. Shilov, 1963. Rasprostranenie stepnoi pishchukhi v Priaral'e [Distribution of steppe pika in cis-Aral area]. *Byull. MOIP, Otd. Biol.*, Vol. 68, Vyp. 4, pp. 37-43.

Gashev, N.S. 1968. O nakhodkakh severnoi pishchukhi na Urale [a localities of northern pika in the Urals]. *Byull. MOIP, Otd. Biol.*, Vol. 73, No. 4.

Gashev, N.S. 1969. Ekologiya ural'skoi severnoi pishchukhi [Ecology of northern pika of the Urals]. Authors' Abstract. Dist. ... Kand. Biol. Nauk, Sverdlovsk.

Gashev, N.S. 1971. Severnaya pishchukha (*Ochotona hyperborea* Pallas, 1811) [Northern pika (*Ochotona hyperborea* Pallas, 1811)]. In: *Mlekopitayushchie Yamala i Polyarnogo Urala*, Sverdlovsk, Vyp. 80, Vol. 1, pp. 4-71.

Gavrilyuk, V.A. 1966. O vzaimosvyazakh zhivotnogo i rastitel'nogo mira v tundrakh Chukotki [On the inter-relationship of the animal and plant world in Chukchi tundra]. *Vopr. Geografii*, Vol. 69, pp. 118-127.

Geptner [Heptner], V.G. 1933. Zaitsy [Hares]. Vneshtorgizdat, Moscow, Leningrad, 51 pp.

Gizenko, A. 1968. Dikii krolik [The wild rabbit]. *Okhota i Okhotnich'e Khoz-vo*, No. 4, pp. 17-19.

Gizenko, A.I. and L.S. Shevchenko. 1973. Perspektivy razvedeniya krolika dikogo (*Oryctolagus cuniculus* L.) na Ukraine [The prospects of breeding wild rabbit (*Oryctolagus cuniculus* L.) in the Ukraine]. *Vestn. Zoologii*, No. 4, pp. 10-15.

Grekov, V.S. 1961. K ekologii zaitsa-rusaka Kzyl-Agachskogo zapovodnika im. S.M. Kirova i prilezhashchikh raionov [On the ecology of European hare of the S.M. Kirov Kzyl-Agach preserve and adjoining areas]. In: *Pervoe Vsesoyuz. Soveshch. po. Mlekopitayushchim. Tez. Dokl.*, Izd-vo MGU, Moscow, pp. 25-26.

Grigor'ev, N.D. 1940. Materialy po razmnozheniyu zaitsev-belyakov (*Lepus timidus* L.) v Kazanskom zooparke [Materials on breeding blue hares (*Lepus timidus* L.) in the Kazan zoological park]. *Tr. O-va Estestvoispytatelei pri Kazan. Un-te*, Vol. 56, Vyp. 3/4, pp. 31-40.

Grigor'ev, N.D. 1956. Gibridy zaitsa-rusaka i zaitsa-belyaka v Kazanskom zooparke [Hybrids of European hare and blue hare in the Kazan zoological park]. *Zool. Zhurn.*, Vol. 35, No. 7.

Gromov, I.M. 1957. Verkhnechetvertichnye gryzuny Samarskoi luki i usloviya zakhoroneniya i nakopleniya ikh ostatkov [The upper Tertiary rodents of Samara bend and conditions of burial and accumulation of their remains]. *Tr. ZIN Akad. Nauk SSSR*, Vol. 22, pp. 122-150.

Grunin, K.Ya. 1962. Fauna SSSR: Nasekomye dvukrylye: Podkozhnye ovoda [Fauna of the USSR: Dipteran Insects: Warble flies]. Izd-vo Akad. Nauk SSSR, Moscow, Leningrad, Vol. 19, Vyp. 4, 238 pp.

Gruzdev, V.V. 1968. Estestvennoe rasselenie zaitsa-rusaka na sever [Natural dispersal of European hare in the north]. *Vestn. MGU Biologiya*, No. 4, pp. 19-24.

Gruzdev, V.V. 1974. Ekologiya zaitsa-rusaka [Ecology of European Hare]. Izd-vo MGU, Moscow, 162 pp. (*Materialy k Poznaniyu Fauny i Flory SSSR*, Novosibirsk, Otd. Zool., Vyp. 48).

Gruzdev, V.V. 1981. Zaitsy-tumaki [Tumak* Hares]. *Okhota i Okhotanich'e Khoz-vo*, No. 9, pp. 24-25.

Gruzdev, V.V. and V.I. Osmolovskaya, 1969. Struktura areala zaitsa-belyaka v evropeiskoi chasti SSSR [Structure of the range of blue hare in the European part of the USSR]. *Biol. Nauki*, No. 11, pp. 27-30.

Gubanov, N.M. 1964. Gel'mintofauna Promyslovykh Mlekopi-tayushchikh Yakutii [Helminth Fauna of Commercial Mammals of Yakutia]. Nauka, Moscow, 154 pp.

Gureev, A.A. 1964. Fauna SSSR: Mlekopitayushchie: Zaitse-obraznye (Lagomorpha) [Fauna of the USSR: Mammals, Lagomorphs (Lagomorpha)]. Leningrad, Vol. 3, Vyp. 10, 275 pp.

Gustavsson, I. 1971. Mitotic and meiotic chromosomes of the variable hare (*Lepus timidus*), the common hare (*Lepus europaeus*) and their hybrids. *Hereditas*, Vol. 67, pp. 27-34.

Guzhevnikov, I.A. and N.S. Tarasov, 1968. Opyt bor'by s daurskoi pishchukhi [Experience of control of Daurian pika]. *Izv. Irkut. N. I. Protivochum. In-ta*, Vol. 27, pp. 413-427.

Gvozdev, E.V. 1948a. K voprosu o vidovom sostave koktsidii zaitsev-peschanikov [On the question of the species composition of coccids in Tolai hares]. *Izv. Akad. Nauk Kaz.SSR, Ser. Parazitol.*, Vyp. 5, pp. 39-41.

Gvozdev, E.V. 1984b. Novyi vid tsestody *Drepanidotaenia fragmentata* sp. nova u samtsa zaitsa-peschanka *Lepus tibetanus* Waterh. [A new cestod species *Drepanidotaenia fragmentata* sp. nova in male cape hare *Lepus tibetanus* Waterh.]. ibid., Vyp. 5, pp. 48-52.

Gvozdev, E.V. 1949. Parazitofauna zaitsa-peschanika *Lepus tibetanus* Waterh. [Parasitic fauna of cape hare *Lepus tibetanus* Waterh.]. ibid., Vyp. 7, pp. 49-54.

Gvozdev, E.V. 1962. Analiz gel'mintofauny pishchukh (Ochotonidae) v svyazi s geograficheskim rasprostraneniem khozyev [Analysis of the helminth fauna of pikas (Ochotonidae) in relation to the geographic distribution of hosts]. *Tr. In-ta Zoologii Akad. Nauk KazSSR*, Vol. 16, pp. 35-41.

Gvozdev, E.V. 1964. Gel'mintologicheskaya otsenka zapadno-kazakhstanskogo zaitsa-rusaka kak ob"ekta akkli-matizatii [Helminthological evaluation of European hare of western Kazakhstan as an object of acclimatization]. *Izv. Akad. Nauk KazSSR, Ser. Biol.*, Vyp. 2, pp. 75-84.

Haga, R. 1960. Observations on the ecology of the Japanese pika. *J. Mammal*, Vol. 41, No. 2, pp. 200-212.

Hall, E.R. 1981. *The Mammals of North America*, N.Y. Wiley, Vol. 1, 90 pp.

Hell, P. and A. Bakos. 1970. Sucasny stav rozserenia a produkcie diveho kralika (*Oryctolagus cuniculus* L., 1750) na Slovensku. *Acta Zootechn. Nitra*, No. 21, pp. 185-199.

Homolka, M. 1985. Spatial activity of hare (*Lepus europaeus*). *Folia Zool.*, Vol. 34, No. 3, pp. 217-226.

Homolka, M. 1986. Daily activity patterns of the European hare (*Lepus europaeus*). ibid., Vol. 35, No. 1, pp. 33-42.

---

*Common name could not be confirmed—Translator.

382

Homolka, M. 1988. Diet of the wild rabbit (*Oryctolaus cuniculus*) in an agrocenosis. ibid., Vol. 37, No. 2.

Honacki, J.N., K.E. Kinman and J.W. Koeppl. 1982. *Mammal Species of the World*. Allen: Laurence, 694 pp.

Hsu, J.C. and K. Benirschke, 1967. *Atlas Mammal. Chrom.* Vol. 1, No. 8.

Ioff, I.G. and O.N. Skalon, 1954. Opredelitel' Blokh Vostochnoi Sibirii, Dal'nego Vostoka i Prilezhash-chikh Raionov [Keys to Fleas of Eastern Siberia, Far East and Bordering Regions]. Medgiz, Moscow, 275 pp.

Ishadov, N. 1974. Zimnee razmnozhenie i chislennost' zaitsa-tolaya v Karakumakh [Winter reproduction and population of tolai hare in the Karakums]. *Ekologiya*, No. 6, pp. 94-96.

Ishunin, G.I. and T.A. Pavlenko, 1966. Materialy po ekologii zhivotnykh pastbishch Kyzylkuma [Materials on the ecology of animals of Kyzylkum pastures]. In: *Pozvonochnye Zhivotnye Srednei Azii*, Fan, Tashkent, pp. 28-66.

Ismagilov, M.I. 1961. Ekologiya gryzunov Betpak-Daiy i Yuzhnogo Pribalkhash'ya [Ecology of rodents of Betpak-Dala and southern Lake Balkhash area]. Alma-Ata, pp. 61-68.

Ivanter, E.V. 1969. Zayats-belyak v Karel'skoi ASSR [Blue hare in Karelian ASSR]. In: *Voprosy Ekologii Zhivotnykh*, Petrozavodsk.

Kapitonov, V.I. 1961. Ekologicheskie nablyudeniya nad pishchukhoi (*Ochotona hyperborea* Pall.) v nizovy'akh Leny [Ecological observations on northern] pika (*Ochotona hyperborea* Pall.) in the lower reaches of the Lena River]. *Zool. Zhurn.*, Vol. 40, No. 6, pp. 922-933.

Karadash, A.I., B.I. Peshkov and S.A. Khamaganov. 1991. K prognozirovaniyu chislennosti daurskoi pishchukhi v Zabaikal'skom prirodnom ochage chumy [On prognostication of the number of Daurian pika in Trans-Baikal natural focus of plague]. In *Ekologiya Pishchukh Fauny SSSR*, Nauka, Moscow, pp. 18-20.

Kawamichi*, T. 1971. Daily activities and social pattern of two Himalayan pikas, *Ochotona macrotis* and *O. roylei*, observed at Mt. Everest. *J. Fac. Sci. Hokkaido Univ., Ser. VI*, Vol. 17, No. 4, pp. 587-609.

Kazarinov, A.P. 1973. Zametki o faune basseina r. Mai [Notes on the fauna of Mai River basin]. In: *Voprosy Geografii Dal'nego Vostoka: Zoogeografiya*, Khabarovsk, Vol. 2, pp. 141-149.

Khlebnikov, A.I. and I.P. Khlebnikova, 1972. Kharakterny'e cherty povedeniya severnoi pishchukhi v raznykh tipakh poselenii [Characteristic features of behavior of northern pika in different types of colonies]. In: *Povedenie Zhivotnykh. Ekologicheskie i Evolyutsionnye Aspekty*, Moscow, pp. 215-217.

Khlebnikov, A.I. and I.P. Khlebnikova, 1991. Chislennost' i metodika ucheta *Ochotona alpina* Pall. v usloviyakh Zapadnogo Sayana [Population and the method of census of *Ochotona alpina* Pall. in conditions of western Sayan]. In: *Ekologiya Pishchukh Fauny SSSR*, Nauka, Moscow, pp. 55-54 [sic.].

Khlebnikova, I.P. 1972. Vliyanie severnoi pishchukhi v vozobnovlenie gornykh kedrovnikov Zapadnogo Sayana [The effect of northern pika on rejuvination

---

*In Russian original, Kawamlehi—Translator.

of cedar forests of western Sayan]. In: *Izuchenie Prirody Lesov Sibiri*, Krasnoyarsk, pp. 77-82.

Khlebnikova, I.P. 1974. Vozdeistvie severnoi pishchukhi na rastitel'nost' taezhnykh biogeotsenozov v uslo-viyakh Zapadnogo Sayana [The impact of northern pika on the vegetation of taiga biogeocenoses in conditions of western Sayan]. In: *Ekologiya Populyatsii Lesnykh Zhivotnykh Sibiri*, Nauka, Novosibirsk, pp. 64-76.

Khlebnikova, I.P. 1976. Kolichestvo fitomassy, potreblyaemoi, zagotavlivaemoi severnoi pishchukhoi na starykh kedrovykh garyakh Zapadnogo Sayana [The quantity of phytomass cured by northern pika in old cedar burned-out forests of Western Sayan]. *Ekologiya*, No. 2, pp . 99-102.

Khlebnikova, I.P. 1978. Severnaya pishchukha v gornykh lesakh Sibiri [Northern pika in the mountain forests of Siberia]. Nauka, Novosibirsk, 118 pp.

Khmelevskaya, N.V. 1961. O biologii altaiskoi pishchukhi [On the biology of Altai pika]. *Zool. Zhurn.*, Vol. 40, No. 10, pp. 1583-1585.

Kim, T.A. 1956. Zametki po ekologii severnoi pishchukhi Vostochonogo i Zapadnogo Sayan [Notes on the ecology of northern pika of eastern and western Sayans]. *Uch. Zap. Krasnoyar. Ped. In-ta*, Vol. 5, pp. 229-232.

Kim, T.A. 1957. K ekologii severnoi pishchukhi Kizyr-Kazyrskogo mezhdurech'ya [On the ecology of northern pika of the Kizyr-Kazyr interfluve]. ibid., vol. 10, pp. 254-256.

Kim. T.A. 1959. K ekologiya pishchukhi Vostochnogo Sayana [On the ecology of pika of eastern Sayans]. ibid., Vol. 15, pp. 207-214.

Kim, T.A. 1962. [The ecology of the northern *Ochotona* in the Sayan mountains]. Problems in ecology *Vyssaya Shkola* (Moscow) Vol. 6, pp. 71-72.

Kirikov, S.V. 1952. Ptitsy i Mlekopitayushchie v us Oviyakh Landshaftov Yuzhnoi Okonechnosti Urala [Birds and Mammals in Terrain Conditions of the Southern End of the Urals]. Izd-vo Akad. Nauk SSSR, Moscow, 412 pp.

Kirikov, S.V. 1955. Ptitsy i mlekopitayushchie Yuzhnoi okrainy Priural'ya [Birds and mammals of the southern edge of Cis-Urals]. *Tr. In-ta Geografii Akad. Nauk SSSR*, Vyp. 56, pp. 5-107.

Kirk, A. 1983. Faktory vliyayushchie na chislennosti zaitsa-rusaka v Estonii [Factors affecting the population of European hare in Estonia]. In: *Lesovedcheskie Issledovaniya*, Tallin, pp. 6-19.

Kirk, A. 1984. Opredelenie vozrasta zaitsa-rusaka po sloistnym strukturam periostal'noi zony kosti [Determination of the age of European hare from the layered structure of periosteal zone of bone]. In: *Registriruyushchie Struktury i Opredelenie Vozrasta Mlekopitayushchikh. Tez. Dokl. Vsesoyuz. Konf.*, Moscow.

Kir'yanov, G.I. 1974. Pishchukhi (*Ochotona* Link) v Altai-skom krae [Pikas (*Ochotona* Link) in Altai Territory]. In: *Teriologiya*, Nauka, Novosibirsk, Vyp. 2, pp. 71-75.

Kishchinskii, A.A. 1969. Severnaya pishchukha (*Ochotona alpina hyperborea* Pall.) v Kol'skom nagor'e [Northern pika (*Ochotona alpina hyperborea* Pall.) in the Kola Uplands]. *Byull. MOIP, Otd. Biol.*, Vol. 74, Vyp. 3, pp. 134-143.

Kokhanovskii, N.A. 1962. Mlekopitayushchie Khakasii [Mammals of Khakasia]. Abakan, 159 pp.

Kokhiya, S.S. 1974. Rasprostranenie i plotnost' nasaleniya zaitsa (*Lepus europaeus* Pall.) na Malom Kavkaze v predelakh Gruzii [Distribution and population density of hare (*Lepus europaeus* Pall.) in Malyi [little] Caucasus in the limits of Georgia]. In: *Materialy k Faune Gruzii*, Vyp. 4, pp. 360-385.

Kolosov, A.M. 1939. Zveri Yugo-Vostochnogo Altaya i smezhnoi oblasti Mongolii [Animals of southeastern Altai and adjoining regions of Mongolia]. *Uch. Zap. MGU*, Vyp. 20, pp. 158-165.

Kolosov, A.M. and N.N. Bakeeva 1947. Biologiya Zaitsa-rusaka [Biology of European Hare]. Izd-vo MOIP, Moscow, 101 pp.

Kolosov, A.M., N.P. Lavrov and S.P. Naumov. 1964. Biologiya Promyslovykh Okhotnich'ikh Zverei SSSR [Biology of Commercial Game Animals of the USSR]. Vysshaya Shkola, Moscow, 416 pp.

Koneva, I.V. 1983. Gryzuny i zaisteobraznye Sibiri i Dal'nego Vostoka (Prostranstvennaya Struktura Naseleniya) [Rodents and Lagomorphs of Siberia and Far East (Spatial Structure of Populations)]. Nauka, Novosibirsk, 216 pp.

Kontrimavichus, V.L. 1959. Gel'mintofauna zaitsev SSSR i opyt ee geograficheskogo analiza [Helminth fauna of hares of the USSR and the experience of its geographic analysis]. *Tr. GELAN*, No.9, pp. 133-144.

Korneev, O.P. 1960. Zayats-rusak na Ukraine [European hare in the Ukraine]. In: *Ekologiya i Puty Ratsional'nogo Ispol'zovaniya*, Kiev, 107 pp.

Koshkin, S.M., L.A. Lazareva, N.I. El'shanskaya and K. Khumarkan, 1978. Nekotorye itogi izucheniya fauny gamazovykh kleshchei melkikh mlekopitayushikh i ptits v ochagakh chumy Gornogo Altaya i Zapadnoi Mongolii [Some results of studying fauna of gamasid ticks of small mammals and birds in plague foci of Gorny Altai and western Mongolia]. In: *Epidemiologiya i Profilaktika Osobo Opasnykh Infektsii v MNR i SSSR*, Ulan-Bator, pp. 140-141.

Kostenko, V.A. 1976. Osobennosti biotopicheskogo raspredeleniya severnoi pishchukhi Ochotona *hyperborea* Pall. (1811) na Dal'nem Vostoke [Peculiarities of habitat distribution of northern pika (*Ochotona hyperborea* Pall. (1811) in Far East]. *Tr. Biol.-Pochv. In-ta DVNTS AN SSSR*, Vol. 37, pp. 70-74.

Kostin, V.P. 1956. Materialy po faune mlekopitayushchikh levoberezh'ya Amu-Dar'ya i Ustyurta i ocherk raspredeleniya vidov [Materials on the mammalian fauna of the left bank area of Amu-Darya and Ustyurt and an outline of species distribution]. *Tr. In-ta Zoologii i Parazitologii*, Tashkent, Vyp. 8, pp. 3-79.

Kovacs, G. and M. Ocseny, 1981. Age structure and survival of an European hare population determined by periosteal growth lines: Preliminary study. *Acta oecol.: Oecol. appl.*, Vol. 2, No. 3, pp. 241-245.

Krivosheev, V.G. 1959. K'biologii zaitsa-peschanika v Severnykh Kyzylkumakh [On the biology of tolai hare in northern Kyzylkums]. *Zool. Zhurn.*, Vol. 38, No. 1.

Krivosheev, V.G. 1964. Biofaunitsicheskie materialy po melkim mlekopitayushchim taigi Kolymskoi nizmennosti [Biofaunistic materials on small mammals of Kolyma lowland taiga]. In: *Issledovaniya po Ekologii, Dinamike Chislennosti i Boleznyam Mlekopitayushchikh: Yakutii*, Moscow.

Krivosheev, V.G. 1971. *Ochotona alpina* Pallas (1773)—altaiskaya ili severnaya pishchukha [*Ochotona alpina* Pallas (1773)—Altai and northern pika]. In: *Mlekopitayushchie Yakutii*, Nauka, Moscow, pp. 115-127.

Krivosheev, V.G. 1989. Issledovanie territorial'noi struk-tury soobshchestva melkikh rastitel'noyadnikh mlekopitayushchikh taezhno-tundrovogo landshafta Okhotsko-Kolymskogo nagor'ya [Study of territorial structure of communities of herbivorous small mammals of taiga-tundra landscapes of Okhotsk-Kolyma uplands]. *Zool. Zhurn.*, Vol. 68, No. 2.

Krivosheev, V.G. and M.V. Krivosheeva, 1991. Voprosy biologii severnoi pishchukhi [Problems in the biology of northern pika]. In: *Ekologiya Pishchukh Fauny SSSR* Nauka, Moscow, pp. 21-24.

Krylova, T.A. 1973. Vozrastnoi sostav populyatsii mongol'skoi pishchukhi (*Ochotona pricei*) [Age composition of the population of Mongolian pika (*Ochotona pricei*)]. *Zool. Zhurn.*, Vol. 52, No. 9, pp. 1422-1425.

Krylova, T.A. 1974. Razmnozhenie mongol'skoi pishchukhi v yugo-zapadnoi chasti Tuvinskoi ASSR [Multiplication of Mongolian pika in southeastern part of Tuva ASSR]. *Biol. Nauki*, No. 11, pp. 31-37.

Kryl'tsov, A.I. 1962. Topografiya lin'ki gryzunov i vozmozhnost' ispol'zovaniya ee v kachestve taksonomi-cheskogo priznaka [Topography of molting of rodents and the possibility of its use as a taxonomic character]. Tr. NII Zashchity Rastenii, Alma-Ata, Vol. 7, pp. 418-451.

Kucheruk, V.V. 1945. Znachenie razlichnykh mlekopitayu-shchikh v chumnykh epizootiyakh i v vozniknovenii lyudskikh zabolevanii v Mongol'sko-Zabaikal'skom endemichenom ochage [The significance of different mammals in plague epizootics and in the appearance of human diseases in Mongolian-Trans-Baikalian endemic focus]. ibid., Vol. 24, No. 5, pp. 309-319.

Kucheruk, V.V., N.V. Tupikova, B.P. Dobrokhotov and others. 1980. Gruppirovki naseleniya melkikh mlekopitayu-shchikh i ikh territorial'noe razmeshchenie v vostochnoi polovine MNR [Population groups of small mammals and their territorial distribution in eastern half of Mongolian Peoples Republic]. In: *Sovremennye Problemy Zoogeografii*.

Kuz'mina, I.E. 1965. Saiga i stepnaya pishchukha v verkhov'yakh Pechory [Saiga and steppe pika in the upper reaches of the Pechora River]. *Zool. Zhurn.*, Vol. 44, No. 2, pp. 307-310.

Kuznetsov, B.A. 1929. Gryzuny Vostochnogo Zabaikal'ya [Rodents of eastern Trans-Baikal area]. *Izv. Assots. N. I. In-tov*, Vol. 2, No. 1, pp. 87-89.

Kuznetsov, B.A. 1932. Gryzunov semipalatinskogo okruga [Rodents of Sempalatinsk district]. *Byull. MOIP. Otd. Biol.*, Vol. 41, Vyp. 1/2.

Kuznetsov, B.A. 1948a. Mlekopitayushchie Kazakhstana [Mammals of Kazakhstan]. Izd-vo MOIP., Moscow, 225 pp.

Kuznetsov, B.A. 1948b. Zveri Kirgizii [Animals of Kirgizia]. Izd-vo MOIP, Moscow, 209 pp.

Kuznetsov, B.A. 1965. Gryzuny [Rodents]. In: *Opredelitel' Mlekopitayushchikh SSSR*, Moscow.

Labutin, Yu.V. 1988. Osobennosti prostranstvennogo raspredeleniya zaitsa-belyaka (*Lepus timidus* L.) v Yakutii kak adaptatsiya vida k usloviyam severa

386

[Peculiarities of spatial distribution of blue hare (*Lepus timidus* L.) in Yakutia as an adaptation of the species to conditions of the north]. *Ekologiya*, No. 2, pp. 40-44.

Labutin, Yu.V. and M.V. Popov. 1960. Mestoobitanie i raspredelenie zaitsa-belyaka po statsiyam [Dwellings and distribution of blue hare according to habitats]. In: *Issledovanie Prichin i Zakonomernostei Dinamiki Chislennosti Zaitsa-Belyaka v Yakutii*, Nauka, Moscow, pp. 17-45.

Labutin, Yu.V., M.V. Popov and Yu.V. Revin. 1974. Osabennosti i vosproizvodstva v nekotorykh populyatsiyakh zaitseobraznykh i gryzunov Yakutii [Peculiarities and reproduction in some populations of lagomorphs and rodents of Yakutia]. In: *Zoologicheskie Issle-dovaniya Sibiri i Dal'nego Vostoka*, Vladivostok, pp. 30-38.

Laptev, I.P. 1958. Mlekopitayushchie Taezhnoi Zony Zapadnoi Sibiri [Mammals of Taiga Zone of Western Siberia]. Tom. Un-t, Tomsk, 284 pp.

Larin, B.A. 1950. Zaitsy [Hares]. Zagotizdat, Moscow, 56 pp.

Larin, N.V. and T.A. Shelkovnikova. 1987. K ekologii severnoi pishchukhi v tsentral'nykh Putoranakh [On the ecology of northern pika in central Putorans). In: *Ekologiya i Okhrana Gornykh Vidov Mlekopitayu-shchikh. Materialy III Vsesoyuz. Shk.*, Moscow, pp. 106-108.

Larin, V.V. and T.A. Shelkovnikova. 1991. Nekotorye osobennosti ekologii severnoi pishchukhi na plato Putorana [Some peculiarities of the ecology of northern pika in Putoran plateau). In: *Ekologiya Pishchukh Fauny SSSR*, Nauka, Moscow, pp. 35-39.

Lazarev, B.V. 1968. K ekologii mongol'skoi pishchukhi v Gornom Altae [On the ecology of Mongolian pika in Gornyi Altai]. In: *Nositeli i Perenoschiki Vozbu-ditelei, Osobo Opasnykh Infektsii Sibiri i Dal'nego Vostoka*, Kyzyl, pp. 17-22.

Lazarev, B.V. 1971. Rasprostranenie i chislennost' mongol'skoi i daurskoi pishchukh na Altae [Distribution and population of Mongolian and Daurian pikas in the Altai]. *Dokl. Irkut. N.-I. Protivochum. In-ta*, Vyp. 9, pp. 194-195.

Lazarev, B.V. 1974. Stroenie nor mongol'skoi i daurskoi pishchuki v Gornom Altae [Construction of burrows of Mongolian and Daurian pikas in Gornyi Altai]. *Probl. Osobo Opasnykh Infektsii*, Vol. 6, No. 40, pp. 46-49.

Leont'ev, A.N. 1968. K razmnozheniyu daurskoi pishchukhi v svyazi s kolebaniyami ee chislennosti [On the multiplication of Daurian pika in relation to fluctuations in its population]. *Izv. Irkut. N.-I. Protivochum. In-ta*, Vol. 27, pp. 23-28.

Letov, G.S. 1960. Rasprostranenie i mesta obitaniya dauriskoi pishchukhi v Tuve i prigranichnoi polose Severo-Zapadnoi Mongolii [Distribution and dwellings of Daurian pika in Tuva and bordering belt of northwestern Mongolia]. In: *Biologicheskii Sbornik*, Irkutsk, pp. 107-115.

Letov, G.S. and G.I. Letova, 1971. Rasprostranenie pishchukh i ikh e'ktoparazitov v Tuve v svyazi s epizootologicheskom znacheniem [Distribution of pikas and their ectoparasites in Tuva in relation to epizootiological significance]. In: *Problemy Osobo Opesnykh Infektsii*, Vyp. 2, pp. 125-111 [sic.].

Li, W. and Y. Ma. 1986. A new species of *Ochotona*, Ochotonidae, Lagomorpha. *Acta Zool. Sin.*, Vol. 32, No. 4.

Lipaev, V.M. and P.P. Tarasov. 1952. Materialy po pitaniyu khishchnykh ptits v Yugo-Vostochnom Zabaikal'e po dannym analiza pogadok [Materials on the feeding of predatory birds in southeastern trans-Baikal area according to the data of analysis of baits]. *Izv. Irkut. N.I. Protivochum In-ta,* Vol. 10, pp. 103-110.

Loskutov, R.I. 1966. Severnaya pishchukha-vreditel' kudra [Northern Pika—Pest of cedar]. *Zool. Zhurn.,* Vol. 45, No. 12, p. 1887.

Loshkarev, G.A. 1970. Biologicheskie gruppirovki zaitsa-rusaka na Severnom Kavakaze [Biological groups of European hare in the northern Caucasus]. In: *Populyatsionnaya Struktura u Mlekopitayushchikh,* Moscow, pp. 50-51.

Machul'skii, S.N. 1949. K voprosu o kokrsidoze gryzunov yuzhnykh raionov Buryat-Mongolskoi ASSR [On the problem of coccidosis of rodents in the southern areas of Buryat-Mongolian ASSR]. *Tr. Buryat. Mon-gol. Zoovet. In-ta,* Vyp. 5, pp. 31-39.

Makridin, V.P. 1956. Stai belyakov v tundre [Colonies of European hares in tundra]. *Okhota i Okhotnich'e Khoz-vo,* No. 10, pp. 21-23.

Mamaev, Yu.F. 1979. Zayats-belyak i zashchitnye svsistva lesa [European hare and protective properties of the forest]. In: *Voprosy Lesnogo Okhotovedeniya,* VNIILM, Moscow.

Marin, Yu.F. 1984. Materialy po ekologii altaiskoi pishchukhi v Priteletskoi chasti Altaiskogo zapovednika [Materials on the ecology of Altai pika in Teletsk part of Altai preserve]. In: *Melkie Mlekopitayu-shchie Zapovednykh Territorii,* Moscow, pp. 71-79.

Mel'nikov, V.K. 1974. K ekologii severnoi pishchukhi (*Ochotona alpina* Pall.) v Zapadnom Sayane [On the ecology of northern pika (*Ochotona alpina* Pall.) in western Sayans]. *Byull. MOIP, Otd. Biol.,* Vol. 79, Vyp. 6, pp. 141-143.

Mel'nikov, V.A. and M.P. Tarasov, 1971. Ob ekologii severnoi pishchukhi* (*Ochotona alpina*) v svyazi s ee znacheniem v pitanie soblya v Verkhnelenskoi taiga [On the ecology of northern pika (*Ochotona alpina*) in relation to its significance in the feeding of sable in the upper Lena taiga]. *Zool. Zhurn.,* Vol. 50, No. 4, pp. 602-604.

Mertts, P.A. 1953. Volk v Voronezhskoi oblasti (Ekologiya khishchnika, organizatsiya bor'by) [Fox in Voronezh Region (Ecology of the predator, and organization of control)]. In: *Preobrazovanie Fauny Nashei Strany,* Moscow, pp. 11-21.

Mikulin, M.A. 1956. Materialy po faune blokh Srednei Azii. 3. Blokhi Tsentral'nogo Kazakhstana [Materials on the flea fauna of Soviet Central Asia. 3. Fleas of central Kazakhstan]. *Tr. Sredneaz. N.-I. Protivochum, In-ta,* Vyp. 2, pp. 109-126.

Minin, N.V. 1938. Ekologo-geograficheskii Ocherk Gryzunov Srednei Azii [Ecological and Geographical Outline of Rodents of Soviet Central Asia]. Izd-vo LGU, Leningrad, 184 pp.

Mokeeva, T.M. and M.N. Meier, 1969. Vliyanie sel'skokhoz-yaistvennogo osvoeniya tselinnykh zemel'na gryzun i pishchukh v Tuvinskoi ASSR [The effect of agricultural development of virgin lands on rodents and pikas in Tuva ASSR]. *Tr. Vsesouz. NII Zashchity Rastenii,* Vyp. 30, pp. 105-139.

Möller, D. 1976. Die Fertelitat der Feldhasen populati-tionen. *Ecology and Management of European Hare Population,* Warszawa, pp. 69-74.

388

Moskovskii, A.A. 1936. K biologii pishchukhi [On the biology of pika]. *Izv. Irkutsk. N.-I. Protivochum. In-ta* Vol. 4, pp. 17-19.

Mykytowycz, R. 1966. Observation on odoriferous and other glands in the Australian wild rabbit *Oryctolagus cuniculus* L. and the hare *Lepus europaeus* Pallas. 1. The anal gland. 34. Harder's lachrimal, and sub-mandibular glands. *CSIRO Wildlife Res.*, Vol. 11, No. 1, pp. 11-29, 65-90.

Nasimovich, A.A. 1949. Zametki po biologii dnevnykh khishchnvkh ptits Zabailkal'ya [Notes on the biology of diurnal predatory birds of Trans-Baikal area]. *Byull. MOIP. Otd. Biol.*, Vol. 65, No. 3, pp. 31-38.

Naumov, N.P. 1934. Mlekopitayushchie Tunguskogo okr ga [Mammals of Tungusk district]. *Tr. Polyar. Komiss. AN SSSR*, Vyp. 17.

Naumov, R.L. 1974. Ekologiya gornoi pishchukhi (*Ochotona alpina*) v Zapadnom Sayane [Ecology of Altai pika (*Ochotona alpina*) in western Sayan]. ibid., Vol. 53, No. 10.

Naumov, R.L. and V.V. Labzin. 1980. Pishchukhi [Pikas]. In: *Itogi Mecheniy Mlekopitayushchikh*, Nauka, Moscow, pp. 98-107.

Naumov, R.L. and A.A. Lur'e. 1971. Mechenie gornoi pish-chukhi radioaktivnym kobal'tom [Tagging Altai pika with radioactive cobalt]. *Zool. Zhurn.*, Vol. 50, No. 11, pp. 1728-1731.

Naumov, S.P. 1947. Ekologiya Zaitsa-belyaka [Ecology of Blue Hare). MOIP, Moscow, 207 pp.

Naumov, S.P. 1960. Issledovaniya Prichin i Zakonomernosti Dinemiki Chislennosti Zaitsa-belyaka v Yakutii [Studies on the Reasons and Regularities of Population Dynamics of Blue Hare in Yakutia]. Izd-vo Akad. Nauk SSSR, Moscow, 270 pp.

Naumov, S.P, and S.P. Shatalova. 1974. Vidovye pokazateli razmnozheniya mlekopitayushchikh i osobennosti ikh geo-graficheskogo varirovaniya (na primere semeistva zayach'ikh) [Species indices of reproduction of mammals and the peculiarities of geographic variation (on the example of the family Leporidae)]. *Zool. Zhurn.*, Vol. 53, No. 2.

Naumova, E.I. 1981. Funktsional'naya morfologiya pishche-varitel'noi sistemy gryzonov i zaitseobraznykh [Functional morphology of the digestive system of rodents and lagomorphs].*

Nekipelov, N.V. 1954. Izmenenie chislennosti daurskoi pishchukhi v Yugo-zapadnom Zabaikal'e [Change of population of Daurian pika in southwestern trans-Baikal area]. *Izv. Irkut. N.-I. Protivochum. In-ta*, Vol. 12, pp. 171-180.

Nekipelov, N.V. 1959a. Pitanie daurskoi pishchukhi [Feeding of Daurian pika]. ibid., Vol. 21, pp. 292-297.

Nekipelov, N.V. 1959b. Epizootiologiya chumy v Mongol'skoi Narodnoi Respublike [Epizootiology of plague in Mongolian People's Republic]. ibid., Vol. 20b.

Nekipelov, N.V. 1961a. Rasprostranenie mlekopitayushchikh v Yugo-Vostochnom Zabaikal'e i chislennost' nekotorykh vidov [Distribution of mammals in southeastern Trans-Baikal area and population of some species]. In: *Biologicheskii Sbornik*, Irkutsk, pp. 3-48.

---

*Citation incomplete—Translator.

Nekipelov, N.V. 1961b. Mlekopitayushchie vysokogornykh raionov Sibiri i Mongolii [Mammals of high-mountain regions of Siberia and Mongolia]. *Dokl. Irkut. N.-I. Protivochum. In-ta*, Vyp. 2, pp. 102-106.

Wikol'skii, A.A. 1984. Zvukovye Signaly Mlekopitayushchikh v Evolyutsionnom Protsesse [Some Signals of Mammals in the Evolutionary Process]. Nauka, Moscow.

Nikol'skii, A.A., N.P. Guricheva and P.P. Dnitrev. 1984. Zimnie zapasy daurskoi pishchukhi na stepnykh past-bishchakh [Winter reserves of Daurian pika in steppe pastures]. *Byull. MOIP, Otd. Biol.*, Vol. 89, Vyp. 6, pp. 9-22.

Nikol'skii, A.A. and E.B. Srebrodol'skaya, 1989. Zvuko-vaya aktivnost' severnoi pishchukhi (*Ochotona hyperborea*) v period zapasaniya korma [vocal activity of northern pika (*Ochotona hyperborea*) during storage of food]. ibid., Vol. 94, Vyp. 2, pp. 22-29.

Novikov, G.A. and E.K. Timofeeva, 1965. K ekologii zaitsa-belyaka na severo-vostoke Leningradskoi oblasti [On the ecology of European hare in the northeast of Leningrad region]. In: *Okhotnich'e-Promyslovye Zveri*, Rossel'khozizdat, pp. 178-196.

Novikov, G.A., A.E. Airapetyants, Yu.B. Pukinskii and others, 1970. Zvery Leningradskoi Oblasti (Fauna, Ekologiya i Prakticheskoe Znachenle) [Animals of Leningrad Region (Fauna, Ecology and Practical Significance)]. Izd-vo LGU, Leningrad, 345 pp.

Nurgel'dyev, O.N. 1969. Ekologiya Mlekopitayushchikh Ravninnoi Turkmenii [Ecology of Mammals of the Plains of Turkmenia]. Ylym, Ashkhabad, 259 pp.

Obukhov, P.A. 1973. Zayats-tolai Tuvy [Tolai hare of Tuva]. In: *Redkie Vidy Mlekopitayushchikh Fauny SSSR i Ikh Okhrana*, Nauka, Moscow.

Ochirov, Yu.D. and K.A. Bashanov, 1975. Mlekopitayushchie Tuvy [Mammals of Tuva]. Kyzyl, 138 pp.

Odinashoev, A. 1987. Zaitseobraznye i Gryzuny Pamira [Lagomorphs and Rodents of the Pamirs]. Donish, Dushanbe, 172 pp.

Ognev, S.I. 1940. Zveri SSSR i Prilezhashchikh Stran. T. 4. Gryzuny [Animals of the USSR and Adjoining Countries. Vol. 4, Rodents]. Izd-vo Akad. Nauk SSSR, Moscow, Leningrad, 615 pp.

Okunev, L.P. 1971. Zimnie nablyudeniya za aktivnost'yu mongol'skikh pishchukh [Winter observations on the activity of Mongolian pikas]. *Dokl. Irkut. N. I. Protivochum. In-ta*, Vyp. 9, pp. 197-199.

Okunev, L.P. 1975. Nablyudenie za aktivnost'yu pishchukh v Gorno-Altaiskom ochage chumy [Observation on activity of pikas in Gorno-Altaisk plague focus]. In: *Mezhdunarodnye i Natsional'nye Aspekty Epidnadzora pri Chume*, Irkutsk, Pt. 2, pp. 10-11.

Okunev, L.P. and G.B. Zonov. 1980. Ekologicheskie adapta-tsii mongol'skikh pishchukh k zhizni v gorno-step-nykh landshaftakh [Ecological adaptation of Mongolian pikas to life in mountain-steppe terrain]. *Ekologiya*, No. 6, pp. 61-66.

Ol'kova, N.V. 1954. Arealy semi dauriskoi pishchukhi [Family territories of Daurian pika]. Izv. *Irkut. N. I. Protivochum. In-ta*, Vol. 12.

Ondar, S.O. 1989. Raspredelenie melkikh mlekopitayushchikh v rastitel'nykh assotsiatsiyakh Ubsnurskoi kotlo-viny [Distribution of small mammals in plant associations of the Ubsnur depression]. In: *Sovetsko-Mongol 'skoi E'ksperiment "Ubsnur" (mnogostoron Soveshch, Stran-Chlenov SE'V, Kyzyl. 1-10 Aug. 1989)* Pushchino, pp. 54-58.

Orlov, G.I. 1983. Sheinaya zheleza pishchukh (Lagomyidae, Ochotonidae) i svyazannoe s ee funktsionirovaniem markirovochnoe povedenie altaiskoi pishchukhi (*Ocho tona alpina*) [Cervical glands of pikas (Lagomyidae, Ochotonidae) and related to it functioning of marking behavior of Altai pika (*Ochotona alpina*)]. *Zool. Zhurn.*, Vol. 62, No. 11.

Orlov, G.I. and G.I. Makushin, 1984. K vozrastnoi struk-ture populyatsii altaiskoi pishchukhi [On the age structure of population of Altai pika]. In: *Registriruvushchie Struktury i Opredelenie Vozrasta Mlekopitayushchikh (Tez. Dokl. Vsesoyuz. Konf.)*, Moscow, pp. 48-49.

Ostapenko, M.M. 1963. Biologiya zaitsa-peshchanika v Uzbekistane [Biology of tolai hare in Uzbekistan]. In: *Okhotnich'e Promyslovye Zhivotnye Uzbekistana*, Tashkent, pp. 77-89.

Pakizh, V.I. 1969. Nekotorye dannye po ekologii stepnoi pishchukhi (na severo-vostoke Kazakhstana v svyazi s izmeneniem ee areala) [Some data on the ecology of steppe pika (in the northeast of Kazakhstan in relation to change of its range). *Zool. Zhurn.*, Vol. 48, No. 8, pp. 1214-1220.

Palvaniyazov, M. 1974. Khishchnye Zveri Pustyn' Srednei Azii [Predatory Animals of Deserts of Soviet Central Asia]. Karakalpakstan, Nukus, 318 pp.

Paramonova, A.N., L.F. Vasil'eva and L.A. Guvva. 1958. Gryzuny i ikh e'ktoparazity gornoi sistemy Bolshoi Balkhan [Rodents and their ecotoparasites in mountain system of the Greater Balkhan). *Tr. Turkmen. Protivochum. Stantsii*, Ashkhabad, Vol. 1, pp. 135-144.

Pavlenko, T.A. and S.T. Gubaidulina, 1970. Zaitseobraznye [Lagomorphs]. In: *Ekologiya Pozvonochnykh Zhivot-nykh Khrebta Nurtau*, Fan, Tashkent, pp. 101-102.

Pavlinin, V.N. 1971. Zayats-belyak (*Lepus timidus* L. 1758) [European hare (*Lepus timidus* L. 1758)]. *Tr. In-ta Ekologii Rastenii i Zhivotnykh UNTs. Akad. Nauk SSSR*, Vyp. 80, Vol. 1, pp. 75-106.

Pavlinin, V.N. and S.S. Shvarts, 1957. K voprosu o grani-tsakh rasprostraneniya nekotorykh vidov gryzunov na Urale [On the problem of boundaries of distribution of some species of rodents in the Urals]. *Tr. In-ta Biologii Ural. Fil. Akad. Nauk SSSR*, Vol. 8, pp. 25-29.

Pavlov, M.A., A. Shulyat'ev and A. Gineev, 1984. Dikii krolik [Wild rabbit]. *Okhota i Okhopinich'e Khozvo*, No. 9, pp. 18-20.

Perevalov, A.A. 1953. Zayats-peshchanik: (Ekologiya i promysel v Kazakhstane) [Tolai hare (Ecology and hunting in Kazakhstan]. Author's Abstract, Diss. ... Kand. Biol. Nauk, Alma-Ata, 23 pp.

Perevalov, A.A. 1956a. Materialy po biologii razmnozheniya zaitsa-peshchanika [Materials on the reproduction biology of tolai hare]. *Zool. Zhurn.*, Vol. 35, No. 1.

Perevalov, A.A. 1956b. Stroenie mekha i lin'ka zaitsa-peschanika [Structure of fur and molting of tolai hare]. *Uch. Zap. Tiraspol. Ped. In-ta*, Vyp. 1.

Perevalov, A.A. 1958. Pitanie zaitsa-peschanika (*Lepus tolai lehmani* Sev. 1873)

[Feeding of tolai hare (*Lepus tolai lehmani* Sev. 1873)]. ibid., Vyp. 7.

Peshkov, B.I. 1957. Dannye po chislennosti i pitaniyu pernatykh khishchnikov yugo-vostochnogo Zabaikal'ya [Data on the population and feeding of perenating predators of southeastern Trans-Baikal area]. *Izv. Irkut. N. I. Protivochum. In-ta*, Vol. 16, pp. 143-153.

Petrov, P. 1963. Faktory, vliyayushchie na koe'ffitsient real'nogo prirosta pogolovya zaitsev [Factors influencing the coefficient of actual rise of numbers of hares]. *Izv. In-ta Gorata* (Sofia), Kn. 2, pp. 61-95.

Petrov, P. 1976. Über die Factoren die den realen Zurwachs des Hasen bestimmen. In: *Ecology and Management of European Hare Population*. Warsazawa, pp. 1-3.

Petrusewicz. 1970. Dynamics and production of the hare population in Poland. *Acta Theriol.*, Vol. 15, No. 24/31.

Petter, F. 1961. Elements d'une revision des Lievres europeens et asiatiques du sousgenre *Lepus. Ztschr. Sangetierk*, Bd. 26, H. 1, pp. 30-40.

Pielowski, Z. 1971. Badania nad zajacem (2). *Lowiec pol.* No. 5 (1392), pp. 79-88.

Pielowski, Z. 1972. Home range and degree of residence of the European hare. *Acta Theriol.*, Vol. 17, No. 12.

Pielowski, Z. 1976. Number of young born and dynamics of the European hare population: In: *Ecology and Management of European Hare Population*. Warszawa, pp. 76-77.

Ponomarev, B.A. 1935. Lovlya zaitsev kapkanami [Catching hares by traps]. *Okhotnik Sibiri*, No. 1, pp. 15-19.

Ponomarev, G. 1934. Za zaitsem-belyakom po sledu [Following European hare from its sign]. ibid. No. 11/12, pp. 9-15.

Popov, M.V. 1960. Kormovye usloviya i ikh znachenie dlya dinamiki chislennosti [Feeding conditions and their significance for population dynamics]. In: *Issledo-vaniya Prichin i Zakonomernosti Dinamiki Chislen-nosti Zaitsa-belyaka v Yakutii*, Izd-vo Akad. Nauk SSSR, Moscow, pp. 69-107.

Popov, V.A. 1960. Mlekopitayushchie Volzhsko-Kamskogo Kraya: Nasekomoyadnye, rukokrylatye, gryzuny [Mammals of Volga-Kama Territory: Insectivores, Chirop-tera, Rodents]. Kazan, 468 pp.

Portenko, L.A. 1941. Fauna Anadyrskogo kraya. Ch. 3. Mle-kopitayushchie [Fauna of Anadyr Territory. Part 3. Mammals]. Moscow, Leningrad, 93 pp. (*Tr. NII Pol-yar Zemledeliya, Zhivotnovodstva i Promyslovogo Khoz-va*, Vyp. 14).

Portenko, L.A., A.A. Kishchinskii and F.B. Chernyavskii, 1963. Mlekopitayushchie Koryatskogo nagorlya [Mammals of Koryatsk Upland]. Izd-vo Akad. Nauk SSSR, Moscow, Leningrad, 130 pp.

Potapkina, A.F. 1967. K parazitofaune altaiskoi pishchukhi (*Ochotona alpina* Pall.) [On the parasite fauna of Altai pika (*Ochotona alpina* Pall.)]. In: *Priroda Ochagov Kleshchevogo Entsefalata na Altae (severo-vostochnaya chast')*. Nauka, Novosibirsk, pp. 46-48.

Potapkin, A.F. 1971. Zametki o geograficheskom rasprede-lenii i ekologii pishchukh na Altae [Notes on the geographic distribution and ecology of pikas in the Altai]. In: *Priroda i Prirodnye Resursy Gornogo Altaya*, Gorno-Altaisk, pp. 284-289.

Potapkina, A.F. 1975. Rasprostranenie i biologiya pishchukh (*Oohotona*) Yuga Zapadnoi Sibiri [Distribution and biology of pikas (*Ochotona*) in the south

392

of western Siberia]. In: *Sistematika, Fauna, Zoogeografiya Mlekopitayushchikh i Ikh Parazitov*, Nauka, Novosibirsk, pp. 92-103.
Priklonskii, S.G. and E.N. Teplova. 1965. Pervyi Opyt Vserossiiskogo Zimnego Marshrutnogo Ucheta Okhotni-ch'ikh Zverei [The First Experiment of All-Russian Winter Route, Estimate of Game Animals]. Moscow 51 pp.
Prokop'ev, V.N. 1957. K ekologii daurskoi i mongol'skoi pishchukh [On the ecology of Daurian and Mongolian pikas]. *Izv. Irkut. N.-I. Protivochum. In-ta*, Vol. 16, pp. 110-113.
Proskurina, N.S. 1991. O pesnyakh i trel'yakh daurskoi pishchukhi (*Ochotona daurica*) [On the songs and trills of Daurian pika (*Ochotona daurica*)]. In: *Ekologiya Pishchukh Fauny SSSR*, Nauka, Moscow, pp. 40-48.
Proskurina, N.S. and V.M. Smirin, 1987. Forma vnutrivido-vykh vzaimodeistvii daurskoi pishcukhi [The form of interspecific interactions of Daurian pikas]. *Byull. MOIP, Otd. Biol.*, Vol. 92, Vyp. 4, pp. 12-21.
Proskurina, N.S., N.A. Formozova and D.G. Derviz, 1985. Sravnitel'nyi analiz prostranstvennoi struktury poselenii trekh form pishchukh: *Ochotona daurica, O. pallasi pallasi, O. pallasi pricei* (Lagomorpha, Lagomyidae) [Comparative analysis of spatial structure of habitations of three forms of pikas: *Ochotona daurica, O. pallasi pallasi, O. pallasi pricei* (Lagomorpha, Lagomyidae)]. *Zool. Zhurn.*, Vol. 64, No. 2, pp. 1695-1701.
Pshennikov, A.E., V.G. Alekseev I.I. Koryakin and D.Yu. Gnutov, 1990. Koprofagiya v severnoi pishchukhi (*Ochotona hyperborea*) v Yakutii [Coprophagy in northern pika (*Ochotona hyperborea*) in Yakutia]. ibid., Vol. 69, No. 12, pp. 106-114.
Pshennikov, A.E., Z.Z. Borisov and I.S. Vasil'ev, 1988. Koprofagiya i ee ritmika u zaitsa-belyaka (*Lepus timidus*) v Tsentral'noi Yakutii [Coprophagy and its rhythm in European hare (*Lepus timidus*) in Central Yakutia]. ibid., Vol. 67, No. 9, pp. 1357-1362.
Rausch, R.L. 1963. A review of the distribution of Holarctic recent mammals. *Pacific Basin Biogeography*, London, Bishop Museum Press, pp. 29-43.
Reimers, N.F. 1960. Burunduk i severnaya pishchukha v kedrovoi taige Pribaikal'ya [Siberian chipmunk and northern pika in cedar-taiga of Cis-Baikal area]. *Tr. 3 Vost.-Sib. Fil. SO Akad. Nauk SSSR, Ser. Zool.*, Vyp. 23, pp. 101-106.
Reimers, N.F. 1966. Ptitsy i mlekopitayushchie yuzhnoi taigi Srednei Sibiri [Birds and Mammals of Southern Taiga of Central Siberia]. Nauka, Moscow, 420 pp.
Reimov, R. 1985. Mlekopitayushchie Yuzhnogo Priaral'ya (Ekologiya, okhrana i ispol'zovanie) [Mammals of Southern Cis-Aral Area (Ecology, Conservation and Exploitation)]. FAN, Tashken.
Revin, Yu.V. 1968. O biologii severnoi pishchukhi (*Ochotona alpina* Pall.) na Olekmo-Charskom nagor'e [On the biology of Altai pika (*Ochotona alpina* Pall.) in Olekmo-Charsk Upland]. *Zool. Zhurn.*, Vol. 47, No. 7, pp. 1075-1082.
Revin, Yu.V. 1989. Mlekopitayushchie Yuzhnoi Yakutii [Mammals of Southern Yakutia]. Nauka, Novosibirsk, 320 pp.
Revin, Yu.V., V.M. Safronov, Ya.L. Vol'pert and A.L. Popov. 1988. Ekologiya i

dinamika chislennosti mlekopita-yushchikh Predverkhoyan'ya [Ecology and Population Dynamics of Mammals of Cis-Verkhoyan Area]. Nauka, Novosibirsk, 199 pp.

Rieck, W. 1956. Untersuchungen über die Vermchrung des Feldhasen. Ztschr. Zagdwiss., Bd. 2, No. 2.

Romanov, A.A. 1941. Pushnye zveri Lensko-Khatangskogo kraya i ikh promysel [Fur Animals of Lena-Khatanga Territory and their Trade]. Izd-vo Gavsevmorputi, Leningrad.

Rossolimo, O.L. 1979. Ocherk geograficheskoi izmenchivosti cherepa zaitsa-belyaka (Lepus timidus L.) [Outline of geographic variation in the skull of European hare (Lepus timidus L.)]. Tr. Zool. Muzeya MGU, Vol. 18, pp. 215-240.

Rozanov, M.P. 1935. Mlekopitayushchie Pamira [Mammals of the Pamirs]. In: Materialy po Mlekopitayushchim i Ptitsam Pamira: Tr. Tadzh. Kompleks. Ekspeditsii 1932 g., Vyp. 32, pp. 17-27.

Sadovskaya, N.P. 1955. O sostave fauny gel'mintov zaitsev Primorskogo kraya [On the composition of helminth fauna of hares of Primorsk Territory]. Soobshch. DVF Akad. Nauk SSSR, No. 5, pp. 57-60.

Salikhbaev, Kh.S., V.P. Karpenko, D.Yu. Kashkarov and others. 1967. Ekologiya, mery okhrany i ratsional'-noe ispol'zovanie pozvonochnykh zhivotnykh Karshinskoi stepi [Ecology, Conservation Measures and Rational Exploitation of Vertebrate Animals of Karshinsk Steppe]. Fan, Tashkent, 172 pp.

Samusev, F.F. and I.F. Samusev. 1972. Materialy po pitaniyu severnoi pishchukhi (Ochotona alpina Pall.) na Altae [Materials on the feeding of northern [Altai] pika- (Ochotona alpina Pall.) in the Altai]. Teriologiya, Nauka, Novosibirsk, No. 1, pp. 358-361.

Sapozhenkov, Yu.F. 1964. Ekologiya zaitsa-tolaya (Lepus tolai Pall.) v peschanykh Karakumakh [Ecology of tolai hare (Lepus tolai Pall.) in sandy Karakums]. Zool. Zhurn., Vol. 44, No. 9, pp. 1382-1387.

Sapragel'dyev, M. 1987. Ekologiya ryzhevatoi pishchukhi v Turkmenistane [Ecology of Afghan Pika in Turkmenistan]. Ylym, Ashkhabad, 141 pp.

Savinov, V.A. and A.N. Lobanov, 1958. Zveri Vologodskoi oblasti [Animals of Vologda Region]. Kn. Izd-vo Vologda, 208 pp.

Selevin, V.A. 1937. Perechen' mlekopitayushchikh okrest-nostei Semipalatinska [Checklist of mammals in the vicinity of Semipalatinsk]. Byull. Saratov. Un-ta, No. 36, Vyp. 22, pp. 21-29.

Semenov, P.V. and A.F. Potapkina, 1975. Podkozhnye ovody (Hypodermatidae) melkikh mlekopitayushchikh gor yuga Sibiri [Warble flies (Hypodermatidae) of small mammals of Siberian mountains in south]. In: Siste-matika Fauna, Zoogeografiya Mlekopitayushchikh i Ikh Parazitov, Nauka, Novosibirsk.

Serzhanin, I.N. 1955. Mlekopitayushchie Belorussii [Mammals of Belorussia]. Izd-vo Akad. Nauk BSSR, Minsk, 310 pp.

Shatalova, S.P. 1970. Morfologiya i fenologiya polovogo tsikla zaitsa-belyaka Yakutskoi taigi [Morphology and phenology of sexual cycle of blue hare in Yaku-tian taiga]. Uch. Zap. Mosk. Ped. In-ta im. V. I. Lenina, No. 272, pp. 37-41.

Shevchenko, L.S. 1986. Morfologicheskaya kharakteristika dikogo krolika na

Ukraine [Morphological characteristics of wild rabbit in the Ukraine]. *Vestn. Zoologii*, No. 5.

Shishkin, A.S. 1988. Zayats-belyak Srednei Sibiri [Blue Hare of Central Siberia]. Krasnoyarsk, 177 pp.

Shnitnikov, V.N. 1936. Mlekopitayushchie Semirech'ya [Mammals of Semirech'e]. Izd-vo Akad. Nauk SSSR, Moscow, 323 pp.

Shtil'mark, F.R. 1963. Novye dannye po ekologii *Apodemus mystacinus i Ochotona hyperborea* v sayanskoi taige [New data on the ecology of *Apodemus mystacinus* and *Ochotona hyperborea* in Sayan taiga]. *Byull. MOIP, Otd. Biol.*, Vol. 68, Vyp. 5, pp. 137-138.

Shubin, I.G. 1956. Vliyanie geograficheskoi sredy na razm-nozhenie mongol'skoi pishchukhi [The effect of geographic environment on the reproduction of Mongolian pika]. *Tr. In-ta Zoologii Akad. Nauk KazSSR*, Vol. 6, pp. 61-77.

Shubin, I.G. 1958. O zarazhennosti mongol'skikh pishchukh lichinkami ovodov *Oestromia falax* Grunin, 1949 [On the infectivity of Mongolian pikas with larvae of the flea *Oestroemia falax* Grunin; 1949]. ibid., Vol. 9, p. 242.

Shubin, I.G. 1959. K ekologii mongol'skoi pishchukhi v Kazakhskom nagor'e [On the ecology of Mongolian pika in the Kazakh uplands]. ibid., Vol. 10, pp. 114-115.

Shubin, I.G. 1962a. Novye dannye po rasprestraneniyu nekotorykh gryzunov Tselinogradskoi oblasti [New data on the distribution of some rodents in Tselinograd Region]. ibid., Vol. 17, pp. 242-245.

Shubin, I.G. 1962b. K pitaniyu khishchnykh zverei i ptits Kazakhskogo nagor'ya [On the feeding of predatory animals and birds of the Kazakh uplands]. ibid., Vol. 17, pp. 183-191.

Shubin, I.G. 1963. Areal maloi pishchukhi i faktory ego opredelyayushchie [The range of steppe pika and the factors determining it]. In: *Zoogeografiya Sushi*, Tashkent, pp. 365-366.

Shubin, I.G. 1965. Razmnozhenie maloi pishchukhi (Reproduction of steppe pika]. *Zool. Zhurn.*, Vol. 44, No. 5, pp. 917-924.

Shubin, I.G. 1966. Dinamika vozrastnogo sostava maloi i pallasovoi pishchukh [Dynamics of age structure of steppe and Pallas's pikas]. *Byull. MOIP, Otd. Biol* Vol. 71, Vyp. 4, pp. 27-33.

Shubin, I.G. 1972. Lin'ka maloi i mongol'skoi pishchukh i ee osobennosti [Molting in steppe and Mongolian pikas and its peculiarities]. In: *Teriologiya*, Nauka Novosibirsk, Vol. 1, pp. 365-367.

Shubin, I.G. 1975. Osobennosti gnezdovogo povedeniya maloi i mongol'skoi pishchukh [Peculiarities of nesting behavior of steppe and Mongolian pikas]. In: *Tr. II Vsesoyuz. Soveshch. po Mlekopitayushchim*, Izd-vo MGU, Moscow, pp. 200-202.

Shubin, N.G. 1963. Severnaya pishcukha Kuznetskogo Alatau [Northern pika of Kuznetskii Alatau]. In: *Dokl. VI Nauch. Konf. po Biol. Naukam. Novokuznetskogo Ped. In-ta*, Novokuznetsk, pp. 69-74.

Shubin, N.G. 1967. Materialy po ekologii rysi v Zapadnoi Sibiri [Materials on the ecology of lynx in western Siberia]. In: *Problemy Ekologii*, Tomsk, Vol. 1.

Shubin, N.G. 1971. Ob ekologii severnoi pishchukhi Kuznet-skogo Alatau [On the ecology of northern pika of Kuznetsk Alatau]. ibid., Vol. 2.

Shukurov, G. Shch. 1962. Fauna Pozvonochnykh Zhivotnykh gof Bolshie Balkhany [Fauna of Vertebrate Animals of the Greater Balkhan Mountains]. Ashkhabad, 110 pp.

Shvetsov, Yu.G. and N.I. Litvinov, 1967. Mlekopitayushchie bassina rechki Nizhnii Kachergat (Yugo-Vostochnoe Prebaikal'e) [Mammals of the Nizhni Kochergat River basin (southeastern Cis-Baikal area)]. *Izv. Irkut. S.-Kh. Inta*, Vyp. 25, pp. 209-223.

Shvedov, Yu.G. and A.A. Moskoskii, 1961. Rasprostranenie i chislennost' osnovnykh vidov gryzunov v pograni-chnoi polose Yugo-Zapadnogo Zabaikal'ya [Distribution and population of main species of rodents in the bordering belt of southwestern Trans-Baikal area]. In: *Biologicheskii Sbornik*, Irkutsk, pp. 96-106.

Shvedov, Yu.G., M.N. Smirnov and G.I. Monakhov, 1984. Mle-kopitayushchie basseina ozera Baikal [Mammals of the Lake Baikal basin]. Nauka, Novosibirsk, 258 pp.

Sludskii, A.A. 1953. Zaitsy [Hares]. In: *Zveri Kazakhstana*, Alma-Ata, pp. 99-112.

Sludskii, A.A. 1962. Vzaimootnosheniya khishchnikov i dobychi [Interrelationships of predators and prey]. In: *Tr. In-ta Zoologii Akad. Nauk KazSSR*, Vol. 17, p. 24.

Sludskii, A.A. and Yu.G. Afanas'ev, 1964. Itogi i perspek-tivy akklimatizatsii okhotnich'e-promyslovykh zhivotnykh v Kazakhstane [Results and prospects of acclimatization of game and commercial animals in Kazakhstan]. ibid., Vol. 23, pp. 5-74.

Sludskii, A.A., Yu.G. Afanas'ev, A. Bekenov and others. 1982. Mlekopitayushchie Kazakhstana: Khishchnye (kuni, koshki) [Mammals of Kazakhstan: Carnivores (martens, cats)]. Nauka, Alma-Ata, Vol. 3, pt. 2, 263 pp.

Sludskii, A.A., B.I. Badamshin, A. Bekenov and others. 1981. Mlekopitayushchie Kazakhstana: Khishchnye (sobach'i, enotovye, lastonogie (nastoyashchie tyu-leni)) [Mammals of Kazakhstan: Carnivores (dogs, raccoons, pinnipeds (True Seals))]. Nauka, Alma-Ata, Vol. 3, Pt. 1, 232 pp.

Sludskii, A.A., A.D. Bernshtein, I.G. Shubin, and others. 1980. Mlekopitayushchie Kazakhstana: Zaitseobraznye [Mammals of Kazakhstan: Lagomorphs]. Nauka, Alma-Ata, Vol. 2, 238 pp.

Smirnov, V.M. 1967. Chislennost' i raspredelenie altai-skoi pishchukhi (*Ochotona alpina* Pall:) v severo-vostochnom Altae [Population and distribution of Altai pika (*Ochotona alpina* Pall.) in northeastern Altai]. In: *Priroda Ochagov Kleshchevidnogo E'ntse-falita na Altae*, Novosibirsk, pp. 60-65.

Smirnov, P.K. 1972. Opyt soderzhaniya i razvedeniya v nevole mongol'skoi pishchukhi (*Ochotona pricei* Thom.) [Experiment of captive rearing and breeding of Mongolian pika (*Ochotona pricei* Thom.)]. *Vestn. LGU*, No. 15, pp. 20-26.

Smirnov, P.K. 1974. Statsial'noe raspredelenie i terri-torial'nya vzaimootnosheniya maloi i mongol'skoi pishchukhi v zone simpatrii i ikh arealov [Habitat distribution and territorial interactions of steppe and Mongolian pikas in sympatric zone and their ranges]. *Byull. MOIP, Otd. Biol.*, Vol. 79, Vyp. 5, pp. 72-81.

Smirnov, P.K. 1976. Nablyudeniya za povedeniem pishchukh [Observations on the behavior of pikas]. In: *Grup-povoe Povedenie Zhivotnykh*, Nauka, Moscow.

Smirnov, P.K. 1985. Mestoobitanie Mongol'skoi pishchukhi v Kazakhskoi skladchatoi strane [Habitat of Mongolian pika in Kazakh folded country]. *Vestn. LUG* No. 10, pp. 117-120.

Smirnov, P.K. 1987. Adaptivnye osobennosti povedeniya pishchukh [Adaptive peculiarities of behavior of pikas]. In: *Ekologiya i Okhrana Gornykh Vidov Mlekopitayushchikh*, Moscow, pp. 153-155.

Smirnov, P.K. 1988. Prostranstvennye osobennosti kommuni-kativnogo povedeniya pishchukh [Spatial peculiarities of communicational behavior of pikas]. In: *Kommunikativnye Mekhanizmy Regulirovaniya Populya-tsionnoi Struktury u Mlekopitayushchikh*, Moscow, pp. 157-159.

Smolina, L.L. 1958. Ob osobennostyakh razmnozheniya daur-skoi pishchukhi [On the peculiarities of reproduction of Daurian pika]. *Izv. Irkut. N. I. Protivochum. In-ta*, Vol. 19, pp. 105-109.

Soares, M.J. and M. Diamond, 1982. Pregnancy and chin marking in the rabbit *Oryctolagus cuniculus*. *Anim. Behav.*, Vol. 30, No. 3, pp. 941-943.

Sokolov, V.E. 1977. Sistematika mlekopitayushchikh (Otrydy: Zaitsobraznykh, gryzunov) [Systematics of Mammals (Orders: Lagomorpha, Rodentia)]. Vysshaya Shkola, Moscow.

Sokolov, V.E. and A.N. Terekhova, 1978. Zapakhovaya marki-rovka territorii u gryzunov i zaitsoobraznykh [Scent marking of territory in rodents and lagomorphs]. *Uspekhi Sovrem. Biologii*, Vol. 86, No. 2, pp. 240-246.

Sokolov, G.A. 1965. Vliyanie sbora kedrovykh orekhov na pitanie i chislennost' sobolya i belki [The effect of collection of cedar nuts on the feeding and population of sable and squirrel]. In: *Fauna Kedrovykh Lesov Sibiri i Ee Ispol'zovanie*, Nauka, Moscow, pp. 53-91.

Solomatin, A.O. 1969. Zayats-rusak na Turgaiskom plato [European hare in Turgai plateau]. *Byull. MOIP, Otd. Biol.*, Vol. 74, Vyp. 6. pp. 5-18.

Solomatin, A.O. 1975. Territorial'noe raspredelenie zaitse-obraznykh na Turgaiskom plato [Territorial distribution of lagomorphs in Turgai Plateau]. In: *Prirodnoe Sel'skokhozyaistvennoe Raionirovanie SSSR*, Moscow, pp. 105-107.

Solomonov, N.G. 1973. Ocherki populyatsionnoi ekologii gryzunov i zaitsa-belyaka v Tsentral'noi Yakutii [Outline of Population Ecology of Rodents and Lagomorphs in Central Yakutia]. Yakutsk, 247 pp.

Sterba, E. 1982. Hlavani priciny ztrat zajece zvere v letech 1975-1979. *Polovn. zb.*, No. 12, pp. 239-259.

Strelkov, A.P. 1989. K ekologii krasnoi pishchukhi *Ochotona rutila* (Mammalia, Lagomorpha) [On the ecology of red pika (*Ochotona rutila* (Mammalia, Lagomorpha)). *Zool. Zhurn.*, Vol. 68, Vyp. 1, pp. 153-155.

Stroganov, S.U. and A.S. Stroganova, 1944. Materialy po biologii zaitsa-peschanika (*Lepus tibetana bucharensis* Ogn.) po nablyudeniyam v Yuzhnom Tadzhikistane [Materials on the biology of Tibetan cape hare (*Lepus tibetana bucharensis* Ogn.) from observations in southern Tadzhikistan). *Izv.Tadzh. Fil. Akad. Nauk SSSR*, No. 5, pp. 11-31.

Svanbaev, S.K. 1958. K poznaniyu fauny kaktsidii gryzu-nov Tsentral'nogo Kazakhstana [On the knowledge of coccid fauna of rodents of central

Kazakhstan]. *Tr. In-ta Zoologii Akad. Nauk KazSSR, Ser. Parazitol.*, Vol. 9, pp. 183-186.

Tarasov, P.P. 1950. K ekologii Mongol'skoi pishchukhi [On the ecology of Mongolian pika]. *Byull. MOIP, Otd. Biol.*, Vol. 55, Vyp. 6, pp. 35-42.

Tarasov, P.P. 1958. Gryzuny yugo-vostochnoi chasti Mongol'skogo Altaya i prilezhashchei Gobi [Rodents of southeastern part of Mongolian Altai and adjoining Gobi]. *Izv. Irkut. N.-I. Protivochum. In-ta*, Vol. 19, pp. 60-71.

Tarasov, P.P. 1959. Nekotorye osobennoti vnutrividovykh otnoshenii u stenotopnykh gryzunov [Some peculiarities of intraspecific relationships of stenotopic rodents]. *Tr. Sredneaz. N.-I. Protivochum. In-ta*, Vyp. 5, pp. 161-176.

Tarasov, P.P. 1962. Statsial'noe razmeshchenie i otnosi-tel'naya chislennost' massovykh vidov gryzunov Zapadnogo Khamar-Dabana [Habitat dispersal and relative population of large species of rodents of western Khamar-Daban]. *Izv. Irkut. N.-I. Protivo-chum. In-ta*, Vol. 24, pp. 248-260.

Tavrovskii, V.A., O.V. Egorov, V.G. Krivosheev and others. 1971. Mlekopitayushchie Yakutii [Mammals of Yakutia]. Nauka, Moscow, 115 pp.

Terent'ev, P.V., V.B. Dubinin and G.A. Novikov. 1952. Krolik [The Rabbit]. Moscow, 362 pp.

Tikhvinskaya, M.V. and P.K. Gorshkov, 1975. Biotopicheskie osobennosti pitaniya lisitsy v Tatarskoi ASSR [Habitat peculiarities of feeding of foxes in Tatar ASSR]. In: *Tr. II Vsesoyuz. Soveshch. po Mlekopita-yushchim.*, Izd-vo MGU, Moscow, pp. 108-110.

Timofeeva, A.A. 1962. Nekotorye nablyudeniya po statsial'-nomu raspredeleniya gryzunov na yuge o. Sakhalin [Some observations on habitat distribution of rodents in the south of Sakhalin Island]. *Izv. Irkut. N.-I. Protivochum. In-ta*, Vol. 24, pp. 261-271.

Ushakova, G.V. 1956. Kleshchi inodsemeistva Ixoidea pustyn Bet-Pak-Dala i prilegayushchikh k nei raionom [Ticks of the family Ixoidea in Bet Pak-Dala deserts and regions adjoining them]. *Tr. In-ta Zoologii Akad. Nauk KazSSR*, Vol. 5, pp. 129-151.

Ushakova, G.V. and N.N. Buslaeva, 1962. Materialy po ikso-dovym kleshcham polupustyn Kargandinskoi oblasti [Materials on ixodid ticks of semideserts of Karaganda Region]. In *Parazity Sel'skokhozyaistvennykh Zhivotnykh Kazakhstana*, Alma-Ata, pp. 31-39.

Ustyuzhin, Yu.A. 1971. Osobennosti vozrastnogo i polovogo sostava populyatsii mongol'skoi pishchukhi tuvinskogo prirodnogo ochaga [Peculiarities of age and sex composition of population of Mongolian pika in Tuva natural focus]. *Dokl. Irkut. N.-I. Protivo-chum. In-ta*, Vyp. 9, pp. 193-194.

Ustyuzhin, Yu.A. 1972. Rasprostranenie i Kolebanie chis-lennosti mongol'skoi pishchukhi v Tuve [On the distribution and fluctuation of population of Mongolian pika in Tuva]. In: *Zoologicheskie Problemy Sibirii* Novosibirsk, pp. 483-484.

Ustyuzhina, I.M. and Yu.A. Ustyuzhin, 1971. O chislennosti i kharaktere raspredeleniya blokh mongol'skoi pish-chukhi na razlichnykh territoriyakh [On the number and nature of distribution of fleas of Mongolian pika in different territories]. *Izv. Irkut. N.-I. Protivochum. In-ta*, Vol. 9, pp. 221-222.

Utinov, S.P. 1979. Klimaticheskie faktory chislennosti zaitsa-belyaka v lesostepnoi zone Severnogo Kazakhstana [Climatic factors of the population of European hare in forest steppe zone of northern Kazakhstan]. In: *Materialy po Ekologii i Fiziologii Zhivotnykh*, Alma-Ata, pp. 74-75.

Valenticic-Stane, S.I. 1956. Resultate zweijahriger beo-bachtungen und Studien uber den idealen Zuwach beim Feldhasen auf der Insel "Biserni otok". *Ztschr. Jagdwiss.*, Bd. 2, No. 3.

Vashchenok, V.S. 1962. O chumnykh epizootiyakh sredi mongol'skikh pishchukh na severo-zapade Mongol'skoi Narodnoi Respubliki [On plague epizootic among Mongolian pikas in northwest of Mongolian People's Republic]. *Zool. Zhurn.*, Vol. 41, No. 10, pp. 1548-1555.

Vasil'ev, G.I. and L.A. Lazareva. 1968. K parazitologi-cheskoi kharakteristike Gorno-Altaiskogo chumovogo ochaga [On parasitological description of Gorno Altai plague focus]. *Izv. Irkut. N. I. Protivochum. In-ta*, Vol. 27, pp. 279-287.

Vasil'ev, N., A. Pankratov and E. Panov. 1965. Zapovednik "Kedrovaya pad" [The "Kedrovaya Pad" Preserve]. Vladivostok, 48 pp.

Vereshchagin, N.K. 1953. K istorii pleistotsenovoi i golot-senovoi faun mlekopitayushchikh v raione srednego techeniya r. Urala [On the history of Pleistocene and Holocene faunas of mammals in the area of middle reaches of the Ural River]. *Byull. Komis. po Izuch. Chetvertich. Perioda*, No. 18, pp. 37-41.

Vinogradov, B.S. 1952. Mlekopitayushchie Krasnovodskogo raiona Zapadnoi Turkmenii [Mammals of Krasnovodsk Region of western Turkmenia]. *Tr. ZIN Akad. Nauk SSSR*, Vol. 10, pp. 1-44.

Vladimirskaya, M.K. 1955. K biologii zaitsa-belyaka na Kol'skom poluostrove [On the biology of blue hare in Kola Peninsula]. *Zool. Zhurn.*, Vol. 34, No. 3.

Vlasenko, G.S. 1954. Gryzuny yuzhnykh raionov Tuvy [Rodents of southern areas of Tuva]. *Izv. Irkut. N.-I. Protivochum. In-ta*, Vol. 12.

Volodin, I.A. 1974. Vliyanie nekotorykh faktorov na dinamiku chislennosti zaitsa-rusaka i zaitsa-belyaka v Bryanskoi oblasti [The effect of some factors on the population dynamics of European and blue hares in Bryansk Region]. In: *Prirodnye Resursy Zapadnykh Oblasti RSFSR i Ikh Rastional'noe Ispol'-zovanie*, Smolensk, pp. 191-194.

Volozheninov, N.N. 1972. K nekotorym voprosam ekologii zaitsa-peshchanika *Lepus tolai* Pallas na yuge Uzbekistana [On some questions in the ecology of Tolai hare *Lepus tolai* Pallas in the south of Uzbekistan]. In: *Ekologiya i Biologiya Zhivotnykh Uzbekistana*, Tashkent, pp. 322-127.

Volozheinov, N.N. 1980. Fenologiya link'i gryzunov i zai-tsev na yuge Uzbekistana [Phenology of molting of rodents and hares in the south of Uzbekistan]. *Uzb. Biol. Zhurn.*, No. 4.

Volozheninov, N.N. 1983. Ekologiya mlekopitayushchikh Zapovednika Aral-Paigambar [Ecology of Mammals of the Aral-Paigambar Preserve]. Fan, Tashkent, 133 pp.

Vol'pert, Ya.L., V.I. Pozdnyakov and N.I. Germogenov. 1988. Territorial'noe raspredelenie i vidovoi sostav mlekopitayushchikh nizov'ev Leny [Territorial distribution and species composition of mammals in the lower reaches of

Lena River]. In: *Zoogeograficheskie i Ekologicheskie Issledovaniya Teriofauny Yakutii*, Yakutsk, pp. 96-106.

Voronov, V.G. 1974. Mlekopitayushchie Kuril'skikh ostro-vov [Mammals of Kuril Islands]. Nauka, Leningrad, 162 pp.

Voronov, G.A. 1964. O biologii severnoi pishchuki verkhne-lenskoi taigi [On the biology of northern pika of upper Lena taiga]. *Zool. Zhurn.*, Vol. 43, No. 4.

Vorontsov, N.N. and E.Yu. Ivanitskaya, 1969. Opisanie khromosomnogo nabora zaitsa-tolaya iz Kyzylkumov (Leporidae, Lagomorpha) [Description of the chromosomal set of Tolai hare from Kyzyl Kums (Leporidae, Lagomorpha)]. In: *Mlekopitayushchie (Evolyutsiya, Kariologiya, Sistematika, Faunistika)*, Novosibirsk.

Vorontsov, N.N. and E.Yu. Ivanitskaya, 1973. Sravnitel'naya kariologiya pishchukh (Lagomorpha, Ochotonidae) Severnoi Palearktiki [Comparative karyology of pikas (Lagomorpha, Ochotonidae) of the northern Palearctic]. *Zool. Zhurn.*, Vol. 42, No. 4, pp. 584-588.

Wallage-Drees, J.M. 1983. Effects of food on onset of breeding in rabbits, *Oryctolagus cuniculus* (L.), in a sand dune habitat. *Acta Zool. Fenn.*, No. 174, pp. 57-59.

Yanushevich, A.I. 1952. Fauna pozvonochnykh Tuvinskoi oblasti [Vertebrate Fauna of Tuva Region] . Novosibirsk, 142 pp.

Yanushevich, A.I., B.M. Aizin, A.K. Kydyraliev and others. 1972. Mlekopitayushchie Kirgizii [Mammals of Kirgizia]. Ilim, Frunze, 462 pp.

Yudakov, A.G. and I.G. Nikolaev, 1974. Nekotorye dannye po biologii manchzhurskogo zaitsa-*Caprolagus (Allolagus) mandshuricus* Radde [Some data on the biology of Manchurian hare—*Caprolagus (Allolagus) mandshuricus* Radde]. In: *Fauna i Ekologiya Nazemnykh Pozvonochnykh Yuga Dal'nego Vostoka SSSR*, Vladivostok, pp. 65-74.

Yudin, B.S., L.I. Galkina and A.F. Potapkina, 1979. Mlekopitayushchie Altai-Sayanskoi gornoi strany [Mammals of the Altai-Sayan Montane area]. Nauka, Novosibirsk, pp. 91.101.

Yudin, B.S., V.G. Krivosheev and V.G. Belyaev, 1976. Melkie mlekopitayushchie severa Dal'nego Vostoka [Small mammals of the north of the Far East]. Nauka, Novosibirsk, pp. 54-70.

Yurgenson, P.B.* Materialy k poznaniyu mlekopitayushchikh Prilteletskogo uchastka Altaiskogo gos. zapovednika [Materials on study of mammals of the Teletsk part of the Altaisk State preserve]. In: *Tr. Altaiskogo Gos. Zapovednika*, Moscow, Vyp. 1, p.**

Yurgenson, P.B. 1939. K ekologii senostavki *Ochotona alpina* Pallas n Vostochnom Altae [On the ecology of hay stacking of *Ochotona alpina* Pallas in eastern Altai]. In: *Nauch.-Metod. Zap. Gl. Upr. po Zapovednikam*, Vyp. 5, p.**

Zagniborodova, E.N. 1960. K faune i ekologii blokh Zapadnoi Turkmenii [On the fauna and ecology of fleas of western Turkmenia]. In: *Voprosy Prirodnoi Ochagovosti i Epizootologii Chumy v Turkmenii*, Ashkhabad, pp. 32-334.

---

*Year of citation missing in the original—Translator.
**Page reference missing in the original—Translator.

400

Zaletaev, V.S. and M.S. Sapargel'dyev, 1975. Ekologicheskie osobennosti dvukh populyatsii ryzhavatykh pishchukh Turkmenii (Kopetdag i Bolshoi Balkhan) [Ecological peculiarities of two populations of Afghan pikas of Turkmenia (Kopetdag and Bolshoi Balkhan)]. *Tr. II Vsesouz. Soveshch. po Mlekopitayushchim.*, Izd-vo MGU, Moscow, pp. 131-137.

Zimina, R.P. 1962. Ekologiya bolsheukhoi pishchukhi (*Ochotona macrotis* Gunther) v Khrebte Terskoi Alatau [Ecology of long-eared pika (*Ochotona macrotis* Gunther) in Terskei Alatau Range]. *Byull. MOIP, Otd. Biol.*, Vol. 67, Vyp. 3, pp. 5-12.

Zimina, R.P. 1964. Zakonomernosti vertikal'nogo rasprostraneniya mlekopitayushchikh [Regularities of Vertical Distribution of Mammals]. Nauka, Moscow, 157 pp.

Zonov, G.B. 1974. Geograficheskie osobennosti zimnego obitaniya mlekopitayushchikh i ptits v nekotorykh gornykh sistemakh Vostochnoi Sibiri [Geographical peculiarities of winter habitation of mammals and birds in some mountain systems of Eastern Siberia]. In: *Voprosy Zoogeografii Sibiri*, Irkutsk, pp. 21-32.

Zonov, G.B. and G.K. Evteev, 1972. K ekologii mongol'skoi pishchukhi (*Ochotona pricei* Thomas) v Tuve [On the ecology of Mongolian pika (*Ochotona pricei* Thomas) in Tuva]. In: *Teriologiya*, Nauka, Novosibirsk, Vol. 1, pp. 363-365.

Zonov, G.B. and L.P. Okunev, 1991. Zimnaya aktivnost' daurskoi pishchukhi [Winter activity of Daurian pika]. In: *Ekologiya Pishchukh Fauny SSSR*, Nauka, Moscow, pp. 14-17.

Zonov, G.B., L.P. Okunev, and I.K. Mashkovskii, 1983. Ispol'zovanie otdushin v snegu melkimi mlekopitayu-shchimi [Use of airholes in the snow by small mammals]. *Zool. Zhurn.*, Vol. 62, No. 12, pp. 1863-1867.

Zörner, H. 1981. Dutherstadt: Der Feldhase. Wittenberg; Ziemen 172 pp.